P9-EEC-550

MASONRY

MASONRY
Materials, Design, Construction

R.C. Smith
T.L. Honkala
C.K. Andres

Reston Publishing Company, Inc.
A Prentice-Hall Company
Reston, Virginia

Library of Congress Cataloging in Publication Data

Smith, Ronald C
Masonry: materials, design, construction.

Includes index.
1. Masonry. I. Honkala, T.L., joint author.
II. Andres, C.K., joint author. III. Title.
TA670.S47 624'.183 78-27178
ISBN 0-8359-4247-3

A Prentice-Hall Company
Reston, Virginia 22091

10 9 8 7 6 5 4 3 2 1

Printed in the United States of America

CONTENTS

TABLES

PREFACE

The art of building with masonry materials has its roots in antiquity. The use of stone and brick goes back to the earliest building efforts of civilized man. The technique of molding hollow clay units—clay tile—was developed at a later date and the production of concrete masonry came much later, in the nineteenth century.

Although lime mortar was used as the binding agent for many hundreds of years, the development of Portland cement and its use in cement mortar and, more recently, the development of organic cements has resulted in greatly improved methods of construction with unit masonry.

Concepts regarding masonry construction have undergone radical changes over the years. Whereas stone and brick were, at one stage, the primary structural materials, resulting in heavy buildings with thick bearing walls and great arched roofs, the modern trend favors lighter construction, using structural frameworks and masonry closure walls or reinforced masonry walls. The result has been that stone is now used almost entirely as a facing material, while brick is used for closure walls, bearing walls in some cases, and as a facing material.

The trend to lighter buildings was enhanced by the development of hollow masonry units—first structural clay tile and later hollow concrete masonry. In fact, the production of concrete masonry units on such a vast scale as has been the case during the past fifty years has led to a great reduction in the use of structural clay tile. However, clay tile in other forms still enjoys wide popularity.

It is the purpose of this book to deal with masonry materials and construction in the modern sense. The study is accomplished in three stages, beginning with a study of the materials themselves—their origin, manufacture, physical characteristics and limitations, the mortars used to bind them together and the products used to protect and decorate them.

Next, the engineering aspects of the materials are investigated, with a look at testing methods, masonry systems and elements of masonry design. Finally, the planning, construction and maintenance of buildings using these masonry units is discussed in some detail.

Although the book is aimed primarily at technical students and those engaged in the pursuit of the masonry trades, a sincere attempt has been made to keep the discussions as easily understandable as possible, in order to make the book useful to the interested layman as well.

Finally, since the decision has been made in North America to adopt an internationally accepted, up-to-date metric system of measurement—the SI (Systeme International) system—this book has been written using that system and giving equivalents in the imperial system.

All three of us wish to express our sincere gratitude for the assistance and cooperation that has been available from so many sources, without which the book would not have been possible. The extent of that assistance is attested to by the following list of contributors.

ACKNOWLEDGMENTS

The authors wish to express their indebtedness to the following organizations for the information and materials they provided for use in this book.

Alberta Masonry Institute (H. Morstead)
Alberta Workers' Compensation Board
American Society for Testing and Materials
Besser Technical Center
The Bonnot Company
Brick Institute of America
Brick and Tile Manufacturing Association of Canada
Building Stone Institute
Canadian Pittsburgh Industries
Canadian Standards Association
Canadian Structural Clay Association
Chambers Brothers Company
Clay Brick Association of Canada
Clayburn Brick Products
Dow Chemical of Canada Ltd.
Edmonton Concrete Block Company
Garson Limestone Company
Gillis Quarries Ltd.
Indiana Limestone Company
International Masonry Institute
I.X.L. Industries Ltd.
J.C. Steele & Sons, Inc.
Jeffrey Manufacturing Company
Morgen Manufacturing Company
National Concrete Masonry Association
National Concrete Producers Association
National Research Council of Canada
Ontario Reinforcing Manufacturers' Association
Portland Cement Association
Precast Concrete Institute
Soiltest Inc.
Southern Alberta Institute of Technology
University of Alberta (Engineering, M. Hatzinkolas)
Vermette Machine Company
Whiteman Manufacturing Company

Chapter 1

BRICK

Early historical records indicate that brick was being used as an important building material even before the Christian Era. The first true arch of sun-dried brick was constructed in the ancient city of Ur about 2000 B.C. and a relatively short time later—certainly by 1500 B.C.—burned brick was being used. The Great Wall of China, built between 300 and 200 B.C., was made of both sun-dried and burned brick. The early Romans used burned brick in the reconstruction of the Pantheon (123 A.D.) and in the construction of the Colosseum, completed in 82 A.D. The art of glazing brick was known to the Babylonians and Assyrians as early as 600 B.C.

Those early construction materials are the ancestors of a wide range of clay building products in use today, and manufacture of these products still follows the same basic procedures of the past. Technological advances of the last seventy-five years, however, have resulted in more advanced procedures and more efficient modern brick plants. A more complete knowledge of the raw materials involved, more effective firing methods and more and better machinery have all contributed to the development of a highly efficient industry.

BRICK INGREDIENTS

The basic ingredient of brick is *clay.* Clays are complex materials but basically they are composed of *silica* and *alumina,* with varying amounts of *metallic oxides* and other ingredients. In general, clays may be classified as *calcareous* and *noncalcareous,* depending on their basic composition.

The calcareous clays contain about 15 percent *calcium carbonate* and burn to a yellowish color. Noncalcareous clays are composed of *silicate of alumina,* from 2 to 10 percent *iron oxide* and *feldspar* and burn red, buff or salmon, depending largely on the iron oxide content.

In order to be suitable for making brick, clay must have some very specific properties. It must have *plasticity* when mixed with water so it can be molded, it must have enough *tensile strength* to maintain its shape after forming and the clay particles must be able to *fuse* together when heated to high temperatures.

Clay occurs in three principal forms, all of which are similar chemically but different in their physical characteristics. Those three forms are: *surface clays, fireclays* and *shales.* Surface clays are of recent sedimentary formation or they are exposed older deposits. Fireclays are found at deeper levels and usually have more uniform physical and chemical properties, their most important characteristic being their ability to withstand high temperatures. Shales are clays that have been subjected to high pressure until they have become relatively hard.

MANUFACTURING PROCESS

The manufacturing process involved in brick making includes six main phases: *mining and storage* of the raw material, *preparation of the raw material* for manufacture, *forming* of units, *drying* the formed units, *burning and cooling* the units and *drawing and storing* the finished products.

Mining and Storage

In most cases, clays are mined by the *open-pit* method, though some fireclays are obtained by *underground mining.* In either case, the material is transported to a plant site by truck or rail and stored in bins to await further processing. If the clay is in large lumps, it undergoes preliminary crushing before being stored. Also, at this stage the clays are blended to minimize variations in physical and chemical properties.

Fig. 1-1. Reversible clay crusher. (*Courtesy Jeffrey Mfg. Co.*)

Preparing Raw Materials

Clay is taken from the storage bins and transferred, usually by belt, to crushers that reduce it in size to 50 mm (2 in) or less and remove the stones. (See Fig. 1-1.) It then goes to grinders (see Figs. 1-2 and 1-3) that grind the clay to a fine powder and mix it thoroughly. The ground clay passes over a screen that removes the coarse particles and returns them to the grinder for further processing.

Forming

In the forming process, the first step is to mix the finely ground clay with water—a process known as *tempering.* This is done in a

Fig. 1-2. Dry pan grinder. (*Courtesy J.C. Steele & Sons, Inc.*)

Fig. 1-3. Roll grinder. (*Courtesy Chambers Bros. Co.*)

pugmill, one type of which is illustrated in Figure 1-4.

The amount of water mixed with the clay depends on the method to be used to form the brick. Two principal methods are in common use: the *stiff-mud* process and the *dry-press* method.

Stiff-Mud Process. In the stiff-mud process, only enough water is added to the clay to produce plasticity, usually from 12 to 15 percent by weight. Then the plastic clay is forced through a *de-airing machine* that removes air pockets and bubbles and increases workability.

From there it is forced by auger through an *extruder*—a die—from which issues a continuous column of clay of the required size and shape. Figure 1-5 illustrates a machine that combines a pugmill, a de-airing machine and an extruder. Notice the *die head* on the right; it will produce a column of clay with three holes in the core.

Textures are applied to the surface before the column passes to an *automatic cutter* that cuts off units of the proper length. (See Fig. 1-6.)

The units are then carried by belt to an inspection area where the properly formed ones

Fig. 1-4. Pugmill. (*Courtesy The Bonnot Co.*)

Fig. 1-5. Combination pugmill, deairing machine, and extruder. (*Courtesy J.C. Steele & Sons, Inc.*)

are unloaded onto drier cars and the imperfect units are returned to the pugmill for recycling.

A great deal of the brick used in industry and all of the structural clay tile is manufactured by the stiff-mud process.

Dry-Press Process. The dry-press method uses less water for tempering than the stiff-mud—the maximum being about 10 percent. The relatively dry mix is fed to huge presses, similar to the one shown in Figure 1-7, that form the brick in steel molds under high pressure. Most bricks produced by this method are made with three or more round, hollow cores. (See Fig. 1-8.)

In both of these forming systems, allowance is made for the fact that the units shrink when dried. Accordingly, the brick are made large enough so that, when they are dried and burned, the finished product will be the proper size.

Fig. 1-6. Automatic brick cutter. (*Courtesy The Bonnot Co.*)

Fig. 1-7. Dry-press molding machine.

Drying

The formed brick, as they come from the extruder or press, contain from 7 to 30 percent moisture, most of which must be removed before they are fired. The drying is done in *drier kilns,* having temperatures ranging from 38°C to 205°C (100°F to 400°F), with the heat usually being provided by the exhaust from burning kilns. In all cases, the heat and humidity are carefully regulated to avoid too rapid shrinkage which would cause excessive cracking.

When brick are to be glazed, the glazing compound is usually applied at the end of the drying period. The operation consists of spraying a mixture of mineral ingredients in liquid form over one or more surfaces of the brick. The glaze melts and fuses to the brick at a given temperature, producing a glasslike coating in various colors.

Burning and Cooling

Burning is a very important step in the manufacture of brick. It is done in a number of different types of *kiln,* two of the most commonly used being *tunnel kilns* and *periodic kilns.* A tunnel kiln consists of a long shaft, containing a number of temperature zones. The dried brick pass through the various zones on special cars that enter the tunnel at one end and move slowly to the other. In periodic kilns, the brick cars are stationary, and the temperature is varied periodically (raised in stages and then lowered) until the burning is complete. In either case, fuel may be natural gas, oil or coal. Also, in either case, the brick must be stacked on the kiln cars in such a way that the hot kiln gases can circulate freely through them.

In the kiln, the brick goes through six general stages: *water smoking, dehydration, oxidation, vitrification, flashing* and *cooling.* Although temperatures vary, depending on the type of material, each step occurs at a definite temperature range. The time required for the entire process varies from 40 to 150 hours, depending on the *type of kiln,* the *type of clay, type of glaze,* if any, *size* of the units and other variables.

Water smoking (the evaporation of free water) takes place at temperatures up to about 205°C (400°F); dehydration at temperatures from about 150°C to 980°C (300°F to 1800°F); oxidation at from 535°C to 980°C (1000°F to 1800°F); and vitrification at from 870°C to 1315°C (1600°F to 2400°F).

Near the end of the burning process, the brick may be *flashed* to produce color variations. This is done by injecting extra natural gas at the appropriate time or place. When this extra fuel burns, patches and variations in color are formed at random throughout the stack of brick.

Cooling takes from 48 to 72 hours, depending on the type of kiln. This cooling process must be carefully controlled because the rate of cooling has a direct effect on the color of the brick and because too rapid cooling will cause checking and cracking in the brick.

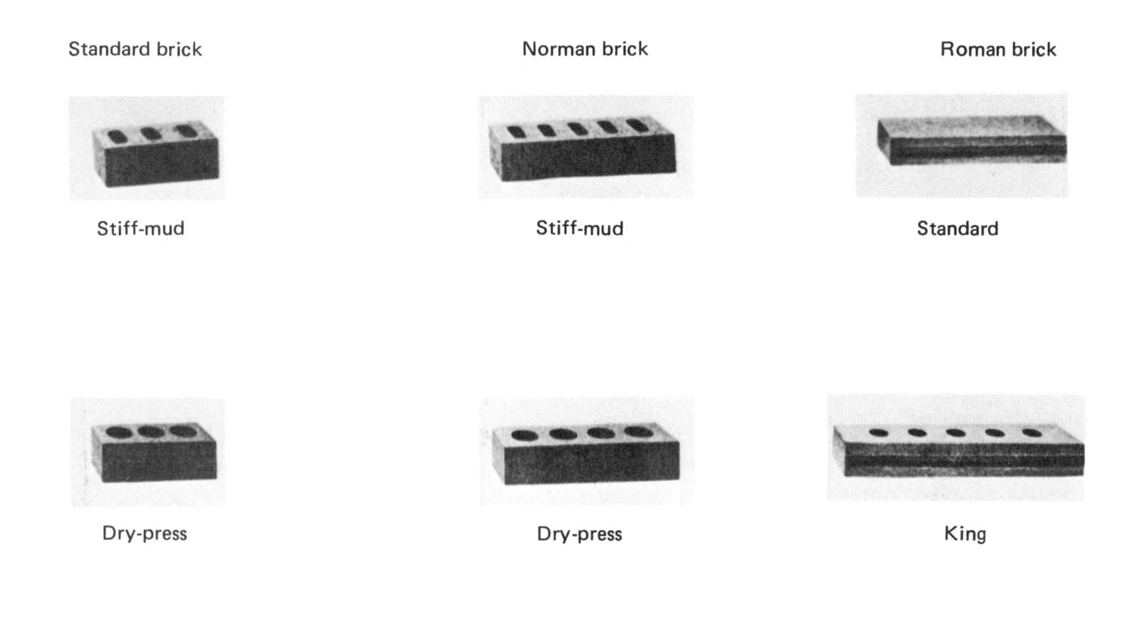

Fig. 1-8. Brick types.

Drawing and Storage

Drawing is the process of removing the cars from the kiln after the brick are cool and unloading them. At this time the brick are sorted, graded, packaged and taken to storage or loaded on rail cars or trucks for shipment.

BRICK SHAPES AND SIZES

Brick Shapes

Beginning with the unit universally known as a *common brick* and recognized as a type or shape with a generally accepted set of dimensions, a number of other shapes have been developed over the years, some designed for a special purpose, others based on an historical shape and still others made for reasons of economy. Among the better known ones are *Roman, Norman, giant,* and *titan,* or *TTW* (through-the-wall). (See Fig. 1-8.) In addition, there are *monarch, Saxon* (see Fig. 1-9) and *hollow* (see Fig. 1-10), along with several variations of these. Each shape has a range of widths in which it may normally be produced.

Special Brick Shapes. In order to provide greater flexibility in design, a number of special brick shapes are made by a number of manufacturers. They include *special corner units, angled brick, inside* and *outside radial brick, fluted brick, solid brick, bullnose brick*—with either single or double bullnose—and *corner brick,* three courses high.

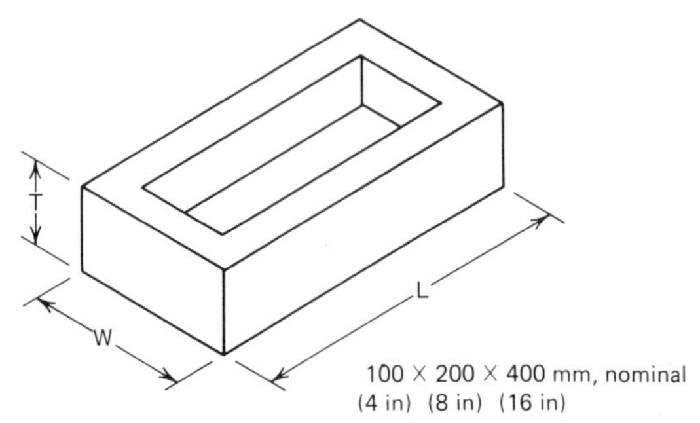

Fig. 1-10. Typical hollow brick.

These units are normally made to order, and the local brick plant should be consulted for items available.

Brick Sizes

In the past, the common brick has been made in a variety of sizes, depending on the locality and the material. But as the need for standardization became apparent, most structural clay products came to be designated by nationally recognized product specifications. With the introduction of the metric (SI) system into North America, that need has become even more acute.

All unit masonry products have been designated by their *nominal* dimensions (the sizes by which they are named but their actual or *modular* dimensions are such that, measured center-to-center of mortar joints of specified thickness, the dimensions will be compatible with a standard planning module.

Under the SI system, the recognized planning modules are 100 mm (4 in), 300 mm (12 in),

75 mm brick

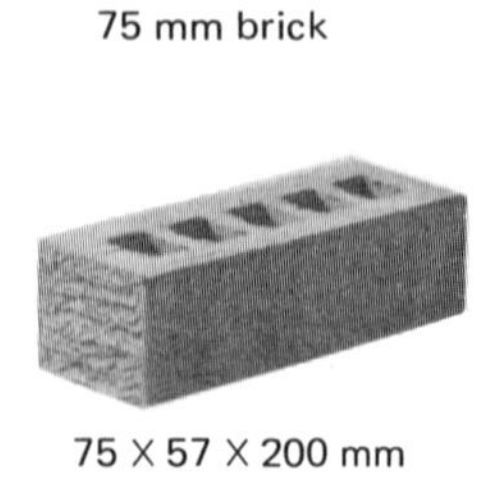

75 × 57 × 200 mm

Monarch

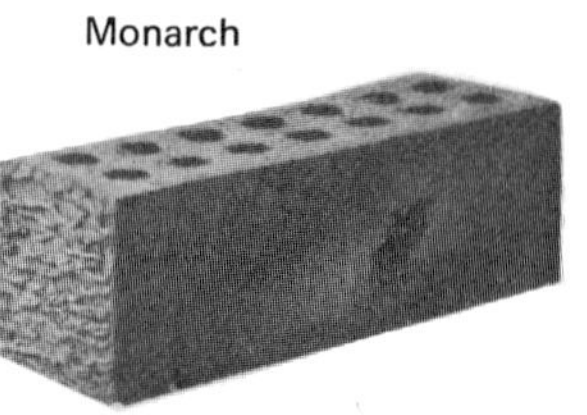

75 × 100 × 400 mm
(Nominal dimensions)

Saxon

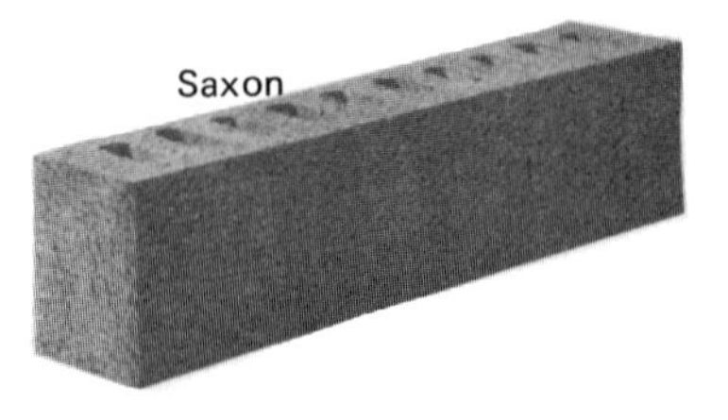

100 × 100 × 300

Fig. 1-9. Unconventional brick sizes.

and 600 mm (24 in), with the internationally recognized *planning grid* being a square 600 X 600 mm (24 X 24 in).

Under this system the recognized module for brick is 100 mm (4 in), with the result that the nominal size for a common brick will be 100 X 67 X 200 mm (4 X $2\frac{2}{3}$ X 8 in). (See Fig. 1-11.) This means that, on the basis of standard 10 mm ($\frac{3}{8}$ in) mortar joints, the modular size of common brick will be 90 X 57 X 190 mm ($3\frac{5}{8}$ X $2\frac{1}{4}$ X $7\frac{5}{8}$ in). Thus one brick length, two brick widths, or three brick thicknesses plus mortar joints will equal 200 mm (2 modules).

In relation to the 600-by-600-mm planning grid, three common brick modular lengths or nine thicknesses fit into the grid. (See Fig. 1-12.)

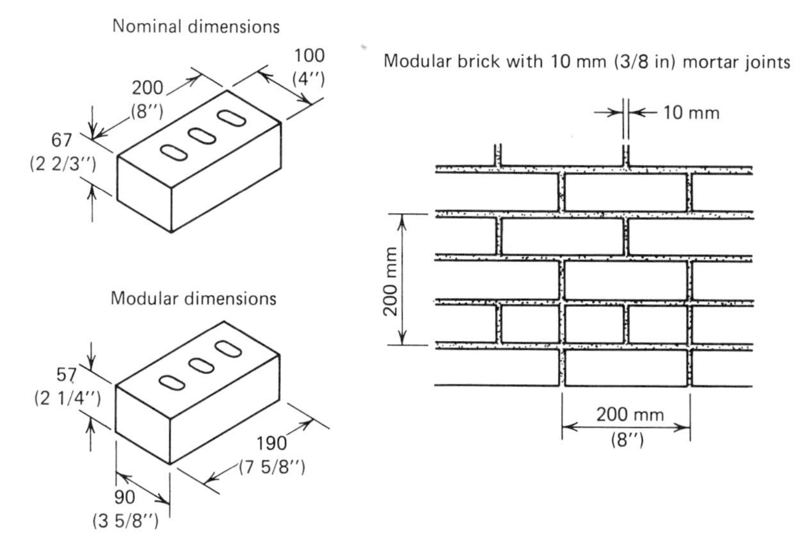

Fig. 1-11. Nominal and modular brick sizes.

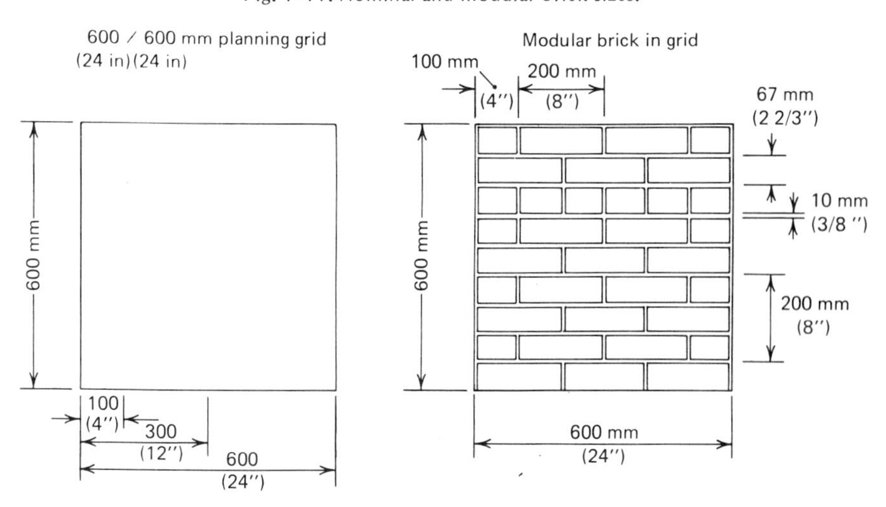

Fig. 1-12. Modular planning grid.

Table 1-1. Nominal and Modular Brick Sizes

Brick Shape	Nominal Size	Modular Size	Nominal Size (in)
Common	100 X 67 X 200	90 X 57 X 190	4 X 2-2/3 X 8
Norman	100 X 67 X 300	90 X 57 X 290	4 X 2-2/3 X 12
Roman			
Standard	100 X 50 X 300	90 X 40 X 290	4 X 2 X 12
King	100 X 67 X 400	90 X 57 X 390	4 X 2-2/3 X 16
Giant			
100 mm	100 X 100 X 400	90 X 90 X 390	4 X 4 X 16
150 mm	150 X 100 X 400	140 X 90 X 390	6 X 4 X 16
200 mm	200 X 100 X 400	190 X 90 X 390	8 X 4 X 16
300 mm	300 X 100 X 400	290 X 90 X 390	12 X 4 X 16
TTW			
150 mm	150 X 67 X 300	140 X 57 X 290	6 X 2-2/3 X 12
200 mm	200 X 67 X 300	190 X 57 X 290	8 X 2-2/3 X 12
75-mm brick	75 X 67 X 200	75 X 57 X 190	3 X 2-2/3 X 8
Monarch	75 X 100 X 400	75 X 90 X 390	3 X 4 X 16
Saxon	100 X 100 X 300	90 X 90 X 290	4 X 4 X 12
Hollow	100 X 200 X 400	90 X 190 X 390	4 X 8 X 16

The sizes that will allow the conventional brick shown in Figure 1-8 to fit into the international planning grid are given in Table 1-1, along with the sizes of the unconventional brick illustrated in Figures 1-9 and 1-10.

PHYSICAL PROPERTIES OF BRICK

All properties of brick are affected by the composition of the raw materials used and the manufacturing process involved. Those properties include *color, texture, size, strength* and *absorption.*

Color

The color of a burned brick depends on its chemical composition, the heat of the kiln and the method used to control the burning.

All clays containing iron will burn red if exposed to an oxidizing fire. The same clay, burned in a reducing fire, will take on a purple tint, due to the ferrous silicate content. If the same clay is underburned, salmon colors are produced. Overburning produces dark red brick. Calcareous clays produce the buff, tan, and brown bricks, depending on the temperature of burning.

Colors may also be changed by adding suitable materials to the plastic clay or by the application of a glazing material to the brick surface before burning. As previously noted, brick may also be flashed (i.e., have patches of color added to their surface at random) by burning excess gas in the kiln at the proper time.

As a result of these techniques, the brick manufacturer is able to provide a wide range of colors in the various brick products.

Textures

The surface texture of brick is as important as color in the design of modern brick structures. A variety of surface textures is available, including *smooth-face, stipple, brush, rug, matte, bark, indented matte, sanded indented* and *split-face.* (See Fig. 1-13.)

The application of a texture is relatively

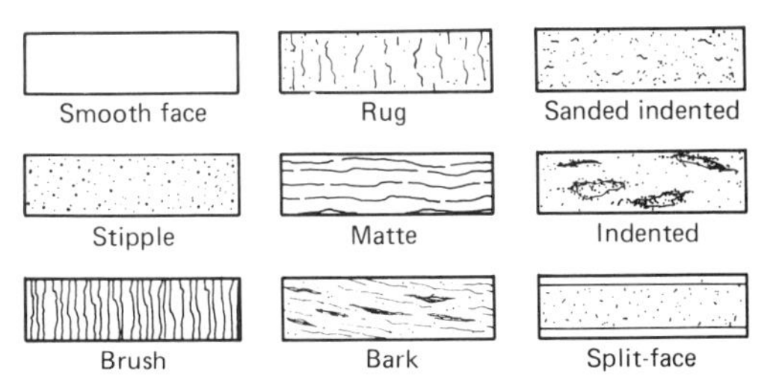

Fig. 1-13. Brick textures.

simple on brick formed by the extrusion process. As the column of clay is extruded from the die, it can be brushed, scratched, indented or otherwise marked to produce a particular texture on the brick face or faces. Smooth-face is the surface produced by the pressure of the clay against the sides of the steel die, while split-face is the result of making brick double width and breaking them along the center to form two units with the characteristic *split* texture. (See Fig. 1-14.)

Size

Most clays shrink during drying and burning from 4.5 to 15 percent and this is taken into account during the molding process by making the brick oversized. Shrinkage will vary, depending on the composition of the clay, its fineness, the amount of water added and the kiln temperature. Because absolute size uniformity is impossible, specifications normally include permissible variations in size.

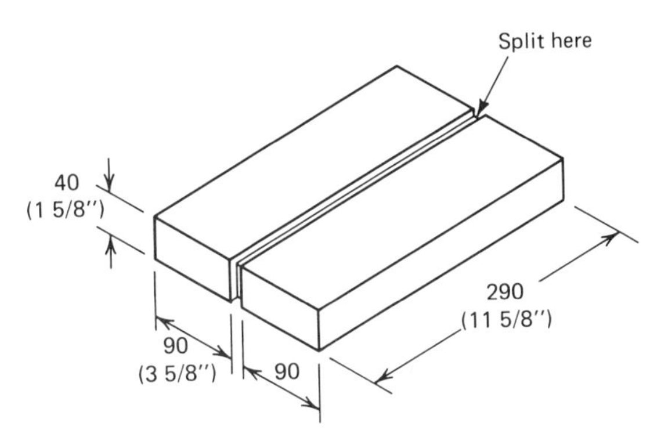

Fig. 1-14. Roman brick before splitting.

Strength

The strength characteristics of brick that must be considered include *compressive, transverse* and *tensile* strengths of the units. All of these will vary with the raw material used and the manufacturing processes.

For example, the compressive strength of brick will vary from 10.5 to 138 MPa (1500 to 20,000 psi). Similarly, transverse strength (i.e., the strength of a brick when it acts as a beam, supported at both ends) varies from 0.8 MPa (100 psi) to an average maximum of 20.5 MPa (3000 psi). The value for transverse strength is usually expressed as the *modulus of rupture.*

Tests indicate that the tensile strength of brick normally runs between 30 and 40 percent of the transverse strength, while punching shear tests on brick indicate shear strengths of from 30 to 40 percent of the net compressive strength.

Absorption

The water absorption of a brick is defined as the weight of water, expressed as a percentage of the dry weight of the brick, that is taken up during a given test period. The water is sucked into the unit through pores that act as capillaries. The initial rate of absorption, or *suction,* of a brick has an important effect on the bond between brick and mortar. Tests indicate that maximum bond strength is obtained when the initial rate of absorption is about 20 grams (0.7 oz) per minute. When brick has a greater suction rate than that, it should be wetted prior to laying to reduce the suction.

Some average physical properties of bricks are given in Table 1-2.

Table 1-2. Average Physical Properties of Brick

Absorption		Modulus of Rupture, MPa (psi)		Compressive Strength, MPa (psi)		Tensile Strength, MPa (psi)	Shear Strength MPa (psi)
5 hr. boil	*48 hr. cold*	*On flat*	*On edge*	*On flat*	*On edge*		
17.4%	14%	7.3 (1060)	7.6 (1100)	32.4 (4700)	37.4 (5424)	2.7 (390)	12.7 (1840)

Glossary of Terms

Alumina Aluminum oxide.

Compressive strength Strength that resists crushing.

Feldspar Aluminum silicates, containing potassium, sodium, calcium, or barium.

Ferrous Pertaining to iron.

Module A standard of measurement.

Oxidizing Combining with oxygen.

Reducing Subjecting to the action of hydrogen.

Shear A stress that tends to cause two adjoining parallel planes in a body to slide in opposite directions.

Tensile strength Resistance to a stress that tends to pull a body apart.

Review Questions

1. Differentiate between *calcareous* and *noncalcareous* clays.
2. What basic qualities must a clay have in order to be suitable for making brick?
3. Explain the reason for blending two or more clays together in the brick manufacturing process.
4. Explain what is meant by:
   (a) tempering clay
   (b) fireclay
   (c) fusion of clay
   (d) de-airing of clay
5. Explain what is meant by each of the following terms used in the brick industry:
   (a) extruding
   (b) firing
   (c) flashing
   (d) glazing
6. Using simple diagrams, illustrate each of the following strength characteristics of brick:
   (a) compressive strength
   (b) transverse strength
   (c) tensile strength

Selected Sources of Information

Alberta Masonry Institute, Calgary, Alberta.

Brick Institute of America, McLean, Virginia.

Clay Brick Association of Canada, Toronto, Ontario.

Clayburn Brick Products, Vancouver, British Columbia.

International Masonry Institute, Washington, D.C.

I.X.L. Industries, Medicine Hat, Alberta.

Chapter 2

CONCRETE MASONRY UNITS

Compared to other masonry building materials, such as brick or stone, concrete masonry units (concrete block, concrete brick, drainage and paving tile) are relative newcomers. The first concrete blocks were molded in the early 1880s and since that time the industry has made great strides in the volume of units produced, in the variety of blocks available and in the quality of the products.

Today, more than two-thirds of the volume of all masonry walls being constructed is made up of concrete block of one kind or another. Blocks are available in a wide range of *types, sizes, shapes,* and *surface textures,* and they are used for a variety of purposes. (Consult manufacturers' brochures for blocks available in your area.) In addition, well-recognized standards have been established covering the physical properties of block; namely, its *solid content, strength, density, water absorption capacity, moisture content* and *linear shrinkage potential.*

MANUFACTURE

The main ingredients of concrete masonry units consist of *graded aggregates,* such as sand, gravel, crushed stone, lightweight aggregates, or cinders, *portland cement* and *water.* In addition, such materials as an *air-entraining agent, coloring agent* and *pozzolanic* or *siliceous materials* may be added when specifications require them. Aggregates are classified as S, N, L_1, or L_2, depending on the type of material from which they are produced. (See Table 2-1 for description of aggregate types.)

The ingredients are mixed as a very dry, no-slump concrete that is fed into a molding machine (see Fig. 2-1). By a combination of pressure and vibration, the machine molds the mix into blocks of the required dimensions. A thousand or more 200 × 200 × 400 mm (8 × 8 × 16 in) blocks, or their equivalent in other sizes, may be produced per hour by such a machine.

The blocks are then cured, usually by some type of accelerated curing process. One such process involves heating the blocks in a steam kiln at atmospheric pressure to temperatures ranging from 49°C to 82°C (120°F to 180°F) for periods of up to 18 hours. In another process, the units are subjected to saturated steam at 163°C to 190°C (325°F to 375°F) for various periods up to 12 hours in autoclaves. (See Fig. 2-2.)

Table 2-1. Standard Concrete Masonry Unit Dimensions and Wall Properties

Dimension or Wall Property	Modular Size in mm (in)				
	100	*150*	*200 (8)*	*250*	*300*
Actual overall width, mm	90	140	190 (7-5/8)	240	290
Min. face-shell thickness, mm	26	26	32 (1-1/4)	35	38
Min. web thickness, mm	26	26	26 (1)	28	32
End flange width, mm	N/A	N/A	50 (2)	50	50
Equivalent thickness, mm	66	80	103 (4)	121	144
Percentage solid	74	57	54	51	50
Approximate mass of wall in place, normal weight, kg/m^2	140	170	215 (44 lb/ft^2)	255	300
Approximate mass of wall in place, lightweight, kg/m^2	110	130	165 (34 lb/ft^2)	195	230
Fire rating, in hours aggregate type S or N	3/4	1	1-1/2	2	3
Fire rating, in hours aggregate type L_1 or L_2	1	1-1/2	2	3	4
Sound transmission class, STC, normal weight, dB	45	47	51	53	56
Sound transmission class, STC, lightweight, dB	42	44	44	49	51

(Courtesy National Concrete Producers Association)

Note: Concrete types

S – is the type in which the coarse aggregate is granite, quartzite, siliceous gravel, or other dense materials containing at least 30 percent quartz, chert, or flint.

N – is the type in which the coarse aggregate is cinders, broken brick, blast furnace slag, limestone, calcareous gravel, trap rock, sandstone, or similar dense material containing not more than 30 percent quartz, chert, or flint.

L_1 – is the type in which all the aggregate is expanded shale.

L_2 – is the type in which all of the aggregate is expanded slag, expanded clay, or pumice.

After curing is complete and before the units are shipped, they go to storage where they dry to reduce the moisture content to the specified limits.

MASONRY UNIT DESIGN

Concrete masonry units are primarily designed as either hollow or solid. (See Fig. 2-3.) A unit is considered to be solid if the net concrete cross-sectional area, parallel to the bearing surface, is more than 75 percent of the gross cross-sectional area. (See Fig. 2-4.) In most cases, solid units have no cores at all (i.e., they are 100 percent solid). (See Fig. 2-5.) In addition to standard solids, concrete brick, paving tile, manhole blocks, septic tank system blocks and some split-block units fall into this category.

Most solid units are used for special purposes; for example, for structures that have very *high design stresses,* for the *bearing course* of load-bearing walls, for increased *fire protection,* for *fillers* in irregular spaces, for *manhole* and *other drainage* purposes, for *paving* and *edging.*

A great majority of the blocks manufactured

Fig. 2-1. Concrete block molding machine. (*Courtesy Besser Co.*)

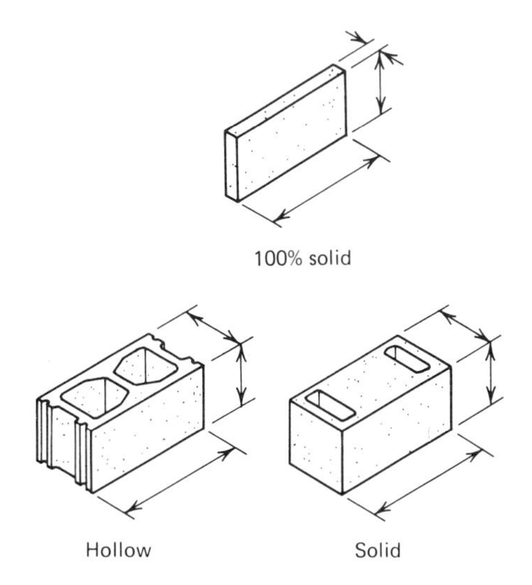

Fig. 2-3. Hollow and solid blocks.

Fig. 2-2. Steam-curing autoclaves. (*Courtesy Besser Co.*)

fall into the hollow block category, of which there is a great variety of sizes and shapes, with some variation from one manufacturer to another. Many blocks are made in half-length as well as in full-length units; half-height blocks are also available in many areas.

CONCRETE MASONRY UNIT SIZES

With the adoption of the metric system in North America, the configuration and dimensions of hollow units have been standardized in order that a common comparable

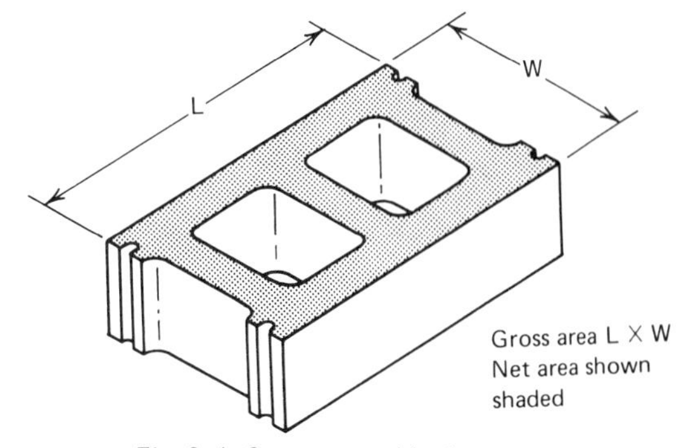

Fig. 2-4. Gross vs. net block area.

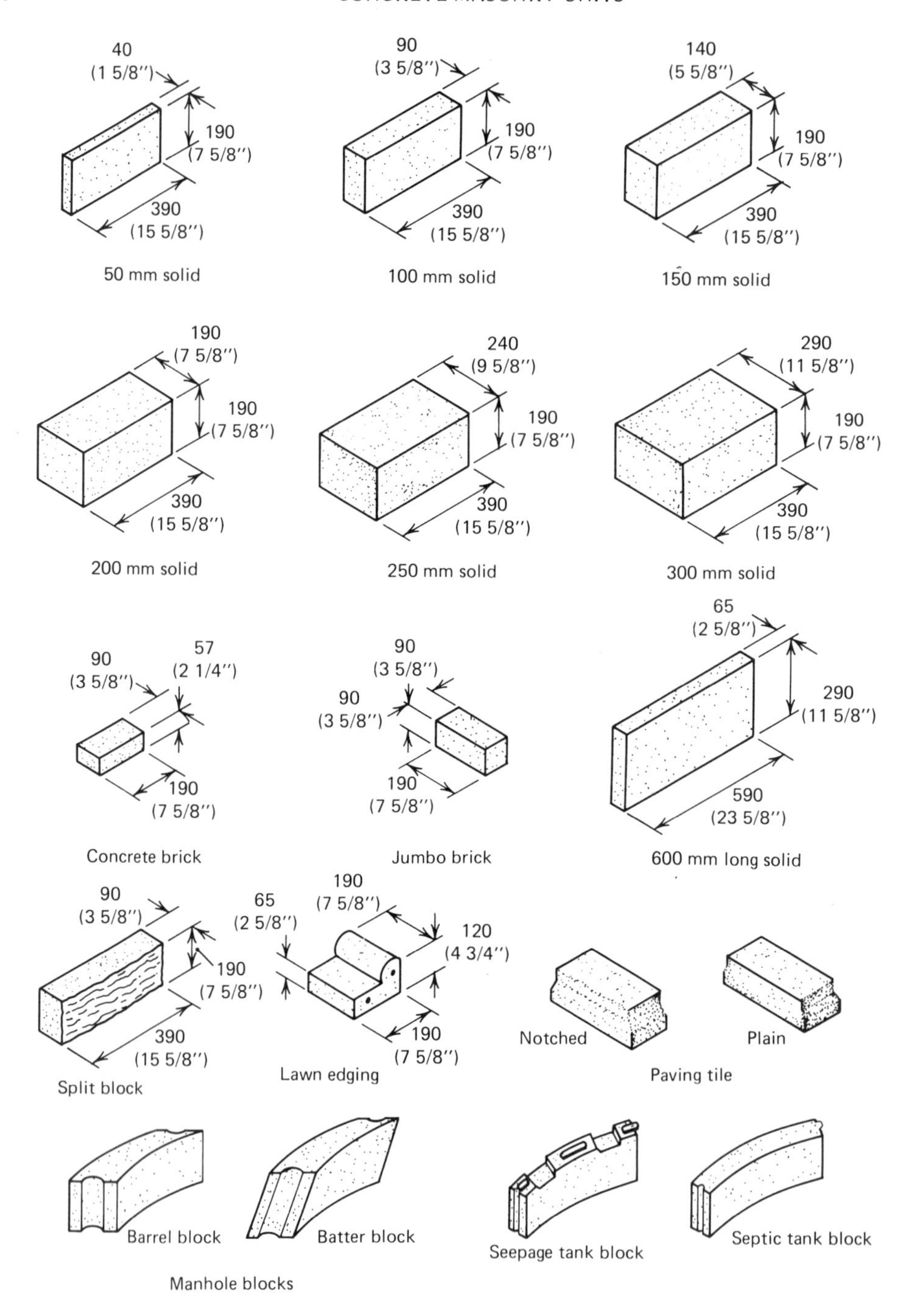

Fig. 2-5. Solid concrete masonry units.

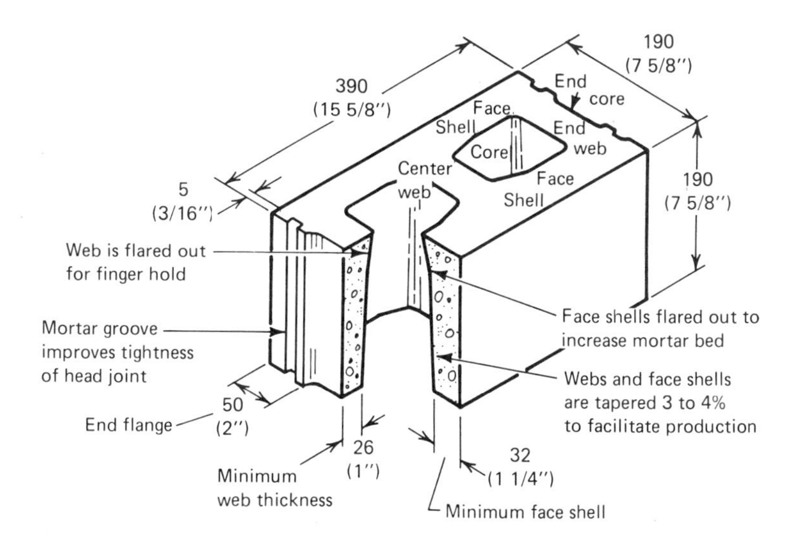

Fig. 2-6. Configuration and dimensions of a typical 200 X 200 X 400-mm concrete masonry hollow unit. (*Courtesy National Concrete Producers Assn.*)

product will be available. (See Fig. 2-6.) This standardization is not intended to limit the design of specific sizes or shapes to satisfy individual requirements but to ensure that the product delivered will conform with the usual design assumptions.

In areas where reinforcing of masonry walls is mandatory, concrete masonry units may be supplied with square-shaped cores. In such a case, face shells will be increased in order to maintain the same net mortar-bedding area and the same percentage of solid.

Dimensions and wall properties of standard concrete masonry units are outlined in Table 2-1.

CONCRETE MASONRY UNIT TYPES

Concrete masonry units may be divided into a number of categories or types, each with a special purpose. Some are *load-bearing,* others *non-load-bearing:* some are intended for *veneer* purposes, some for *architectural* effect; others have a special *structural* use; some are *utility* blocks, designed for special purposes other than building construction. Units designed for architectural purposes may also be load bearing.

Load-Bearing Units

Load-bearing units include a number of shapes, all of which are used essentially in the construction of load-bearing walls. Included in this group are *stretchers, headers, corner blocks, L-corners, joist blocks, sash blocks, control joint blocks, coping blocks, sill blocks* and *corner sash* blocks. (See Fig. 2-7.)

Non-Load-Bearing Units

Units of this type may be made with a lower compressive strength rating and are used largely in the construction of nonbearing partitions. Blocks 100 mm and 150 mm (4 in and 6 in) in size are commonly used for this purpose, but partition block is also made in the 200 mm (8 in) nominal thickness. (See Fig. 2-8.)

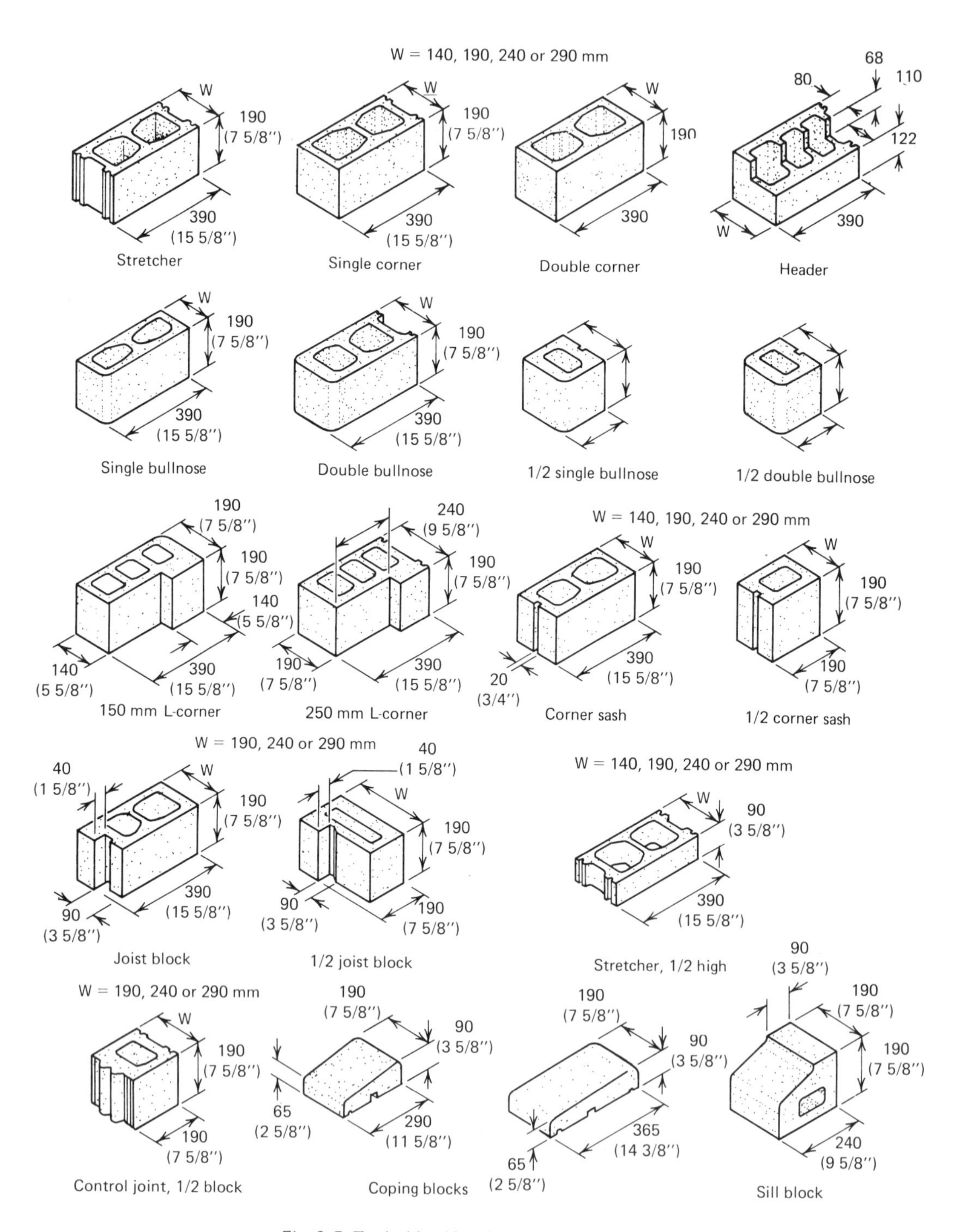

Fig. 2-7. Typical load-bearing concrete masonry units.

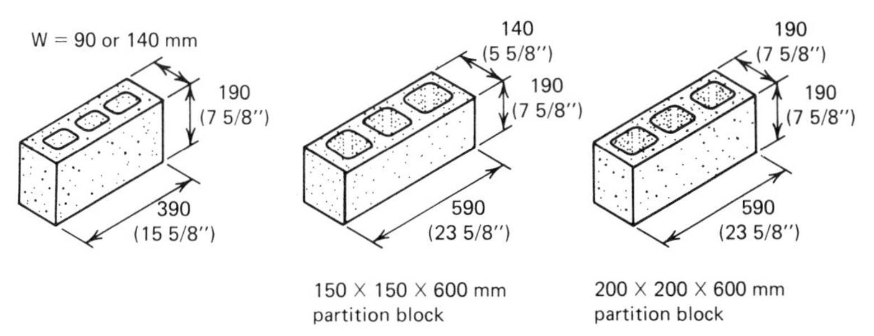

Fig. 2-8. Non-load-bearing concrete masonry units.

Veneer Units

Veneer units are those that are used as a facing material over a plain backup wall. They include concrete brick, split blocks and a variety of units with a designed face (e.g., fluted, ribbed, split-ribbed, etc.) (See Fig. 2-9.)

Architectural Units

A large number of units have been designed that are not only load bearing in quality but have a face that provides some particular architectural effect. Figure 2-10 illustrates some of the blocks available from this category.

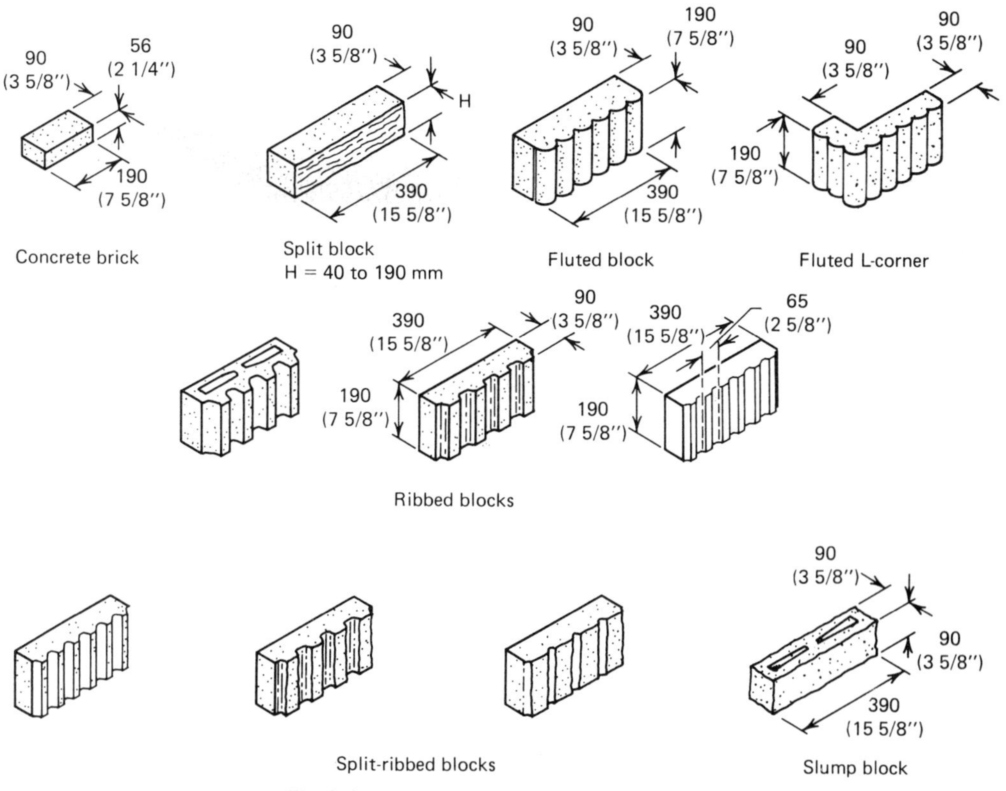

Fig. 2-9. Concrete masonry veneer units.

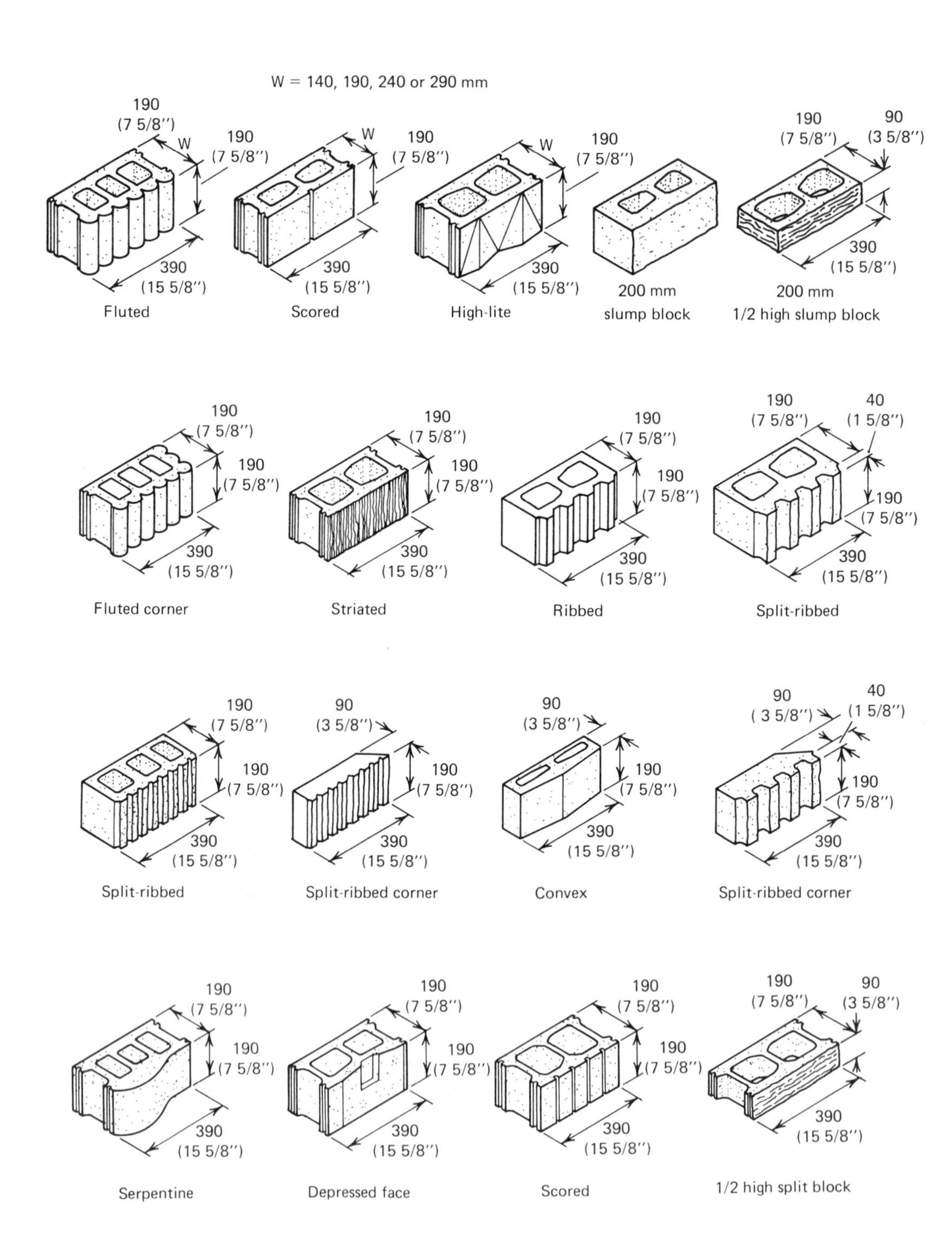

Fig. 2-10. Concrete masonry architectural units.

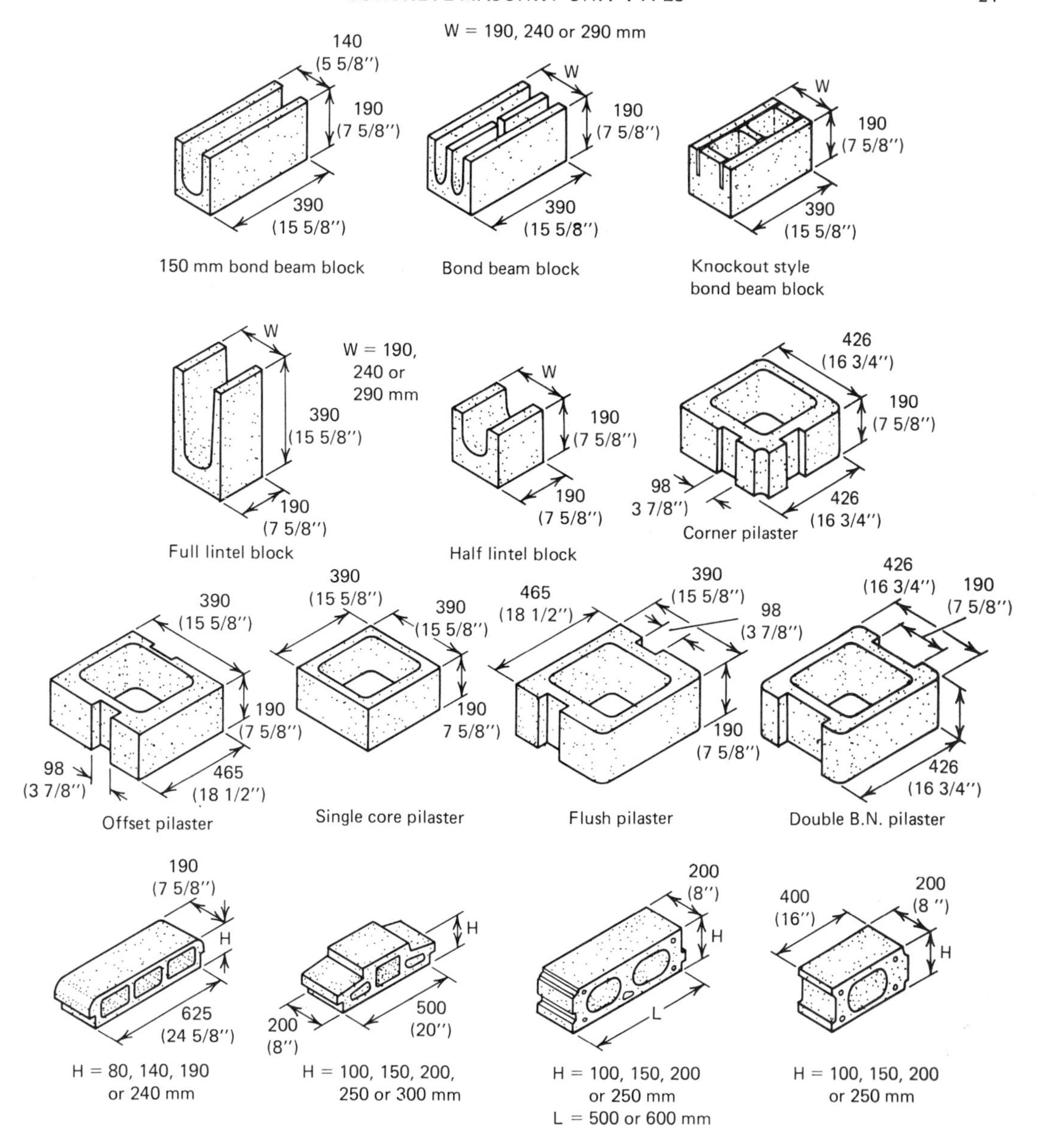

Fig. 2-11. Concrete masonry structural units.

Structural Units

A number of units have been designed to serve special structural purposes. These include *lintel* blocks, used to form a beam over an opening in a wall; *bond beam* blocks, used to form a continuous horizontal beam at specified intervals in a block wall; *pilaster* blocks, used to provide stiffness to a wall and to provide additional bearing area for roof beams, trusses, etc., resting on the top of the wall; and *floor* blocks, used as filler and plank systems for floors. (See Fig. 2-11.)

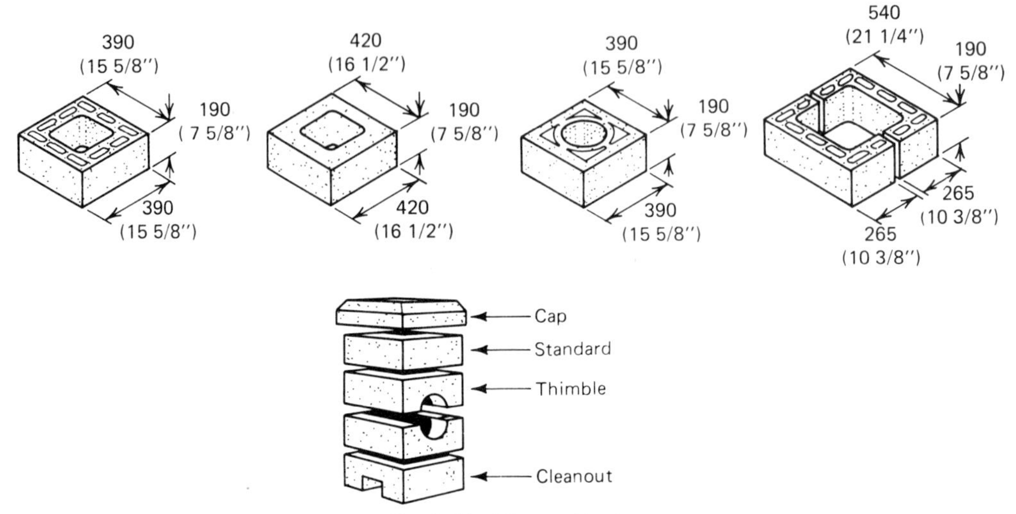

Fig. 2-12. Chimney blocks.

Utility Units

Some concrete masonry units are designed for purposes other than building construction. Among them are *chimney blocks* (see Fig. 2-12), *paving* blocks, *manhole* blocks and *septic tank* system blocks. (See Fig. 2-5.)

MASONRY UNIT SURFACE TEXTURES

Concrete masonry units are produced in a wide variety of textures, either for a specified physical requirement or for the sake of some particular architectural effect. Textures are generally classified as *open, tight, coarse, medium* and *fine.*

An open-textured surface contains numerous closely spaced and relatively large voids between the aggregate particles, while in a tight texture, the voids are well filled with cement paste. (See Fig. 2-13.)

The relative *smoothness* of the texture is described as fine, medium or coarse. A fine texture is made up of small, closely spaced granular particles, providing a smooth surface. A coarse texture will have a large proportion

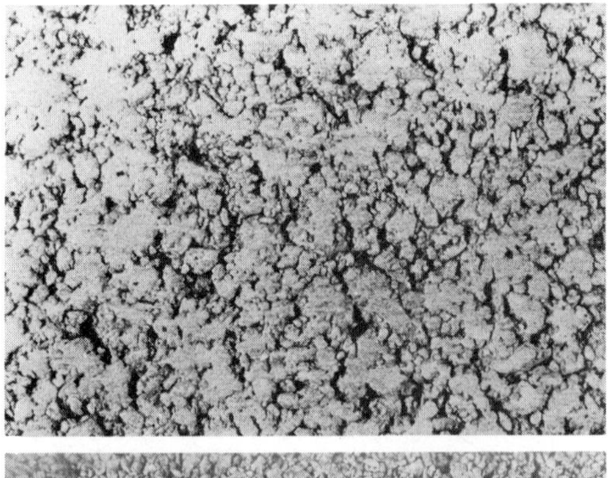

Fig. 2-13. (*Top*) open vs. (*bottom*) tight texture with normal weight aggregate. (*Courtesy Portland Cement Assn.*)

of larger-sized aggregate particles on the surface, with the result that the texture is large grained and rough. A medium texture will be intermediate between fine and coarse. (See Fig. 2-14.)

The texture of concrete masonry surfaces is important from the standpoint of finishing. A coarse texture is desirable as a base for stucco or plaster. Coarse and medium textures also contribute to sound absorption. On the other hand, a fine texture is preferred for easy painting.

All of the texture variations mentioned above may be found on standard block. Other texture variations may be produced for architectural reasons. For example, blocks may be produced with a *striated* face or they may be made double size so that they can subsequently be split lengthwise to produce a roughcast, or *split,* surface. Still another surface texture is created by the use of *slump* blocks–blocks that sag after forming and produce a rough surface. (See Fig. 2-15.)

Another type of surface texture may be produced on concrete masonry units by grinding one or both surfaces of either normal weight or lightweight blocks. From 1 to 3 mm (0.04 to 0.12 in) of concrete is ground from the surface, resulting in a smooth, open-textured surface, which shows aggregate particles of varying color. The use of aggregate of varying size and color and the use of color pigments in the mix will help to enhance the surface characteristics. Sandblasting may be used as an alternative to grinding, either before or after the blocks are laid.

Still another method of providing a unique texture to concrete masonry units is that of *pre-facing* the units. This is done by applying

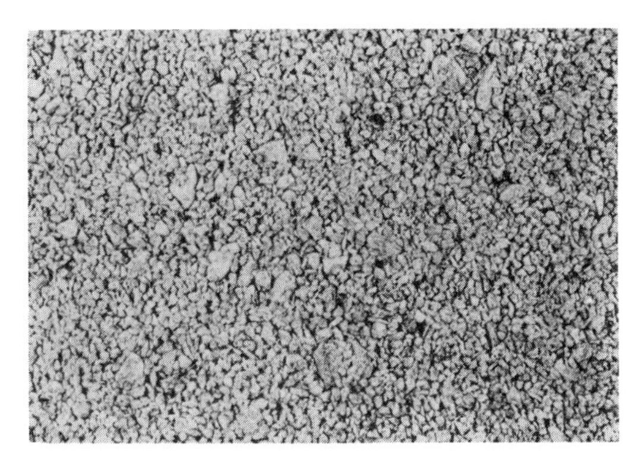

Fig. 2-14. (*Top*) fine vs. (*bottom*) coarse texture with lightweight aggregate. (*Courtesy Portland Cement Assn.*)

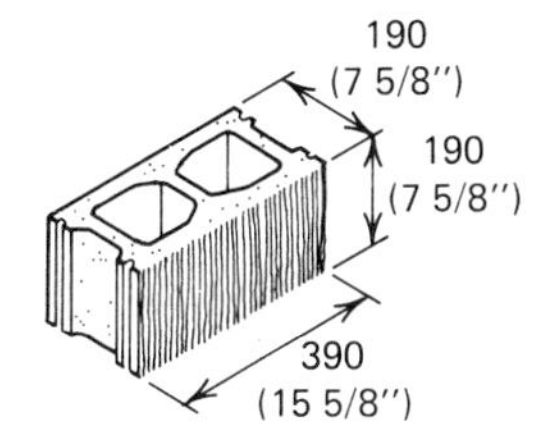

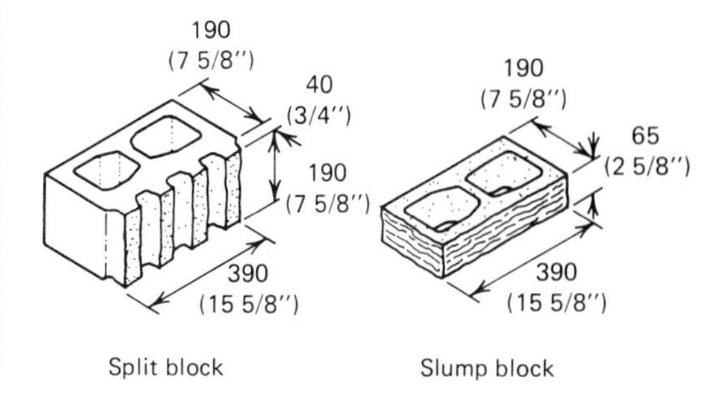

Fig. 2-15. Typical rough-cast surface texture on concrete masonry units.

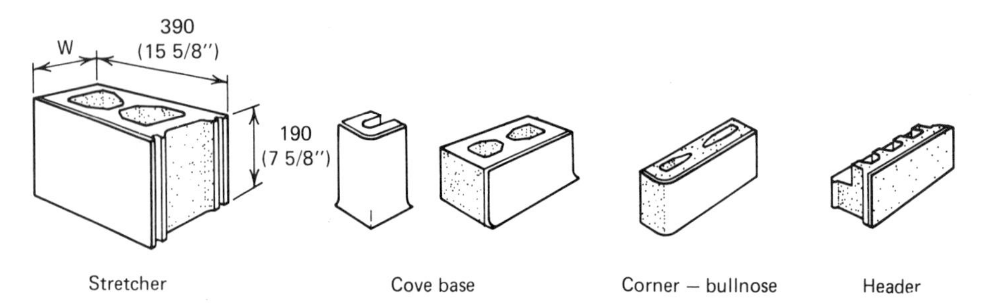

Fig. 2-16. Typical prefaced concrete masonry units.

some type of surfacing material to one or more faces of the blocks. The surfacing material may be a *synthetic resin,* a mixture of resin and fine sand, *portland cement and fine sand, ceramic glaze, porcelain glaze* or *mineral glaze.* It may be applied by *casting* and *heat treating* on a hardened surface or by *integrated casting* of block and face. (See Fig. 2-16.)

These facings are usually hard, dense and resistant to water penetration, abrasion and harsh cleaners and are in demand where a high degree of cleanliness, low maintenance and decoration are important.

SPECIALTY UNITS

In recent years a number of concrete masonry units have been designed for special purposes and, as the need arises, the number of specialty blocks will no doubt increase.

Among the more recent specialty units to be designed are *acoustical block, H-blocks* and *systems blocks.* Each is designed to provide improved performance in a special area, over what was possible using the more conventional types of block.

Acoustical Block

These patented blocks are designed with a slot in the face (see Fig. 2-17) to provide unusually high sound energy absorption. The slotted openings conduct sound waves into the cores, where the waves are effectively damped down. They are particularly effective with sound in the middle and high frequencies.

H-Blocks

In recent years there has been a trend toward more heavily reinforced concrete masonry unit walls. With conventional units, it was necessary to first erect the wall and subsequently insert the reinforcement down through the cores. H-blocks are designed with open ends (see Fig. 2-18) so that reinforcement may be set in place and the units laid around the bars. Then the wall is grouted.

Systems Block

This is another patented block that has been designed to improve the efficiency of block panelization. The unit has double tongue-and-

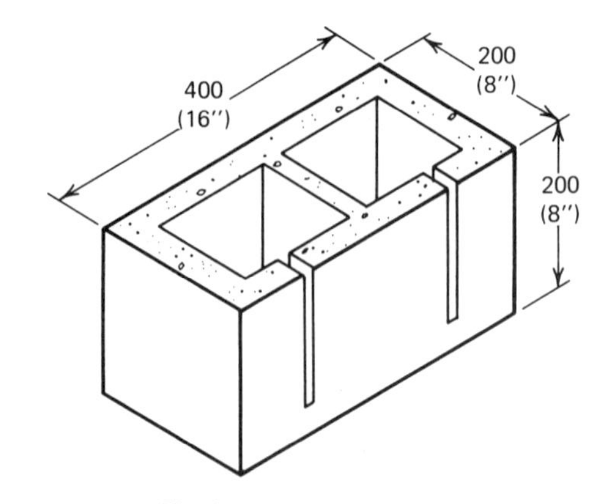

Fig. 2-17. Acoustical block.

Fig. 2-18. H-block and O-block.

Figure 2-19. "Systems" block.

grooved ends (see Fig. 2-19) to increase the ease of alignment and to allow all vertical joints to be laid dry, if desired. Either standard or thin-line, organic mortars may be used for the horizontal joints. A unique, staggered stack method of erection permits laying with no mortar in either vertical or horizontal joints. In such a system, the panel will be reinforced and grouted after fabrication.

PHYSICAL CHARACTERISTICS OF CONCRETE MASONRY UNITS

Concrete Masonry Unit Weights

Concrete masonry units are usually referred to as either *normal weight* or *lightweight*, depending on the kind of aggregate from which they are made. Normal weight units are made from aggregates such as *sand, gravel, crushed stone* and *air-cooled blast furnace slag.* Aggregates such as *expanded shale* or *clay, expanded blast furnace slag, sintered fly ash, coal cinders, scoria* and *pumice* produce lightweight blocks. Aggregates are strictly classified according to the basic nature of the material from which they are produced. (See note following Table 2-1.)

However, concrete masonry units are actually divided into three weight classes, depending on the density of the concrete that they contain. A unit is considered *lightweight* if it has a density of 1700 kg/m^3 (106 lb/ft^3) or less, *medium weight* if its density is between 1700 and 2000 kg/m^3 (106 and 125 lb/ft^3), *normal weight* if its density exceeds 2000 kg/m^3.

In addition to the density of the concrete used, the weight of a concrete masonry unit depends on its design; that is, whether it is solid or hollow and, if hollow, the number of cells that it contains, the thickness of the shells and webs and the design of the face.

The unit weight of concrete made with various kinds of aggregates is given in Table 2-2.

Using these unit weights of concrete, the weight of various concrete masonry units, in kilograms, may be determined by referring to the chart shown in Figure 2-20. For example, the weight of a 200 × 200 × 400 mm (8 × 8 × 16 in), 2-core unit, made with sand and gravel aggregate would be about 18 kg (40 lb) according to the chart, if the average unit weight of concrete made with that type of aggregate is used. On the other hand, a unit of the same size made with expanded shale would weigh approximately 10.75 kg (24 lb).

Table 2-2. Unit Weight of Concrete Made with Various Aggregates

Type of Aggregate	Unit Wt. of Concrete, kg/m^3	(lb/ft^3)
Sand and gravel	2082–2323	130–145
Crushed stone and sand	1922–2242	120–140
Air-cooled slag	1602–2002	100–125
Coal cinders	1282–1682	80–105
Expanded slag	1282–1682	80–105
Expanded clay, shale, slate, and sintered fly ash	1200–1442	75–90
Scoria	1200–1602	75–100
Pumice	960–1361	60–85

(Courtesy Portland Cement Association)

Concrete Masonry Unit Strength

Compressive strength is an important property of concrete masonry units and, in general, the use to which units will be put is directly related to that strength. Thus some units may be used for exterior work, while others will be confined to interiors only. Units are available in a number of strengths and so may be used economically over a wide range of applications.

The method of classification of concrete masonry varies somewhat between Canada and the U.S., but in general, the strength requirements for a specific usage are similar. In Canada, concrete masonry units are manufactured to conform to the requirements of **CSA Standard A165-M 1977** (for concrete block) and **Standard A165.2** (for concrete brick). In the

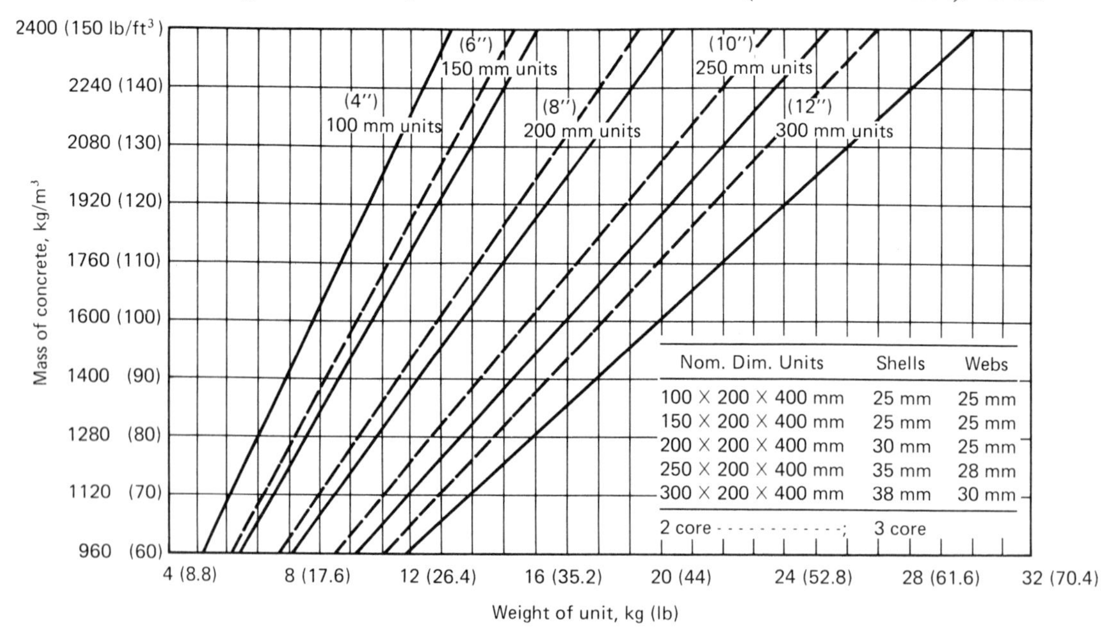

Fig. 2-20. Weights of some hollow units for various concrete densities. (*Courtesy Portland Cement Assn.*)

U.S., requirements are governed by **ASTM C90**, **C145**, and **C129** (for block) and by **ASTM C55** (for concrete brick).

CSA Standard A165.1 classifies all concrete masonry units, except concrete brick, according to their important physical properties, including compressive strength, using a four-facet system, as illustrated in Table 2-3. Under this system, all facets are used in order to designate the properties of a unit and thus to iden-

Table 2-3. Physical Properties of Concrete Masonry Units (Except Concrete Brick)

Facet	Symbol	Property		
		Solid Content		
First	H	Hollow		
	S	Solid		
		Minimum Compressive Strength Calculated on Gross Area MPa (psi)		
		Average of 5 Units	Individual unit	
	2.5	2.5 (363 PSI)	The compressive strength of any individual unit shall not be less than 85% of the specified average strength.	
	5	5.0 (725)		
Second	7.5	7.5 (1,088)		
	12.5	12.5 (1,813)		
	20	20.0 (2,900)		
	27.5	27.5 (3,989)		
		Oven Dry Mass Density of Concrete kg/m^3 (lb/ft^3)	Maximum Water Absorption kg/m^3 (lb/ft^3)	
	A	Over 2,000 (125)	175 (11)	
Third	B	1,700–2,000 (106–125)	225 (14)	
	C	Less than 1,700 (106)	300 (18.7)	
	N*	no limits	no limits	
		Maximum Moisture Content, Percent of Total Absorption-Average of 5 Units		
			Moisture Content	
		Linear Shrinkage	R.H. Over 75 Percent**	R.H. Under 75 Percent**
Fourth		Less than 0.03	45	40
	M	0.03 to 0.045	40	35
		over 0.045	35	30
	O	no limits (where drying shrinkage is not of importance)		

(Courtesy National Concrete Producers Association)

*Only applicable to classification H/2.5

**Average annual climatic relative humidity, in percent, at the point of manufacture.

Table 2-4. Strength and Absorption Requirements for Concrete Brick

Type	Min. Compressive Strength, MPa (psi) on Gross Area		Max. Water Absorption, kg/m³, based on Oven-Dry Unit Wt. of Concrete		
	Average of 5 Units	*Single Unit*	*More than 2000 kg/m³*	*2000 to 1700 kg/m³*	*Less than 1700 kg/m³*
I-3	20 (2900)	17.25	160	208	240
I-4	27.5 (4000)	24.0			
I-5	34.5 (5000)	31.0			
II	17.25 (2500)	13.75	208	240	288

(Courtesy Portland Cement Association)

Note: Type I brick is suitable for use in facing masonry exposed to the weather, while type II brick is intended for use as backup or interior facing masonry and is not suitable for exposure to the weather.

tify it. For example, H/12.5/B/M refers to a *hollow* unit with a *strength* of 12.5 MPa (1800 psi) (average of 5 units), a *density* of 1700-2000 kg/m³ (106-125 lb/ft³) and a *defined moisture content* at the time of shipment.

Under this system, there are six minimum compressive strength ratings for block, namely: 2.5, 5.0, 7.5, 12.5, 20.0 and 27.5 MPa. A compressive strength of 7.5 MPa is considered essential for a block to be designated for exterior use.

As indicated in Table 2-3, the strength of a masonry unit is measured over the gross area of the unit. As a result, when engineering analysis using net area strengths has been used in a design of masonry, the designer should clearly specify the solid content and state whether the strength is based on net or gross area.

Concrete brick are classified according to CSA Standard A165.2, and types and minimum strength requirements are shown in Table 2-4.

According to the requirements of ASTM C90, C145, C55, and C129, all concrete masonry units are classified by grade, according to use, as shown in Table 2-5.

Then, according to the type of unit and its grade, a minimum compressive strength requirement is assigned to each one, as shown in Table 2-6.

Units known as *high-strength* and *extra-high-*

Table 2-5. Grades of Concrete Masonry Units for Various Uses

Grade	ASTM C90 or C145 (Block)	ASTM C55 (Concrete Brick)
N	For general use such as in exterior walls below and above grade that may or may not be exposed to moisture penetration or the weather and for interior walls and backup.	For use as architectural veneer and facing units in exterior walls and for use where high strength and resistance to moisture penetration and to severe frost action are desired.
S	Limited to use above grade in exterior walls with weather protective coatings and in walls not exposed to the weather.	For general use where moderate strength and resistance to frost action and to moisture penetration are required.

(Courtesy Portland Cement Association)

Table 2-6. Strength and Absorption Requirements for Concrete Masonry Units

			Minimum Compressive Strength, MPa (PSI), on Average Gross Area		Maximum Water Absorption, kg/m^3 (lb/ft^3), Based on Oven-Dry Unit Weight			
					Lightweight Concrete			
Type of Unit	*ASTM Designation*	*Grade of Unit*	*Average of 3 Units*	*Single Unit*	*Less than 1360 kg/m^3*	*Less than 1700 kg/m^3 (106 lb/ft^3)*	*Medium Weight Concrete, 1700 to 2000 kg/m^3*	*Normal Weight Concrete, 2000 kg/m^3 or more*
Concrete brick	C55	N	24* (3500)	20.7*	240	240 (15)	208	160
		S	17.25* (2500)	13.8*	288	288 (18)	240	208
Solid load-bearing block	C145	N	12.4 (1800)	10.34	–	288 (18)	240	208
		S	8.27 (1200)	6.89	320	–	–	–
Hollow load-bearing block	C90	N	6.89 (1000)	5.5	–	288 (18)	240	208
		S	4.83 (700)	4.14	320	–	–	–
Hollow, non-load-bearing block	C129	–	2.4 (350)	2.07	–	–	–	–

(Courtesy Portland Cement Association)

*Concrete brick tested flatwise

Table 2-7. Comparative Strengths of Various Strength-Rated Blocks

Type of Block	Gross Area Strength, MPa (psi)	Net Area Strength, (53% Solid Units) MPa (psi)
Regular strength	7.5 (1088)	14.0 (2030)
High strength	12.5 (1813)	24.0 (3480)
Extra-high strength	18.5 (2683)	34.5 (5000)

(Courtesy Portland Cement Association)

strength block are available in some areas. These provide additional strength for certain applications; for example, when it may be necessary to limit wall thickness in buildings over ten stories high. Table 2-7 outlines strength ratings for such units.

Concrete Masonry Water Absorption

The amount of water absorption of concrete masonry units is related to the density of the units, as indicated in Table 2-6. The denser the unit, the less it tends to absorb water. These two properties, density and water absorption, affect construction, insulation, acoustics, porosity, painting, appearance and the quality of mortar required. Units with high absorption rates will need mortar with more water retentivity, to prevent the too rapid withdrawal of water from the mortar, resulting in a lower strength mortar and poor mortar bond. Absorption rate is not a specification requirement for concrete masonry units, as it is for brick, but control is exercised by limiting the maximum water absorption, as indicated in Tables 2-3 and 2-6. However, when concrete masonry units are used in a dry location, such as interior partitions, the maximum water absorption limits need not apply.

Concrete Masonry Moisture Content

The moisture content of concrete masonry units is expressed as a percentage of the total absorption of the unit, and limits are placed on the maximum moisture content allowed. These moisture content requirements are intended to indicate whether the unit is sufficiently dry for use in wall construction. Concrete shrinks slightly with loss of moisture down to an air-

Table 2-8. Moisture Content Requirements for Moisture-Controlled Concrete Brick

Linear Shrinkage, Percent	Moisture Content, Percent Based on Maximum Total Absorption (Average of Five Units)	
	RH over 75	*RH under 75*
Up to 0.03	40	40
0.03 to 0.045	40	35
Over 0.045	35	30

(Courtesy Portland Cement Association)

dry condition and if moist units are placed in a wall and this natural shrinkage is restrained, tensile and shear stresses are developed that may result in cracking in the wall.

Table 2-3 indicates the maximum allowable moisture content of all concrete masonry units except concrete brick by Canadian standards, based on an average climatic relative humidity of either over or under 75 percent. The moisture content requirements for concrete brick are outlined in Table 2-8.

In the U.S., three relative humidity ratings are used. Allowable moisture contents are indicated in Table 2-9.

Concrete Masonry Linear Shrinkage

Concrete masonry units will lose moisture to the surrounding atmosphere until they have reached an air-dry condition; that is, until they are in balance with the existing relative humidity conditions. When moisture is given up, some shrinkage occurs, and if this takes place after the units have been laid up into a structure in such a way that the natural shrinkage is restrained, stresses will be developed that may lead to cracking.

Linear shrinkage limits, in percent, have been established and are indicated in Tables 2-3 and 2-9.

As the concrete masonry industry has grown, a great deal of research has focused on the significance of the physical properties outlined above to the performance of concrete masonry units as building material. Standards have risen to ensure top performance of both exterior and interior block. As a result, concrete masonry has come to occupy an important role in the construction industry and is used in a wide variety of building types, including high-rise structures.

Table 2-9. Moisture Content Requirements for Type 1 Units (U.S.)

	Moisture Content, Percent, Based on Maximum Total Absorption (Average of 3 Units)		
Linear Shrinkage, %	*Humid*	*Intermediate*	*Arid*
0.03 or less	45	40	35
0.03 to 0.045	40	35	30
0.045 to 0.065 (max)	35	30	25

(Courtesy Portland Cement Association)

Glossary of Terms

Autoclave A vessel designed to contain steam under pressure.

Bond beam A solid horizontal member used to strengthen a wall made up of small units.

Ceramic glaze A material used in liquid form to coat earthenware objects.

Configuration Figure, contour, or pattern produced by a set distribution of the parts.

Linear shrinkage Reduction in linear dimensions.

Lintel A solid horizontal member spanning the top of an opening in a wall.

Mortar-bedding area Area available on which mortar may be placed.

Mortar bond The cohesion between mortar and the masonry unit.

Pilaster A solid vertical member used to stiffen or strengthen a wall.

Pozzolanic Referring to a rock of volcanic origin, containing a large proportion of silica.

Pumice A variety of light, volcanic rock.

Relative humidity The ratio of the quantity of water vapor actually present to the greatest amount possible at the given temperature.

Scoria A slag-like lava rock.

Septic system A system of sewage disposal by breakdown in a tank under bacterial action.

Siliceous Pertaining to or containing silica.

Stretcher A standard unit forming the main body of a wall.

Striated Marked with small channels or grooves.

Veneer A thin layer of material overlaid on another.

Review Questions

1. Explain why:
   (a) a vibrating mold is used in the manufacture of concrete masonry units.
   (b) concrete masonry units are usually cured by steam.
2. Explain the reason why (a) an air-entraining agent or (b) a pozzolanic material may be added to a mix used to make concrete masonry units.
3. Give the specific use for each of the following concrete masonry units: (a) stretcher, (b) header, (c) joist block, (d) corner sash block.
4. (a) What is the purpose of a bond beam in a block wall?
   (b) Give two reasons for using pilasters in a block wall.
5. Explain why blocks with open surface texture have better sound-absorbing qualities than smooth-faced blocks.
6. Explain:
   (a) What is meant by the *water retentivity* of mortar.
   (b) Why good water retentivity in mortar used for laying concrete masonry is important.
   (c) How the water retentivity of mortar may be improved.
7. List four situations for which you would specify pre-faced concrete masonry units for the walls.
8. With the aid of a diagram, describe the advantages of using H-blocks for constructing a heavily reinforced concrete block wall.
9. Give the approximate weight of each of the following concrete masonry units:
   (a) a 200 × 200 × 400 mm (8 × 8 × 16 in), 2-core block, made with crushed stone and sand.
   (b) a 200 × 200 × 400 mm, 3-core block, made from coal cinders.
   (c) a 150 × 200 × 400 mm (6 × 8 × 16 in) solid block, made from scoria.
10. Explain why limitations on moisture content and linear shrinkage are important in concrete masonry unit specifications.

Selected Sources of Information

Besser Company, Alpena, Michigan.
Canada Cement Company, Montreal, Quebec.
Edcon Block, Edmonton, Alberta.
National Concrete Masonry Association, Washington, D. C.
National Concrete Producers Association, Downsview, Ontario.
Portland Cement Association, Skokie, Illinois.

Chapter 3

CLAY TILE

Clay tile are made from the same materials and by the same manufacturing process as extruded brick. They may be divided into two general categories: *hollow tile,* used for the same structural purposes as brick or concrete masonry and commonly known as *structural clay tile;* and *solid tile,* which includes such products as *floor tile, ceramic veneer, architectural terra cotta* and *ceramic mosaic.*

STRUCTURAL CLAY TILE

Structural clay tile are used in the construction of load-bearing walls, backup walls and partitions and are called *wall tile* (see Fig. 3-1). Structural clay tile are also used as facing material and called *facing tile.* (See Fig. 3-3)

Wall Tile

Structural clay wall tile are made in two grades–LBX and LB. Grade LBX is suitable for general use in masonry construction and is especially adaptable for use where the masonry is exposed to the weather. Grade LB is considered suitable for use where the masonry is not exposed to frost action or for exposed masonry if it is protected by at least 75 mm (3 in) of stone, brick, facing tile or terra cotta. Such tile may be used in either *side* or *end* construction. (See Fig. 3-2.) The physical requirements of these two grades are given in Table 3-1.

The only specified physical requirement for non-load-bearing tile, such as might be used for partitions, deals with the maximum water absorption allowance, outlined in ASTM Designation C56.

Facing Tile

Facing tile is produced from high-grade, light-burning clay that is suitable for the application of ceramic or salt glaze. (See Fig. 3-3.) Two grades are produced: *Grade S (select),* intended for use with conventional 10 mm ($^{3}/_{8}$ in) mortar joints and *Grade G (ground edge),* used where the variation in face dimension must be very small.

Two types of glazed tile are made, single-faced units (Type I) and units with two opposite faces glazed (Type II). Type I is used to face a backup wall, while Type II is used where the tile is exposed on two opposite faces.

ASTM Designation C126, which covers this type of tile, specifies requirements for compressive strength, absorption rate, number of cells, shell and web thicknesses, tolerances on dimensions and the properties of the ceramic finish.

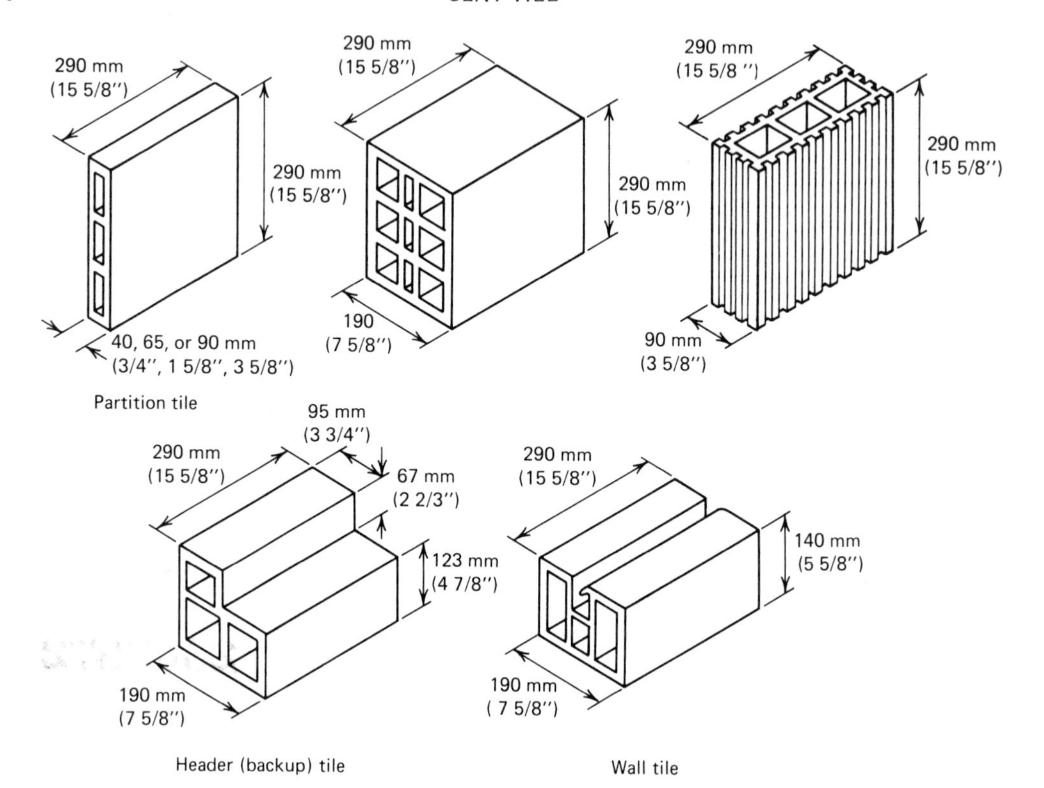

Fig. 3-1. Structural clay tile.

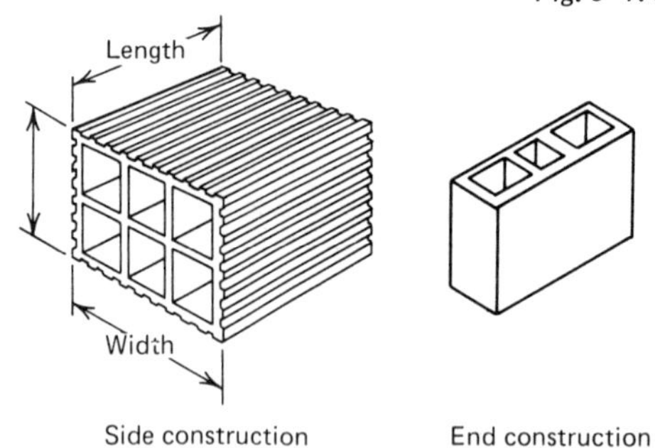

Fig. 3-2. Side vs. end construction with clay tile.

Shapes and sizes are similar to those of unglazed tile, except that ceramic-glazed tile units are available in 50 and 100 mm (2 and 4 in) thicknesses and in 400 mm (16 in) lengths only.

Structural Clay Tile Sizes

The standard length for structural clay wall tile (see Fig. 3-2) is 300 mm (12 in) nominal—290 mm (11⅝ in) actual size—to conform to the metric building module. The standard

Table 3-1. Physical Requirements for Load-Bearing Wall Tile

Grade	Min. Compressive Strength, Avg. of 5 Tests, Based on Gross Area, MPa (psi)		Max. Water Absorption, Avg. of 5 Tests, Percent
	End Const.	*Side Const.*	
LBX	9.65 (1400)	4.85 (700)	16
LB	6.90 (1000)	4.85 (700)	25

Fig. 3-3. Facing tile. (*Courtesy Whiteman Mfg. Co.*)

height is also 300 mm nominal but backup tile may sometimes be produced in heights of 135, 200, and 250 mm ($5\frac{1}{3}$, 8, and 10 in) nominal, as well. Tile thicknesses vary from 50 to 300 mm (2 to 12 in) nominal with usual actual thicknesses being 40, 65, 90, 140, 190, 240 and 290 mm ($1\frac{5}{8}$, $2\frac{5}{8}$, $3\frac{5}{8}$, $5\frac{5}{8}$, $7\frac{5}{8}$, $9\frac{5}{8}$, and $11\frac{5}{8}$ in).

Structural Clay Tile Surfaces

Structural clay tile are produced with three types of surface—*smooth, textured* and *corrugated.* (See Fig. 3-4.) The smooth surface may be glazed or unglazed, while the corrugated tile face is intended as a base for plaster.

Fig. 3-4. Clay tile surfaces: (a) corrugated; (b) smooth; (c) textured.

FLOOR TILE

Clay floor tile are solid units, approximately 25 mm (1 in) in thickness, intended for use as a finish flooring. They are produced in several face dimensions, as illustrated in Figure 3-5. Physical characteristics of clay floor tile are indicated in Table 3-2.

Clay floor tile may be installed in a 25 mm (1 in) mortar bed over a concrete slab, dry set in a well-compacted sand bed, 75 to 100 mm (3 to 4 in) in depth, or over a wood floor in a 20 mm ($\frac{3}{4}$ in) mortar bed, with wire mesh reinforcing suspended on nail heads (See Fig. 3-6.)

The joints are filled by applying grout over

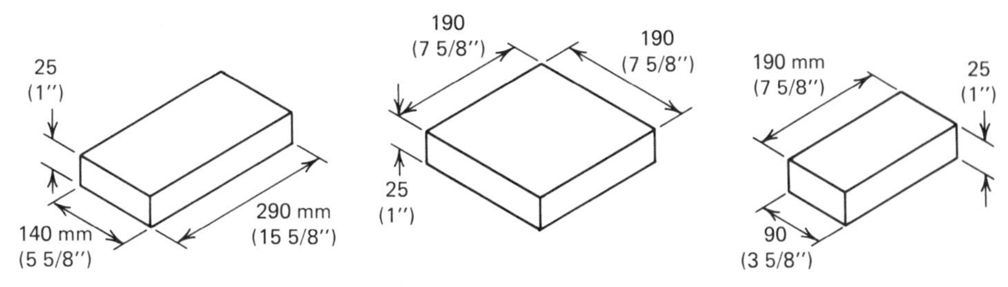

Fig. 3-5. Clay floor tile dimensions.

Table 3-2. Physical Characteristics of Clay Floor Tile

Mass per m^2 (ft^2)	Compressive Strength, MPa (PSI)	Flexural Strength, MPa (PSI)	Max. Water Absorption, 24 Hrs, Cold Water	Loss in mass, 48 Cycles, Freeze-Thaw
44–58.5 kg. (95–125 lb)	46 (6,670)	8.8 (1,280)	8.33%	nil

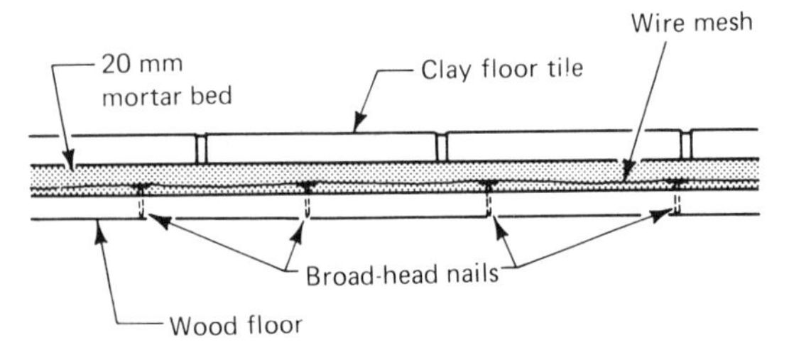

Fig. 3-6. Clay floor tile over wood base floor.

the surface of the tile and rubbing it into the joints with burlap or spreader, as illustrated in Figure 3-7.

Fig. 3-7. Grouting floor tile joints.

CERAMIC VENEER

Ceramic veneer is a solid, flat, machine-extruded product, intended for use as an interior or exterior wall-facing material. One face is highly glazed and produced in a variety of colors.

Application may be by an appropriate organic cement over a solid backing or by adhesion to a mortar coating over various types of backing. (See Fig. 3-8.)

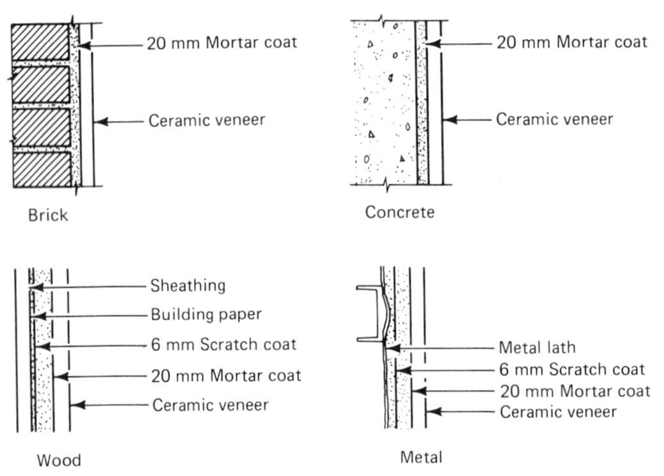

Fig. 3-8. Ceramic veneer over various backings.

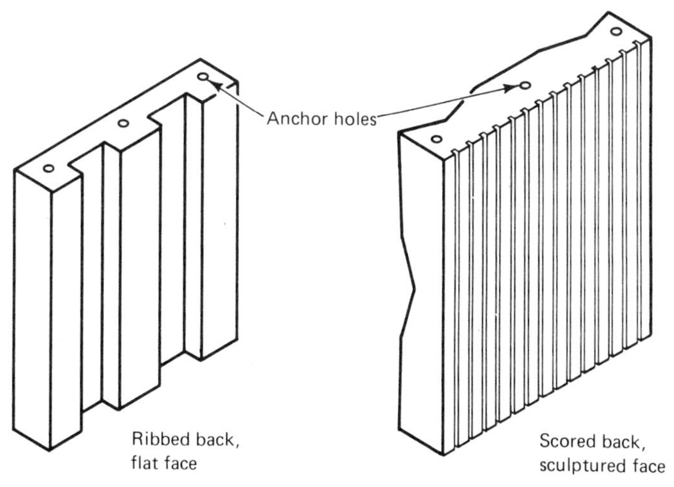

Fig. 3-9. Typical architectural terra cotta units.

Material thickness usually ranges from 6 to 20 mm (¼ to ¾ in), and face dimensions are normally restricted to a maximum width of 600 mm (24 in) and maximum length of 900 mm (36 in).

ARCHITECTURAL TERRA COTTA

Architectural terra cotta is another solid clay product intended for use as an exterior facing material, but unlike the mass-produced ceramic veneer, it is normally made-to-order to fit a specific building. As a result, before such units can be made, detailed drawings must be prepared, showing size, shape, profile, joint sizes, location and type of anchors and hangers and location of expansion joints.

Architectural terra cotta are molded or pressed so that a variety of surfaces may be produced, including *smooth, textured* or *sculptured.* The slab thickness will range from 50 to 60 mm (2 to 2⅜ in), while the face dimensions, within reason, are a matter of choice. The backs of the units are *scored* or *ribbed* (see Fig. 3-9) and anchor holes are provided in the bed edges.

The ends of wire anchors are set in the holes, with the opposite ends being secured to a back-up wall. The space between the back of the units and the backup wall is filled with grout.

CERAMIC MOSAIC

Ceramic mosaic consists of small pieces of glazed clay tile, arranged in patterns (see Fig. 3-10) in a standard sheet unit. In some cases a paper or open mesh fabric backing is used as a

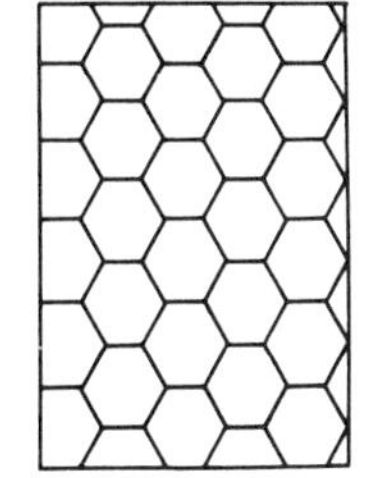
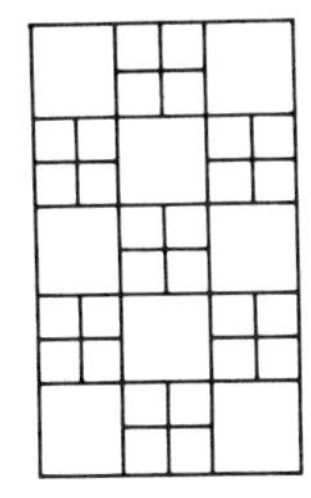
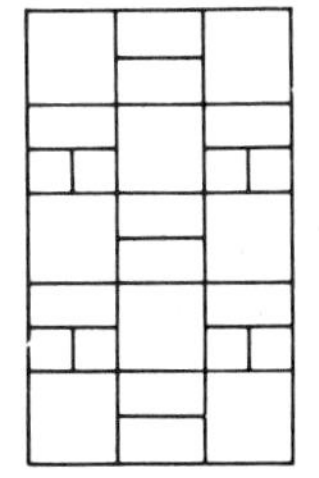
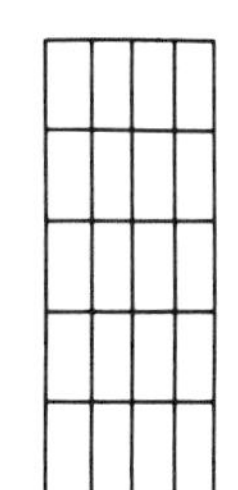
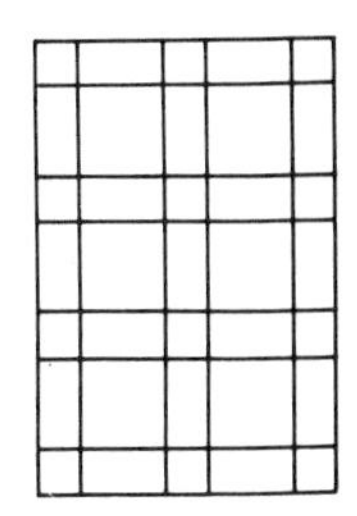

Fig. 3-10. Ceramic mosaic patterns.

base to which to cement the pieces to hold them together, while in others they are bonded to a preformed rubber grid. The latter sheets are normally 225 X 225 mm (9 X 9 in), while the paper or fabric-backed ones are approximately 300 X 600 mm (12 X 24 in). All have a nominal thickness of 6 mm (¼ in).

These small tile are made in a variety of shapes, including *square, rectangular, hexagonal,* and *octagonal,* with a hard, glasslike finish, in a wide variety of colors.

Ceramic mosaic may be used as finish flooring or as wall facing. They may be set in mortar over a concrete base or over a wood or asphaltic surface, using organic adhesives or inorganic bonding coats to hold them in place.

After the tile have been laid and the mortar or adhesive set, grout is poured or plastered over the surface and rubbed into the joints until they are full. Finally, all adhering grout is cleaned from the surfaces of the tiles.

Glossary of Terms

Backup tile Tile used to construct a wall that is to be faced with another material.

Ceramic Pertaining to pottery or earthenware.

Corrugated tile Tile with heavy ribs on the surface.

Mosaic A surface decoration made by using small pieces of glass, stone, or other such material.

Organic cement High strength cement, made from organic materials.

Side construction A method of laying structural tile with the cells horizontal.

Terra cotta Clay products that have their surfaces coated with a glaze; used in the facing of buildings.

Review Questions

1. Outline the basic differences between tile and brick.
2. What are the two basic functions of backup tile?
3. Explain the difference between ceramic veneer and architectural terra cotta.
4. Explain why architectural terra cotta is a relatively expensive type of facing material.
5. What advantages do you see in using structural clay tile as a building material over brick or concrete block?

Selected Sources of Information

Brick and Tile Mfg. Assn. of Canada, Toronto, Ontario.
Division of Building Research, N.R.C., Ottawa, Ontario.
Federal Seaboard Terra Cotta Corp., New York, N.Y.
Structural Clay Products Institute, Washington, D.C.

Chapter 4

STONE

Stone must be included among the oldest building materials known to man. Throughout recorded history, stone has been regarded as the preferred material in the construction of permanent buildings. This is doubtless due to its unique qualities—*beauty, permanence, workability* and *accessibility,* among others. It was, in fact, the predominant material used in the construction industry prior to the turn of the twentieth century.

Since that time, stone has assumed a new role in the construction field. Because the height of buildings continued to increase, it became necessary to look more closely at the *mass* of the materials that went into the basic structures. Stone began to be developed as a facing material, rather than as a basic structural material.

In this new role it is used as a *veneer,* in comparatively thin slabs, over a building frame of reinforced concrete or steel. In this way, the inherent qualities which initially brought stone into prominence—*beauty, permanence, adaptability* and *economy*—are used without having to contend with the great mass imposed by a solid stone structure.

ORIGIN

An understanding of the geological origin of building stones, their structure and composition may be an aid in choosing stone that will fit properly into the regional setting and with the particular architectural design proposed for a specific building. Such an understanding may also be helpful in the selection or rejection of a particular stone proposed for a job and in the correct placing of each selected unit during the stone-laying process.

Rock may be divided into three broad categories depending on its geological origin; namely, *sedimentary, igneous* and *metamorphic.*

Sedimentary rock was formed by the action of water. This action may have resulted either in the depositing of minerals at the bottom of a body of water through sedimentation or in depositing them on the earth's surface. The latter action takes place when water flows to the surface from the interior of the earth, bringing dissolved minerals that are deposited on the surface by evaporation.

Sedimentary rock, like the other groups, has an extensive range of composition and character, but for our purposes it is possible to divide it into four groups. These are: *rudaceous rock*, which includes breccia and conglomerate; *arenaceous rock*, which includes sandstone, arkose and quartzite; *argillaceous rock*, largely shale; and *calcareous rock*, which includes a number of limestones.

Igneous rock was formed by the cooling and consolidation of molten matter that was brought to the earth's surface by volcanic action. It may be divided into three main rock types: *volcanic*, which includes pumice, obsidian, rhyolite, andesite and basalt; *hypabyssal*, which includes felsite, quartz, porphyry and dolerite; and *plutonic*, encompassing granite, diorite and gabbro.

Metamorphic rocks are distinctive new rock types that have been changed from their original igneous or sedimentary structure by the action of extreme pressure, heat, moisture, chemical fluids or various combinations of these forces. There are a great many of these metamorphic rocks, but only two groups have much significance as building stone, namely the *slates* and the *marbles*. A slate is a rock derived from argillaceous sediments or volcanic ash by metamorphism. It has very distinct cleavage lines along planes that are quite independent of the original bedding. A marble is a granular limestone, recrystallized by heat or pressure.

COMPOSITION

Rock may also be classified according to its composition. A great many minerals occur in rock formations throughout the world, but stone used for construction purposes generally comes from rock that falls into one of three classifications: rock that contains mostly *silica*, rock composed largely of *silicates* and rock containing *calcareous minerals*.

Silica, which is the oxide of the element *silicon*, is the most abundant mineral on the earth's surface. It is the chief ingredient of sand and is found in most clays and in a number of the building stones. Free silica, of which crystalline quartz is the most common form, either alone or in combination with other elements, comprises about 60 percent of the earth's crust.

Silicates are minerals that are compounds of silica and one or more other elements and are the main ingredients of igneous rock. The quantity of silica in igneous rock varies from about 30 to 80 percent and is sometimes used as a method of classification of such rock. For example, igneous rock containing more than 66 percent silica is known as **acid** igneous rock, while **basic** igneous rock contains between 30 and 52 percent silica, with no free quartz. **Intermediate** igneous rock has a silica component of between 52 and 66 percent.

Silicate minerals include *feldspar, hornblende, mica,* and *serpentine*. Feldspar is a silicate of alumina in combination with lime or potash. Depending on the combination, colors may be red, pink or clear. Hornblende is a silicate of alumina with lime or iron. It is a tough, strong mineral appearing in green, brown and black crystals. Mica is mostly silicate of alumina but may occur in combination with other minerals such as iron or potash. It appears in soft, usually clear crystals that split easily into flat flakes. Serpentine, which is a silicate of magnesia, often appears in combination with lime. It is light green or yellow in color and has no readily defined planes along which it will split.

Included among the most common calcareous minerals are *calcite*, which is basically carbonate of lime and *dolomite*, a carbonate of lime in combination with varying amounts of magnesia.

PHYSICAL REQUIREMENTS FOR BUILDING STONE

Despite the fact that rock is such an abundant material, relatively few types of stone satisfy the requirements of the building industry. The important physical characteristics required are: *strength, hardness, workability, durability, limited porosity,* proper *color* and *grain* and proper *texture.* In addition, *ease of quarrying* and *accessibility* are important considerations.

Strength

Many stones satisfy the requirements for strength, though most stones will show quite a wide variation in this regard. For example, the compressive strength of granite may range from 124 to 276 MPa (18,000 to 40,000 psi). For most building purposes, a compressive strength of 34.5 MPa (5000 psi) is satisfactory. In a few instances, good shear strength is important and here the range of values may be even greater.

Hardness

Hardness of stone is vitally important where it is to be used for stairs, floors, walks, and so forth, but it also has a bearing on workability. Hardness varies all the way from soft sandstone, which can be easily scratched, to some stones harder than steel. Hardness tests may be conducted according to ASTM D1706.

Workability

Workability is the ease with which the stone may be cut or shaped and is important since the ease of producing the required sizes and shapes has a direct bearing on the cost. Workability varies with the type of stone and its composition.

Durability

Durability is the ability of stone to withstand the effects of rain, spray, wind, dust, frost action, heat, fire and air-borne chemicals. The durability of stone determines, in large part, the maintenance-free life of a stone structure. This will vary from about ten to two hundred years.

Porosity

The porosity of a stone depends on the amount of spaces contained between the particles of solid material. The smaller the ratio of spaces to solid matter, the less porous the stone. The degree of porosity will determine the amount of moisture that may be absorbed into the surface. Porosity has a direct bearing on the ability of the stone to withstand frost action and also on the marking and staining of the surface, caused by the dissolving of some mineral constituents of the stone in water.

Color and Grain

Color is important from the standpoint of aesthetic design and location, but it is also partially a matter of taste and fashion. The grain or surface appearance of stone has an effect on its desirability for decorative purposes.

Texture

Texture of stone is related to the fineness of the grain, which in turn affects workability and therefore cost. For ornamental purposes also, the texture of the surface is important. In general, fine-textured stone splits and dresses more readily than coarse-textured stone.

Ease of Quarrying

Quarrying is the removal of stone from its natural bed and the relative ease with which this may be done is a prime consideration in judging the suitability of the stone from a standpoint of cost. The bedding and joint planes must be such that the stone can be produced in sound blocks of reasonable size. The rock surface should be free from closely spaced joints, cracks and other lines of weakness. Deep and irregular weathering is also undesirable.

Table 4-1. Physical Characteristics of Commercial Building Stones

Stone	Density, kg/m^3 (lb/ft^3)	Compressive Strength, Avg., MPa (psi)*	Tensile Strength, Avg., MPa	Relative Hardness	Durability to Weathering**	Cleavage or Fracture	Porosity, Percent
Argillite	2723 (170)	131 (19,000)	Tensile strength of stone about 1/7 to 1/11 (avg. .117) of the compressive strength.	Soft	Poor	Excellent cleavage	0.29
Granite	2643 (165)	181 (26,250)		Very hard	Excellent	Fracture	2.0
Limestone	2643 (165)	90 (13,050)		Med. to soft	Good	Fracture	4.10
Obsidian	2482 (155)	595 (86,300)		Very hard	Excellent	Fracture	4.0
Quartzite	2627 (163)	252 (36,550)		Hard	Excellent	Fracture	0.60
Travertine	2627 (163)	103 (14,950)		Soft	Poor	Fracture	1.9
Marble	2643 (165)	138 (20,000)		Med. to soft	Fair to poor	Fracture	0.25
Schist	2820 (175)	69 (10,000)		Medium	Poor	Fair cleavage	0.75
Serpentine	2723 (170)	93 (13,500)		Soft	Excellent	Fracture	0.76
Sandstone	2323 (145)	131 (19,000)		Medium	Poor to excellent	Fracture	5.26
Slate	2720 (170)	132 (19,150)		Med. to soft	Fair	Excellent cleavage	0.30

(Courtesy B. Ellison, Petroleum Dept., Southern Alberta Inst. of Technology)

*Compressive strengths may vary by a factor of five from max. to min.

**Resistance to weathering subject to type of climate and air quality conditions—extremes, moisture, industrial gases.

(a) (b)

Fig. 4-1. Stone rubble: (a) rough fieldstone; (b) rough-cut rubble.

Accessibility

The nearness of the stone deposit to the surface is also important from the standpoint of cost. Building stone is seldom obtained from underground; normally it is obtained by stripping whatever *overburden* is present and removing the stone from an open pit or quarry. Accessibility to centers of population also affects cost. Transportation over long distances is particularly expensive because of the masses involved, but in some cases it is a necessity.

BUILDING STONE

Classification

Stone for building purposes may be classified according to the *form* in which it is available commercially. These are: *rubble* (fieldstone), *dimension* (cut stone), *flagstone* (flat slabs) and *crushed rock.*

Rubble includes rough fieldstone that may simply have been broken into pieces of suitable size, or it may include irregular pieces of stone that have been roughly cut to size. (See Fig. 4-1.)

Dimension stone makes up the largest portion of the stone used in building construction. It consists of pieces that have been cut to some specific dimensions and, in general, falls into two categories. The first consists of units of relatively small size, cut to various dimensions, and generally referred to as *ashlar.* This type of cut stone is used as a facing over a structural backup wall. (See Fig. 4-2.) The second category includes units with relatively large face areas but of limited thicknesses, finished to quite specific dimensions and used as *curtain wall panels,* interior and exterior *facings* and various kinds of *trim.* (See Fig. 4-3.)

(a)

(b)

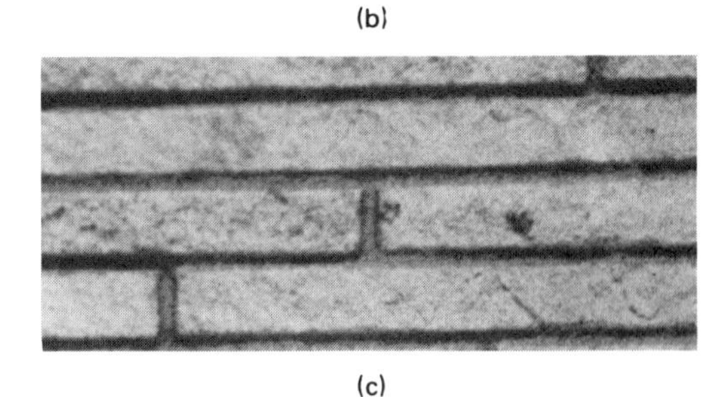

(c)

Fig. 4-2. Stone ashlar: (a) random ashlar; (b) irregular coursed ashlar; (c) regular coursed ashlar.

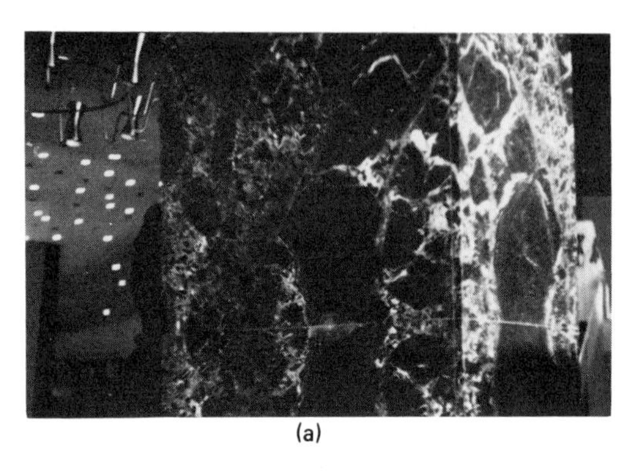

(a) (b)

Fig. 4-3. Stone facings and trim: (a) marble column facings; (b) argillite window sill.

Flagstone consists of thin sheets of stone—12 mm (½ in) and up—that may or may not have their face dimensions cut to some particular size. Such stone is used for flooring, patios, walks, and so forth. (See Fig. 4-4.)

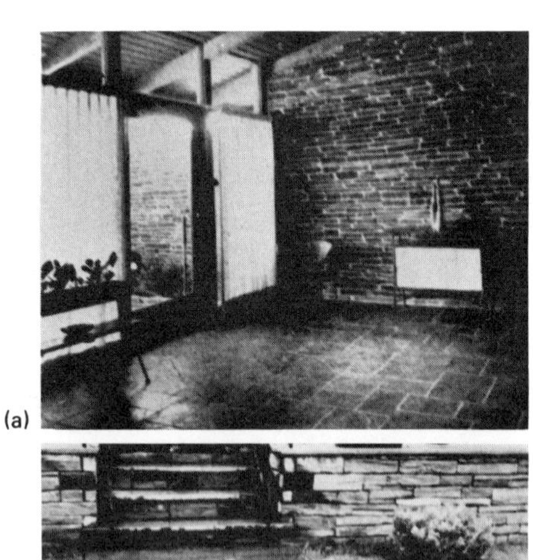

(a)

(b)

Fig. 4-4. Flagstone: (a) flagstone flooring; (b) flagstone walkway.

Crushed stone consists of pieces varying from 10 to 150 mm (⅜ to 6 in) in diameter. It is used as coarse aggregate in the concrete industry and as a facing material for some architectural precast concrete curtain wall units. (See Fig. 4-5.)

Fig. 4-5. Crushed stone used as facing material for precast concrete panel. (*Courtesy Precast Concrete Institute*)

Types of Building Stone

The stones which are most commonly used in modern building construction include: *argillite, granite, limestone, natural lavas, quartzite, travertine, marble, schist, serpentine, sandstone* and *slate.*

Argillite. Argillite is a metamorphic rock, formed from clay, that normally has well-defined cleavage lines. (See Fig. 4–10). It is produced as ashlar and rubble-facing stone, as floor tile and stair treads, wall base, coping stones, window stool and sills. (See Fig. 4-3.) Colors range from deep red to purple; a deep blue color with faint shades of green is common along with seam face colors of gray, buff, tan and russet.

Granite. Granite is of igneous origin, composed of quartz, feldspar, mica and hornblende. It is generally a hard, strong, durable stone, capable of taking a high polish, so that finishes may vary from rough-sawn or natural surface (see Fig. 4-2a) to a mirror-smooth polished surface. Granite is produced as ashlar and rubble, wall panels, flooring tile, stair treads, flagstones, column facings, copings and sills and small regular or natural shapes used as facing for precast concrete facing panels. (See Fig. 4-6.)

Limestone. Limestone, a sedimentary rock, is one of the most popular building stones and has a wide variety of uses. There are three distinct types of limestone. *Oolitic* is a calcite-cemented calcareous rock, formed of shells and shell fragments from small aquatic creatures, and is noncrystalline in nature. (See Fig. 4-7.) It has no cleavage lines and is usually very uniform in structure and composition. *Dolomitic* is rich in magnesium carbonate and frequently somewhat crystalline in character. It has a greater variety of textures and is

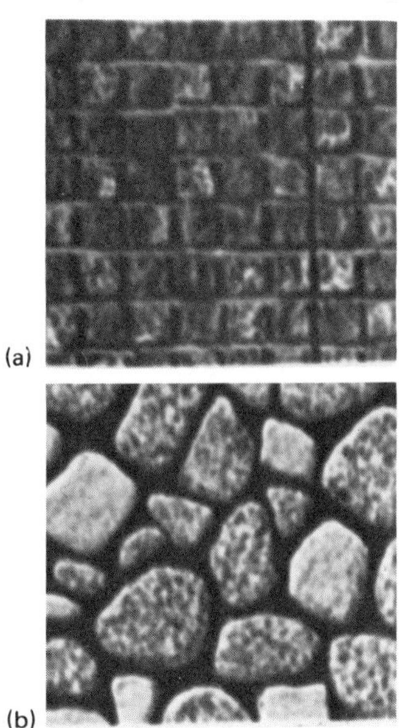

Fig. 4-6. Small granite pieces used for precast panel facing: (a) regular shapes; (b) natural shapes.

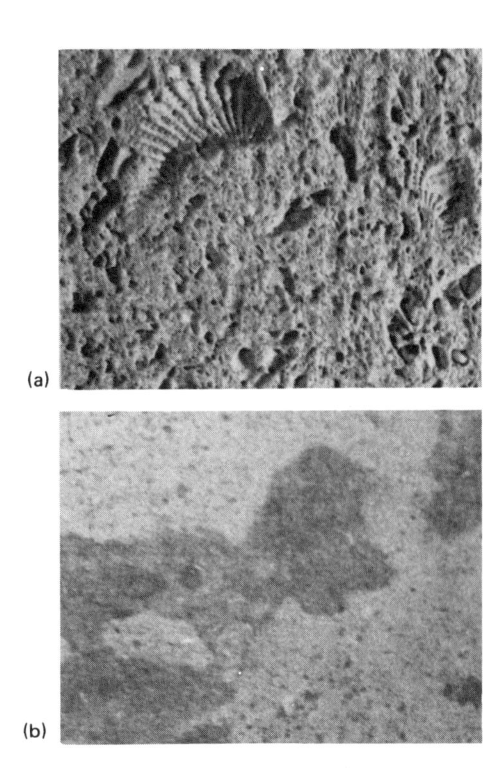

Fig. 4-7. Oolitic limestones: (a) goldenshell limestone; (b) Tyndall-stone.

normally stronger than oolitic limestone in both compression and tension. *Crystalline* is composed largely of calcium carbonate crystals, with high compressive and tensile strength and smooth texture. The color is usually light gray.

Some of the better known limestones are: Carthage and Miami Valley limestone, which are crystalline limestones; Kasota, Mankota, Winona and Niagara stone, which are dolomitic limestones; and Indiana limestone, Tyndallstone, Cordova limestone, shadowvein, cottonwood and goldenshell limestones, which are oolitic limestones. Some of the oolitic limestones are unique in appearance, with the fossil shapes being clearly visible in the face of the cut stone. (See Fig. 4–7.)

Limestone uses include panelling, ashlar, flagstone, window sills and stools, copings, mantles and facings of all kinds.

Natural Lavas. The lavas that flowed from ancient volcanoes have left rock deposits that, in some cases, have proven to be useful as building stone. A unique feature of many of them is their relative lightness; some may weigh only one-fifth as much as granite.

Some of the lightweight stones used in building construction include *obsidian, corkstone* and some *scorias.* Their composition is largely silica, with colors of charcoal, gray and tan.

Production is mainly in the form of rubble veneer, in thicknesses of from 25 to 100 mm (1 to 4 in), in random face sizes.

Quartzite. Quartzite is made up of grains of quartz sand cemented together with silica and is usually distinguishable by its coarse, crystalline appearance. Because of this feature, it is often used where a rustic appearance is desired. Colors include red, brown, gray, tan and ivory, with several colors being found in one piece.

Production is usually in the form of ashlar, commonly in 50 to 100 mm (2 to 4 in) thicknesses.

Travertine. Travertine is a sedimentary rock, made up largely of calcium carbonate. It has been formed at the earth's surface, in most cases, through the evaporation of water from hot springs.

Most of the production is in the form of small slabs that are used as an interior decorative stone because of their pleasing and unique texture.

Marble. Marble is another metamorphic rock, having been formed by the recrystallization of limestone and dolomite. There are a number of well-known types of marble, including *carrara, onyx, numidian, Vermont, parian* and *brecciated* marbles.

Marble has a very wide range of colors due to the presence of various oxides of *iron,* as well as *silica, mica, graphite* and *carbonaceous matter* that are scattered all through the rock in streaks, blotches or grains. Brecciated marbles are made up of small fragments embedded in a cementing material.

Some varieties of marble deteriorate readily when exposed to the weather and are suitable only for interior use. Others are quite stable and are suitable for exterior, as well as interior use.

Much of the marble produced for the building industry is in the form of slabs from 12 to 50 mm (½ to 2 in) in thickness, for use as exterior and interior wall panelling, column facings, flooring, counter tops and sanitary installations.

A variety of finishes are possible, depending on the particular marble being used. They include *polished, honed, sand blasted, bush-hammered, sawn* and *machine-tooled* surfaces. (See Fig. 4–12.)

Schist. Schist is also a metamorphic rock, of *foliated* structure, that splits easily into slabs or sheets. Its composition varies depending on the basic material from which it was formed but, in most cases, the basic component will be about 85 percent silica, with iron and magnesium oxides making up the remainder. Colors include red, gold, green, blue, brown, white and gray.

Rubble veneer and flagstone make up the greatest percentage of schist production; the veneer ranges in thickness from 50 to 155 mm (2 to 6⅛ in) and the flagstone from 20 to 40 mm (¾ to 1⅝ in).

Common uses include exterior and interior wall facings, fireplaces, sidewalks, patios and landscaping.

Serpentine. Serpentine is an igneous rock, so-called because it is largely made up of the mineral *serpentine*—a magnesium silicate. It is dense and homogenous in structure, with a fine grain and no cleavage lines. Colors normally range from olive green to greenish black, but the presence of impurities in the rock may give it other colors.

Some types of serpentine are subject to deterioration due to weathering and are useful for interior work only. Black serpentine is highly resistant to chemical attack and is useful in situations where it is likely to come in contact with moisture-borne chemicals.

Most serpentine is produced in relatively thin slabs of from 20 to 30 mm (¾ to 1³⁄₁₆ in) in thickness, that are used as panelling, window sills and stools, stair treads and landings.

Sandstone. Sandstone is a sedimentary rock, composed of grains of quartz cemented together with silica, iron oxide or clay; the hardness and durability of the particular sandstone depends on the type of cement. Other materials such as lime, feldspar or mica may be present in some sandstones as well, resulting in considerable variation in texture and color.

Textures range from very fine to very coarse, and some sandstones are quite porous, with as much as 30 percent of their volume composed of pores. Bccause of this characteristic, these sandstones lend themselves to textured finishes. (See Figs. 4-8 and 4-13.) Colors include red, russet, purple, buff, brown, copper, gray, beige and pink.

Fig. 4-8. Sandstone textured surface.

Some well-known sandstones are *bereastone, linroc stone, ledgestone, Dunbar stone, Kaibab stone* and *pearl sandstone.* From these are produced rubble, ashlar, dimension stone (see Fig. 4-8) and panels in a wide range of sizes, which vary from one producer to another.

Slate. Slate is a metamorphic rock, formed by the metamorphosis of clays and shales that have been deposited in layers. A unique characteristic of the stone is the relative ease with which it may be separated into thin, tough sheets, known as slates, 6 mm (¼ in) or more in thickness. Colors include red, black, purple, gray and green. (In some cases the color may change after long exposure.)

Slate is commonly used for flooring, flagstones (see Fig. 4-4), window sills and stools, counter tops, interior and exterior wall facing and roofing. (See Fig. 4-9.)

Fig. 4-9. Slate roofing.

(a)

(b)

(c)

Fig. 4-10. Stone quarries: (a) granite quarry, (b) limestone quarry; (c) argillite quarry.

PRODUCTION OF STONE

Quarrying

The removal of stone from its natural bed is carried out by a process known as *quarrying.* The method of quarrying will depend to a considerable extent on the kind of stone being removed and the nature of the particular stone involved. (See Fig. 4-10.)

Some stone deposits have natural horizontal divisions (i.e., they are *stratified* horizontally), and the horizontal lines of demarcation between the strata are known as *bedding planes.* Other types of stone will have more visible vertical separations, which are known as *cleavage lines.* In some stone, both are apparent. These natural planes of separation are an aid in removing the stone from its natural location. (See Fig. 4-10.)

In one method of quarrying, holes are drilled close together at right angles to the bedding planes, cleavage lines, or both and wedges are driven into the holes to split the rock along the drilled line.

The layout for another method that involves quarrying on a large scale is illustrated in Figure 4-11. After the overburden has been removed, a *channel machine* or a *rock saw* is brought in to outline the boundaries of the quarry.

The channel machine is a heavy machine with a pneumatically-operated chisel mounted on one side which cuts a narrow channel into the

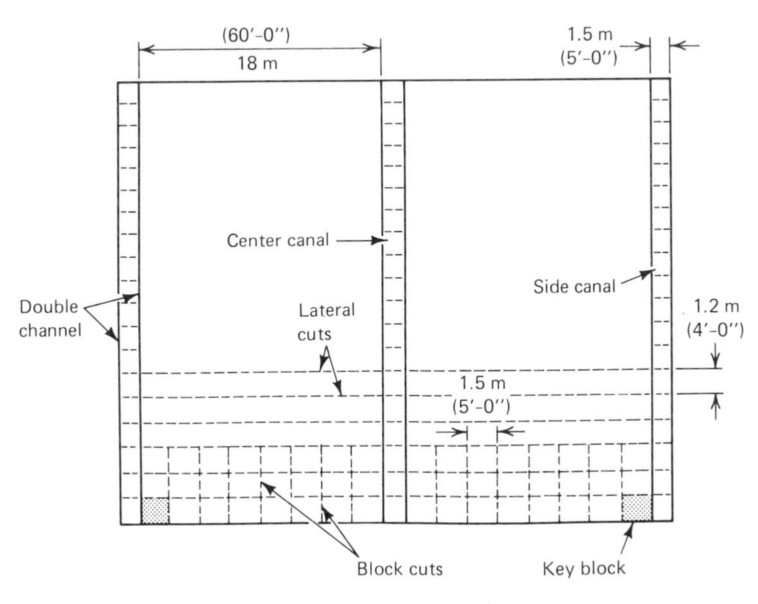

Fig. 4-11. Diagram of quarry layout.

rock, while the rock saw is simply a huge circular saw with diamond-studded teeth. (See Fig. 4-12.) These machines, travelling on temporary tracks, cut a single or double cut, depending on the subsequent method of operation, up to 3 m (10 ft) deep, down each side and down the center of the area of exposed stone.

When a double cut is made, the material between the two cuts is split into blocks and removed to leave a *canal,* the depth of the cuts, down the sides and center of the quarry. (See Fig. 4-13.)

Then, using the channel machine, rock saw or a *wire saw,* lateral cuts are made (see Fig. 4-11), and the long blocks thus produced are cut into convenient lengths and removed from the quarry bed for transportation and further processing. (See Fig. 4-14.)

Fig. 4-12. Rock saw. (*Courtesy Garson Limestone Co.*)

Fig. 4-13. Canals along sides of stone bed. (*Courtesy Indiana Limestone Co.*)

Fig. 4-14. Stone blocks removed from quarry bed. (*Courtesy Indiana Limestone Co.*)

A wire saw is a 6 mm (¼ in) diameter, endless wire, strung between pulleys which are mounted on a frame. The wire transports hard sand that acts as the cutting agent. It is when this method of cutting is used that the canals are required. The wire saw support frames are set in the canals, and the wire saw cuts across the span from side to center canal.

Finishing

The large blocks of stone from the quarry are taken to a mill where they are first passed through a *gang saw* that cuts them into thick slabs. Power saws (*chat* saws, *shot* saws and *diamond* saws) are then used to cut the slabs into pieces of the required dimensions.

Each piece is then given whatever surface treatment is required to produce the texture specified in the order. In some cases, the surface produced by the saw is all that is required. In other cases, *planers, grinders, polishers* or *hammer and chisel* are used to produce the desired texture. Figure 4-15 illustrates some of the surface finishes produced on stone products.

Stones which are to be anchored to the building must be drilled either on the edge or the back, depending on their position. (See Fig. 4-16.) Those that have a special shape are produced by *stone carvers,* working with a hand-sized air hammer and chisel.

Glossary of Terms

Arenaceous Essentially sandy in grain size and texture.

Argillaceous Composed largely of finely divided particles of clay.

Brecciated Composed of angular fragments cemented together.

Calcareous Composed of calcium and magnesium carbonates.

Chat Saw A saw using small pieces of hard steel as the cutting agent.

Coping A cap on a wall to protect it from water penetration.

Foliated Composed of leaflike layers.

Geological Origin Origin related to the history of the earth.

Hypabyssal Partly crystalline in texture, formed at a moderate depth below the earth's surface.

Overburden Material lying on top of a rock bed.

Plutonic Crystalline throughout and formed deep below the earth's surface by the solidification of molten matter.

Porosity The ratio of the volume of the pores in a material to the volume of its mass.

Rudaceous Composed mainly of large fragments of older rock.

Sedimentation The settling of solid matter to the bottom of a body of water.

Shot Saw A saw that uses chilled steel shot as the cutting agent.

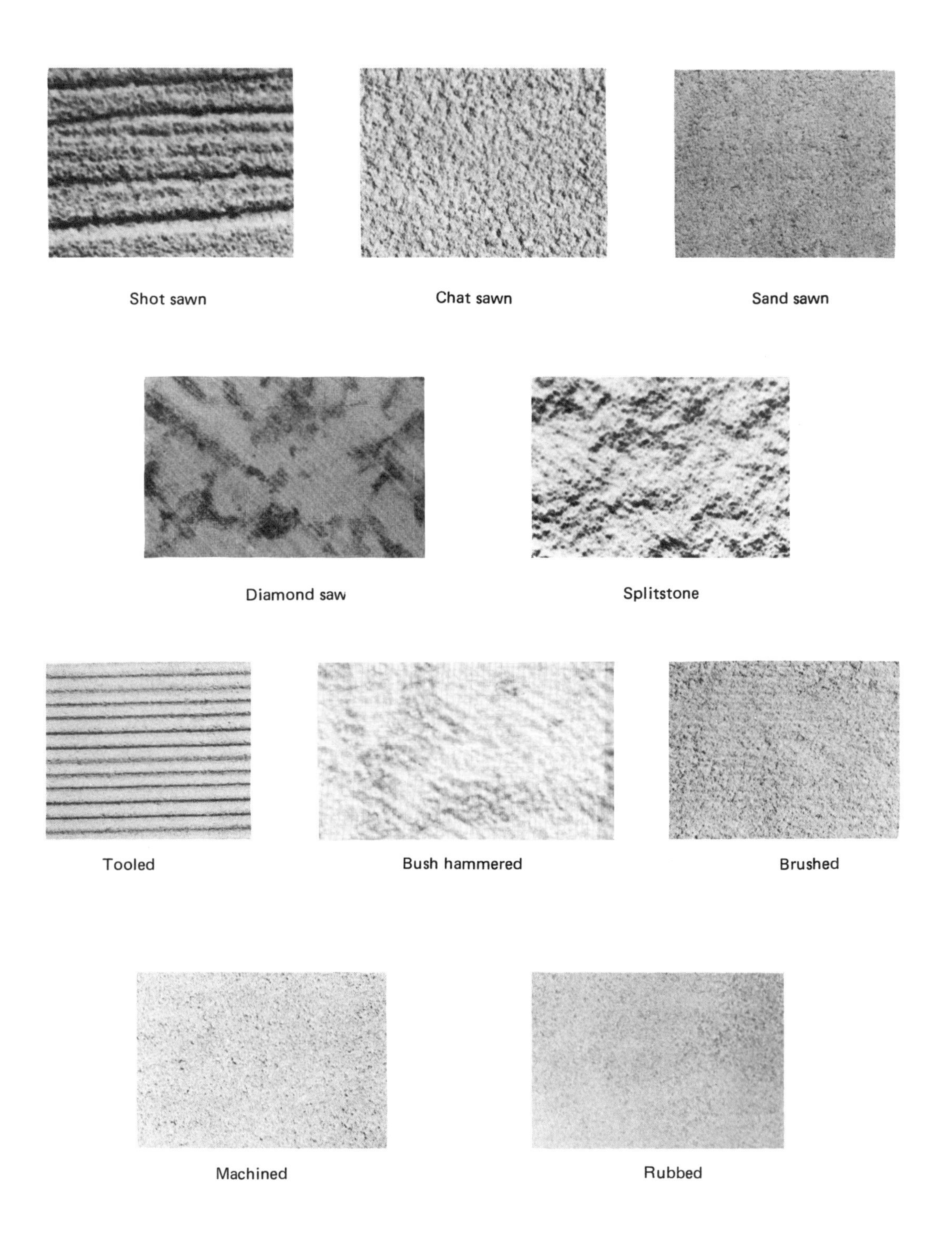

Fig. 4-15. Typical stone finishes.

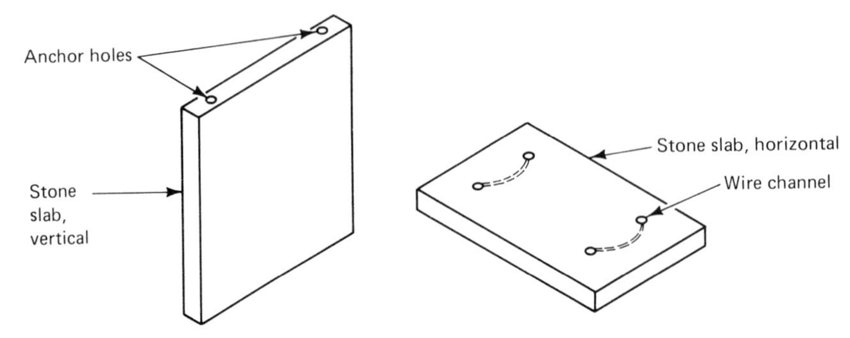

Fig. 4-16. Stones drilled for anchors.

Review Questions

1. Explain what is meant by each of the following terms:
   (a) geological origin
   (b) metamorphic rock
   (c) argillaceous rock
   (d) brecciated marble
2. Explain why each of the following physical characteristics of stone is important from the standpoint of building construction:
   (a) compressive strength
   (b) workability
   (c) porosity
   (d) durability
3. (a) How does oolitic limestone basically differ from other types of limestone?
   (b) What is the chief use of travertine as a building stone?
4. By means of a diagram, illustrate the operation of a wire saw.
5. Explain the difference between:
   (a) cleavage and fracture
   (b) compressive and tensile strength
   (c) density and relative hardness
   (d) bedding plane and cleavage plane
6. Illustrate each of the following by diagram:
   (a) random rubble
   (b) irregular coursed ashlar
   (c) coping stone
   (d) stone base
   (e) stone header

Selected Sources of Information

Briar Hill Stone Company, Glenmont, Ohio.
Building Stone Institute, New York, N.Y.
Chicago Cut Stone Contractors' Assn., Chicago, Illinois.
Cold Springs Granite Company, St. Cloud, Minnesota.
Crab Orchard Stone Company, Inc., Crossville, Tennessee.

Delaware Quarries Inc., Lumberville, Pennsylvania.
Garson Limestone Company Ltd., Winnipeg, Manitoba.
Georgia Marble Company, Atlanta, Georgia.
Gillis Quarries Ltd., Winnipeg, Manitoba.
Indiana Limestone Company, Bedford, Indiana.
Sinclair Cut Stone Company, Hamilton, Ontario.
Vermont Marble Company, Proctor, Vermont.

Chapter 5

MORTAR AND GROUT

Mortar and grout are similar products to the extent that both require a cementatious material (e.g., portland cement) and aggregates (natural or manufactured) for their production. But mortar and grout are not the same product. They are introduced into the wall system differently, are used for different purposes, usually require a different size of aggregate and quite often use a different type of cement.

Mortar is designed for a number of purposes, but it serves primarily to *join masonry units together* in an integral structure. In addition to that, it is also required to *hold the units a specified distance apart;* to produce *tight seals between units* to prevent the passage of air or moisture; to *bond metal ties and anchor bolts to the steel joint reinforcement* in order to integrate them into the masonry structure; to *provide a bed* that will help accommodate variations in the size of units; and to *provide an architectural effect* on exposed masonry walls through various styles of mortar joints.

In order to achieve these purposes, mortar is applied to the *edges of masonry units,* horizontal and vertical, as illustrated in Figure 5-1.

On the other hand, grout is designed primarily to *bond masonry units and steel* together in reinforced masonry walls, so that they act together to resist imposed loads. (See Fig. 5-2.) It is usual to place grout in only those cells containing steel reinforcement (see Fig. 5-3), but in some load-bearing reinforced masonry walls, all cells will be filled with grout. Sometimes the cells of nonreinforced masonry walls are grouted to *give added strength* (see Fig. 5-4), and grout may also be used to *bond together the two wythes of a cavity wall.* (See Fig. 5-5.)

In order to achieve its purposes, grout is introduced into the *cells of the units* or the *cavities between wythes.*

CEMENT MORTAR

Masonry mortar made with portland cement is composed of one or more *cementatious materials,* clean, well-graded *masonry sand* and enough *water* to produce a plastic, workable mixture. In addition to these basic ingredients, *admixtures* may be added for some special purpose.

Fig. 5-1. Mortar applied to edges of masonry units. (*Courtesy Portland Cement Assn.*)

Fig. 5-4. Cells of nonreinforced masonry wall grouted. (*Courtesy Portland Cement Assn.*)

Fig. 5-2. Reinforced masonry wall.

Fig. 5-3. Cells containing reinforcement, grouted.

Fig. 5-5. Grouting cavity wall.

To be useful, mortar must possess a number of important qualities, both in the plastic stage and after it has hardened. In the plastic stage, the important qualities are *workability, water retentivity* and a *consistent rate of hardening.* Hardened mortar must have *good bond, durability, good compressive strength* and *good appearance.*

Workability

Workability is probably the single most important quality of plastic mortar because of the effect that it has on other mortar qualities, in both the plastic and hardened stages. It is a combination of a number of interrelated factors, including the *quality of the aggregate,* the *amount of water* used, the *consistency,* the *ability of the mortar to retain water, setting time, flowability, adhesion* and *mass.*

Mortar with good workability should spread easily on the masonry unit, cling to vertical surfaces (see Fig. 5-6), and extrude readily from joints without dropping as the unit is being placed. (See Fig. 5-7.) The consistency should be such that the unit can be readily aligned but that its weight and the weight of successive courses will not cause further slumping of the mortar.

Fig. 5-6. Good mortar clings to vertical surfaces. (*Courtesy Portland Cement Assn.*)

Fig. 5-7. Good mortar extrudes readily from joints. (*Courtesy Portland Cement Assn.*)

Mortar that has been properly mixed but not used immediately tends to lose its workability through loss of water by evaporation and absorption. *Under these conditons,* workability may be restored by *retempering* (i.e., by adding a small amount of water and thoroughly remixing. (See Fig. 5-8.) The retempered mortar will be slightly lower in compressive strength but not enough to seriously affect the overall strength of the structure.

However, mortar that has stiffened because of an advanced hydration process should not be used. The most practical way of determining whether or not the mortar is suitable for use is on the basis of time. Under normal conditions, the elapsed time between mixing and use should not exceed two and a half hours.

Fig. 5-8. Retempering mortar by hand. (*Courtesy Portland Cement Assn.*)

Water Retentivity

Water retentivity is the property of mortar that allows it to hold on to its water (i.e., to resist loss to the air or to absorptive masonry units). Such a loss will result in premature stiffening of the mortar, reducing the achievement of good bond and watertight joints.

Poor water retentivity may be caused by *poorly graded aggregate, aggregate of too large a size, insufficient mixing* or the *wrong type of cement.* The addition of an air-entraining agent, pozzolanic material or sometimes more water to the mortar and increased mixing may increase its water retentivity.

Consistent Rate of Hardening

The hardening of mortar is due to hydration—the chemical reaction between cement and water—and the rate at which this occurs determines the time required for the mortar to develop resistance to an applied load. If the mortar hardens too quickly, it interferes with the proper laying of the units and finishing of the joints. On the other hand, if hardening is too slow, work may be impeded by having to wait for the mortar to harden sufficiently so that it will not flow from completed joints.

An increase in temperature tends to speed up the hydration process and during hot weather it may be necessary to use colder water or to decrease the length of time that mortar is exposed to the hot atmosphere, or both. Conversely, in cold weather, it may be necessary to use warm mixing water and/or to provide heat for the surrounding atmosphere.

The constant development of the hydration process and consequent consistent rate of mortar hardening depends on intimate contact between molecules of water and the tiny particles of cement taking place all at the same time. Cement particles tend to cling together (i.e., to flocculate), and thus some of them are shielded from contact with water for longer periods than others. This slows down the hydration process and delays the setting time. Increased mixing or the use of a *cement dispersal agent* will alleviate this condition.

Mortar Bond

The *bond* between mortar and a masonry unit is a combination of the degree of contact of the mortar with the masonry unit and the adhesion of the cement paste to the unit surface.

The degree of mortar contact with the unit depends on the *workability of the mortar* and its *water retentivity,* on masonry units having a *medium initial rate of absorption, full mortar joints* and *good workmanship* in the application of the mortar.

The adhesion of the paste to the masonry surface also depends on a number of factors, including *workability* and *water retentivity* of the mortar, the *amount of cement* used in the mix, the *surface texture, moisture content* and *suction* of the masonry units, *good workmanship* in placing the mortar, *temperature* and *relative humidity.*

Cement and water content are of primary importance in the development of bond strength in mortar. Tests show that bond strength increases as the cement content of the mortar increases and that the best bond strength is obtained when the largest amount of water possible, compatible with workability, is used in a mortar mix, even though this will result in decreased compressive strength.

Workability is the other factor of primary importance to the development of good mortar bond. The spread mortar (see Fig. 5-1) should be exposed for a minimum amount of time. The pressure applied to the mortar bed during the placing of the units and the fast and accurate alignment of the units are essential for a good mortar bond.

Finally, freshly laid masonry must be protected from extreme weather conditions if good mortar bond is to develop. It may have to be shielded from the sun, provided with heat against the cold or dampened down to prevent too rapid loss of moisture due to low relative humidity.

Durability

The durability of mortar is its ability to withstand exposure to weather and atmospheric conditions. Polluted air may cause some mortar deterioration, but the major cause is frost action on mortar saturated with water.

The deterioration of mortar in walls above grade due to frost action is not usually a major problem, provided that the mortar has first been allowed to develop its full potential strength. For frost damage to occur, hardened mortar must first be subjected to a high degree of saturation. During the hardening process, the original water content of the mortar is greatly reduced. The mortar does not normally become saturated again unless the masonry is in contact with saturated soil or is exposed to heavy rain, or the formation of horizontal ledges on the wall allows water to be held there. Under any of these conditions, the mortar may again become saturated and subsequent frost action can cause deterioration.

There are a number of steps which may be taken to help prevent such deterioration. *Mortar with high compressive strength* normally has good durability. Mortar containing *air entrainment* will withstand a great many freeze-thaw cycles. Masonry below grade should be protected by a *waterproof coating* on the exterior surface and the *tooling of mortar joints* should not produce ledges which may collect and hold water.

Compressive Strength

The compressive strength of mortar is largely dependent on the type and quantity of cement used in the mortar mix. The compressive strength increases as the amount of cement in the mix is increased in relation to the amount of water used, provided the mix remains workable. An increase in water content, air entrainment or lime content will reduce compressive strength. An increased lime content may be due to the type of cement being used or to the addition of hydrated lime to the mortar mix.

Appearance

The overall appearance of a masonry structure is affected by the appearance of the mortar joints. Factors contributing to the appearance of mortar joints include *uniformity of color and shade, uniformity of texture* and *thickness* and the *workmanship* involved in tooling the joints.

The color and shade of the joints are affected by the *moisture content* of the masonry, *admixtures* in the mortar, *atmospheric conditions, uniformity of proportions* of the mortar mix and the *time of tooling.* Good uniformity of texture depends on careful measurement of batch proportions from batch to batch and also on thorough mixing. Even joint thickness and consistent tooling are both the results of good workmanship.

TYPES OF MORTAR

Mortar types are identified either by the proportions of cementatious materials used per batch or by the compressive strength of representative mortar cubes, tested after seven and twenty-eight days of curing time. When the *proportions* method is being used, the aggregate proportion in each case will not be less than two and one-quarter or more than three times the total volume of cement and lime in the batch. The various types of mortar designated under this method and the proportions of cement and lime used in each case are indicated in Table 5-1.

When mortar is classified according to its compressive strength, the total aggregate will not be less than two and one-fourth or more than three and one-half times the total volume of cement and lime used in the batch. The compressive strength ratings required for each type of mortar are given in Table 5-2.

Table 5-1. Mortar Types by Cement and Lime Proportions

Specification	Mortar Type	Parts by volume: *Portland Cement or Portland Blast Furnace Slag Cement*	*Masonry Cement*	*Hydrated Lime or Lime Putty*
For plain masonry, ASTM C270, CSA A179	M	1	1	–
		1	–	1/4
	S	1/2	1	–
		1	–	over 1/4 to 1/2
	N	–	1	–
		1	–	over 1/2 to 1-1/4
	O	–	1	–
		1	–	over 1-1/4 to 2-1/2
	K	1	–	over 2-1/2 to 4
For reinforced masonry, ASTM C476	PM	1	1	–
	PL	1	–	1/4 to 1/2

(Courtesy Portland Cement Association)

It has been shown that the *compressive strength, water retentivity* and *workability* of cement-lime mortars can be varied over wide ranges by changing the proportions of cement and lime in the mortar.

As noted previously, the cement is the main contributor to the *strength* of mortar. But, at the same time, the cement contributes to *rapid setting, low water retentivity* and *poor workability.* On the other hand, lime in the mix contributes very little towards strength, but it does improve the workability and makes for better water retentivity.

The results of tests made on the strength and

Table 5-2. Compressive Strength of Mortar Cubes

Mortar Type		Minimum Compressive Strength, MPa (PSI) Average of Six 50 mm (2 in) Cubes: *Tested at 7 Days*	*Tested at 28 Days*
Laboratory prepared	M	11 (1600)	17.5 (2540)
	S	7.5 (1088)	12.5 (1813)
	N	3 (435)	5 (725)
	O	1.5 (220)	2.5 (363)
	K	0.3 (45)	0.5 (73)
Job prepared	M	9 (1300)	14 (2030)
	S	6 (870)	10 (1450)
	N	2.5 (363)	4 (580)
	O	1 (145)	2 (290)
	K	0.2 (30)	0.4 (58)

If the mortar does not meet the 7-day strength but meets the 28-day strength, it is acceptable. If it fails to meet the 7-day strength but reaches 2/3 of it, the contractor may continue at his own risk, pending the 28-day results.

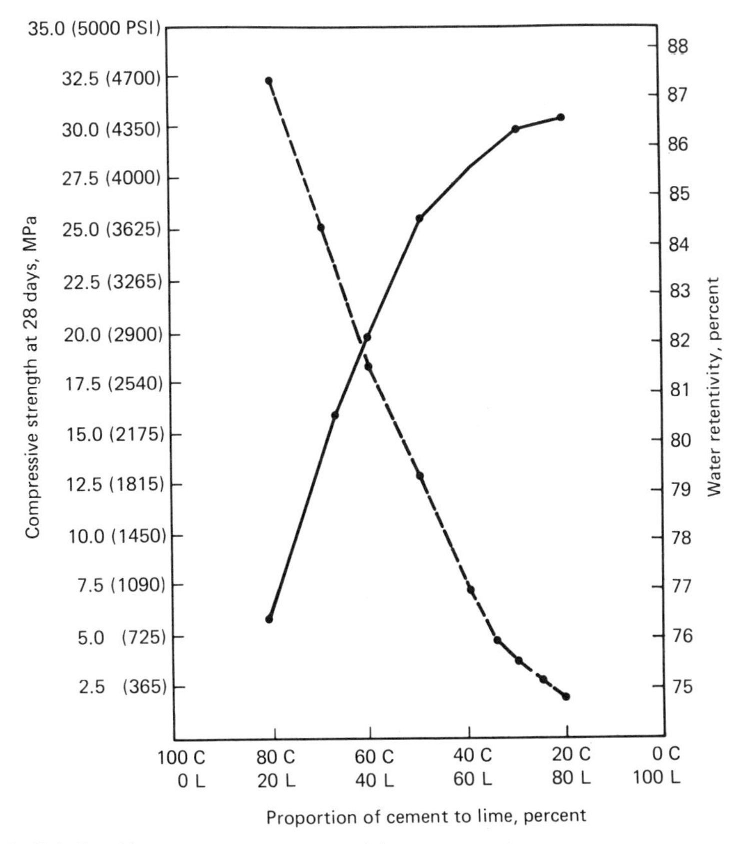

Fig. 5-9. Relationship among mortar composition, compressive strength and water retentivity.

water retentivity of mortar types M, S, N, O and K when various proportions of cement and dry hydrated lime were used in the mix, are indicated in Figure 5-9.

It is evident from Figure 5-9 that the improvement of one quality of a mortar may well mean a decrease in the other. For example, if it is considered necessary to produce mortar with a greater compressive strength—to improve its durability and the strength of the masonry—the proportion of cement in the mix is increased. That can only be done, however, by reducing the proportion of lime, which may result in a lowering of water retentivity. That, in turn, may have an adverse effect on the bonding between mortar and the masonry units and ultimately on the weather resistance of the masonry structure.

For mortar types classified by either method, the amount of water to be used on the job should be the maximum that will produce a workable consistency during construction. Similarly, in either case, the water retention limit is specified. It is measured using a *flow-after-suction* test, described in ASTM C91 and CSA A8-M. For required limits, see Chapter 8.

Once the design loads, the type of structure and the type of masonry to be used have been decided, the mortar type can be selected. In the U.S., for nonreinforced masonry, the mortar type is selected on the basis of Table 5-3. For reinforced masonry, the job specification may simply say that the mortar shall meet the requirements of ASTM C476, which means

Table 5-3. Guide for the Selection of Mortar Type

Kind of Masonry	Types of Mortar
Foundations:	
Footings	M or S
Walls of solid units.	M, S, or N
Walls of hollow units.	M or S
Hollow walls.	M or S
Other than foundation masonry:	
Piers of solid masonry.	M, S, or N
Piers of hollow units.	M or S
Walls of solid masonry.	M, S, N, or O
Walls of solid masonry, other than parapet walls, not less than 300 mm (12 in) thick or more than 10.67 m (3.2 ft) in height, supported laterally at intervals not exceeding 12 times the wall thickness.	M, S, N, O, or K
Walls of hollow units–load-bearing or exterior and hollow walls 300 mm (12 in) or more in thickness.	M, S, or N
Hollow walls, less than 300 mm in thickness, where assumed design wind pressure:	
1. exceeds 958 Pa (20 lb/ft^2).	M or S
2. does not exceed 958 Pa.	M, S, or N
Linings of existing masonry, either above or below grade.	M or S
Masonry other than above.	M, S, or N

(Courtesy Portland Cement Association)

that either PM or PL type may be chosen, depending on circumstances.

In Canada, the selection of mortar for either plain or reinforced masonry is based on whether or not the choice of masonry was made on the basis of engineering analysis of the structural effects of the loads and forces involved. For masonry which is chosen on the basis of such engineering analysis, type M, S or N mortar is permitted. For masonry not based on such an analysis, type M, S, N, O or K is permitted, with two exceptions. Types O and K mortar are not allowed where the masonry is to be in direct contact with the soil or where the masonry is exposed to the weather on all sides.

MORTAR COMPONENTS

As indicated earlier, the basic components of mortar are *cement, sand* and *water.* To these may be added *hydrated lime, pozzolanic materials, admixtures* and *color.* In order to produce good mortar, it is important that the quality of all these ingredients be maintained at as high a level as possible.

Cement

A number of types of cement are used to make mortar, including *portland cement,* type *I, IA, II, IIA, III* or *IIIA,* according to

ASTM specification C150 or type *10,* (normal), type *20* (moderate) or type *30* (high-early strength), according to CSA specification A5-M; *portland blast furnace slag cement,* type *1S* or *1S-A* (ASTM specification C595); *masonry cement,* type *H* or *L* (ASTM specification C91 and CSA specification A8-M; and *portland pozzolan cement,* type *1P* or *1P-A* when fly ash is the pozzolanic material (ASTM specification C595).

Sand

Sand makes up a large part of the volume of mortar and consequently must be of good quality to produce good results. It should be clean, well-graded, meet the requirements of ASTM C144 or CSA A82.56-M and meet the gradation requirements outlined in Table 5-4. (See Fig. 5-10.)

Generally speaking, sand that is too coarse—having less than 5 to 15 percent passing the 300 μm and 150 μm sieves—will produce coarser mortar with poor workability and mortar joints with low resistance to moisture penetration. On the other hand, sand that is finer (specified in Table 5-4) will produce mortar with good workability but poor strength and poor resistance to moisture penetration.

The thickness of the mortar joint will also have some bearing on the fineness of the mortar sand used. If mortar joints are to be less than the standard 10 mm (3/8 in) in thickness, 100 percent of the sand should pass the 2.50 mm (0.1 in) sieve and 95 percent should pass the

Table 5-4. Gradation Percentages for Masonry Sand

	Gradation Specified, Percentage Passing		
	ASTM C144		
Sieve Size	*Natural Sand*	*Mfg Sand*	*CSA A82.56M*
5.00 mm (0.2 in)	100	100	100
2.50 mm (0.1)	95 to 100	95 to 100	95 to 100
1.25 mm (0.05)	70 to 100	70 to 100	60 to 100
600 μm (0.02)	40 to 75	40 to 75	35 to 80
300 μm (0.01)	10 to 35	20 to 40	15 to 50
150 μm (0.006)	2 to 15	10 to 25	2 to 15
75 μm (0.003)	—	0 to 10	—

(Courtesy Portland Cement Association)

Notes:
1. Not more than 50 percent shall be retained between any two sieve sizes and not more than 25 percent between 300 μm and 150 μm sieve sizes.
2. Fine aggregate shall be so graded that neither the proportion of particles finer than a 1.25 mm sieve and coarser than a 600 μm sieve nor the proportion of particles finer than a 600 μm sieve and coarser than a 300 μm sieve exceeds 50 percent.

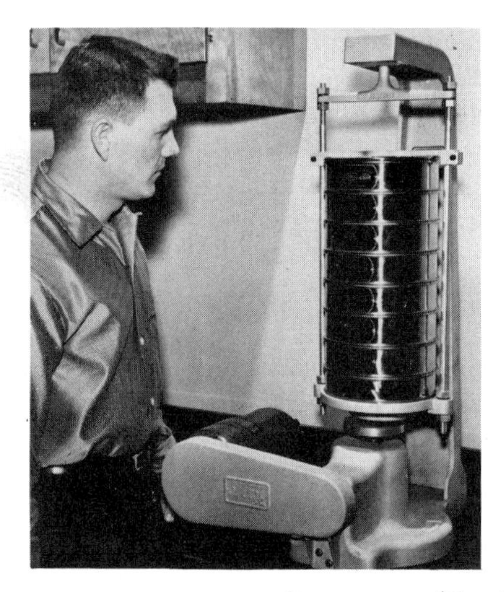

Fig. 5-10. Standard sand-grading screens. (*Courtesy Soiltest Inc.*)

1.25 mm (0.05 in) sieve. If joints are thicker than the standard 10 mm, the gradation of sand should be within the limits of *concrete sand,* as laid down in ASTM C33 or CSA A23.2-M.

Water

Water that is to be used for mixing mortar should be clean and free from injurious amounts of acids, alkalies and organic materials.

Generally, mixing water containing common inorganic acids, such as hydrochloric or sulfuric acid, in concentrations of less than 10,000 parts per million, have no adverse effect on the strength of mortar.

Water containing up to 10,000 ppm sodium or potassium sulfate may be satisfactory as far as strength is concerned but may contribute to efflorescence at the joints later. Sodium chloride may be tolerated in concentrations of up to 20,000 ppm and will not adversely affect mortar strength, but it may cause corrosion to any unprotected reinforcing material.

Organic materials include such things as finely powdered *peat, humus, organic loam, sugar* or *finely powdered coal.* The presence of any of these organic materials may seriously delay the setting and hardening of mortar, and the finely powdered coal, in excessive amounts, may affect the durability of the mortar.

Lime

Lime used in mortar is *hydrated* lime [$Ca(OH)_2$], normally of the nonhydraulic or semihydraulic variety. It may be added dry from the bag, but it is better, where possible, to soak the dry hydrate for 16 hours prior to using it to provide better plasticity and to ensure that the lime is completely *slaked* (i.e., converted to the hydroxide).

Lime added to cement-sand mortar improves its plasticity and its water retentivity.

Pozzolanic Materials

A pozzolan is a finely ground, siliceous material, usually of volcanic origin, used as an additive in concrete and mortar. A number of natural materials such as *diatomaceous earth, opaline cherts* and *shales, tuffs* and *pumicites* and some artificial materials such as *fly ash* are used as pozzolans.

Diatomaceous earth is a soil-like material, composed of the silicified skeletons of microscopic, unicellular animals. Cherts are impure, flintlike rocks, usually dark in color; shale is a sedimentary rock, formed by the consolidation of clay or silt and containing a large proportion of silica; tuffs is a usually stratified rock composed of fine volcanic debris; and pumicites are stones composed of volcanic glass, rich in silica.

Fly ash is a fine residue that results from the combustion of powdered coal and may contain various amounts of carbon, silica, sulfur, alkalies and other ingredients.

Pozzolans may be added to mortar mixes to improve *workability, resistance to water penetration* and *resistance to chemical attack.* Some pozzolans also have been found to reduce the expansion caused by alkali-aggregate reaction in mortar. The expansion is caused by the interaction of alkali in portland cement and certain siliceous types of sand. The result of this action is excessive expansion of the mortar, leading to increased permeability at the mortar joints.

Admixtures

Admixtures are materials that are added to concrete or mortar as it is being mixed in order to produce some specific result, either in the fresh or hardened state. They include such products as *air-entraining agents, retarders, accelerators, water reducers, cement dispersal agents* and products to *improve compressive, tensile* and *bond strength* of mortar.

Air-Entraining Agents. Materials used as air-entraining agents include a number of *natural wood resins,* various *sulfonated compounds* and *some fats* and *oils.* They have the property of reducing the surface tension of water, thus enabling water to trap and hold microscopic bubbles of air, approximately 0.025 mm (0.001 in) in diameter, which are relatively stable.

They help to improve the plasticity of mortar by acting like tiny ball bearings, on which the fresh mortar moves easily. They also reduce permeability because each bubble of air is surrounded by a film of cement paste that, when set, breaks up the pore structure of mortar and thus reduces capillary action. These products may have an adverse effect on the strength of mortar, however, and great care should be taken in their use. Air content of mortar mixes should not exceed about 14 percent by volume.

Retarders. Retarders are materials that delay or extend the setting time of the cement paste and are not very commonly used in mortar. The more commonly known retarders are *carbohydrate derivatives* and *calcium lignosulfonate,* used in fractions of one percent by mass of the cement. Retarders tend to reduce strength development and increase the likelihood of efflorescence.

Accelerators. An accelerator is an admixture that is used to speed up the initial set of cement paste. Its use in mortar is normally confined to cold weather when otherwise low temperatures might seriously retard the hardening of mortar. Among the materials used as accelerators are *calcium chloride,* some of the *soluble carbonates, silicates, fluosilicates* and *triethanolamine.* Manufacturers' instructions should be carefully observed in the use of any of these products.

Water Reducers. A water reducer is a material that is used to decrease the amount of water needed to produce a required workability of mortar. A reduction in the amount of water used per unit of cement will increase the strength of cement paste. However, some water-reducing admixtures will retard the setting time of mortar and some may increase the drying shrinkage. Testing should be carried out to determine the effect of any given product on mortar. Typical water-reducing agents are made from the *metallic salts of lignosulfonic acids.*

Cement Dispersal Agents. As mentioned previously, cement tends to *flocculate* (i.e., gather into small, relatively compact masses) with the result that cement particles in the center of the mass may not come into contact with water and, therefore, will not undergo the hydration process. This, in turn, means that the cement paste is not able to reach its full potential strength.

A cement dispersal agent causes cement particles to separate by imparting *like* electrostatic charges to them. A typical agent of this type is *calcium lignosulfonate.*

Strength-Improving Materials. Among the more recently developed admixtures is one that is designed to improve the strength of masonry

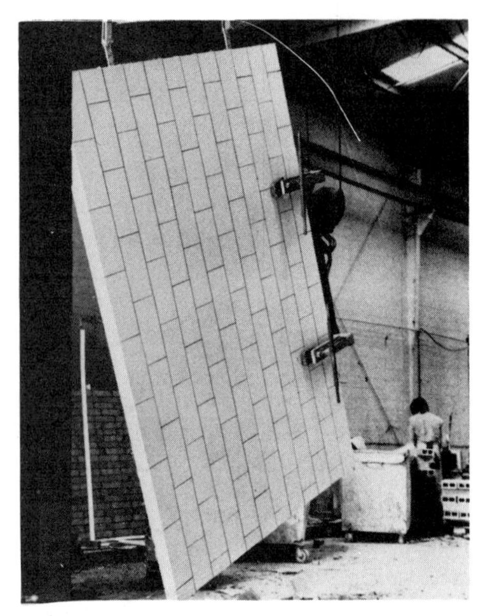

Fig. 5-11. Single-wythe prefabricated wall panel being lifted. (*Courtesy Dow Chemical of Canada, Ltd.*)

mortar. It is a *synthetic polymeric emulsion,* and tests have shown that its addition to mortar has resulted in very marked increases in compressive, tensile and bond strengths. Its use has facilitated the handling of prefabricated masonry panels. (See Fig. 5-11.)

Colors in Mortar

White or colored mortars are sometimes used to produce some particular architectural effect. White mortar is made with *white masonry cement* or *white portland cement* and *lime* and *white sand.* Colored mortar is normally produced by the introduction of color pigments into the mortar mix.

These pigments should be *inactive* mineral oxides, such as *iron, manganese, cobalt, chromium oxides* and *carbon black.* Red, yellow, brown or black mortar may be produced by the use of various iron oxides; chromium oxide produces a green color; cobalt oxide will produce blues and carbon black will produce grays or black. The latter should be limited to a maximum of three percent by weight of the cement.

MEASURING MORTAR MATERIALS

It is important that the materials being used to make mortar be accurately measured in order to ensure that workability, strength, yield and mortar color be maintained from batch to batch.

Although the aggregate for mortar is very often measured by volume, this is not always a satisfactory method, due to the phenomenon known as *bulking* in sand. When small amounts of water are added to dry, rodded sand, the volume increases quite dramatically, the amount of increase depending on the amount of water added and on the fineness of the sand. As indicated in Figure 5-12, the greatest volume increase occurs with the addition of about five percent water by mass.

Because of this bulking, measurement of sand by mass provides a much more positive control over the actual amount of aggregate being added to the mix.

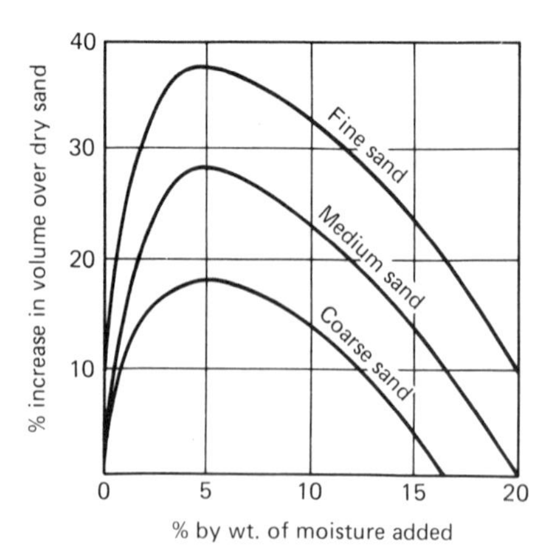

Fig. 5-12. Bulking in sand.

MIXING OF MORTAR

In order to obtain good workability and allow development of the maximum strength possible, mortar ingredients must be thoroughly mixed. Whenever possible, the mixing should be done by machine.

A typical mortar mixer (a paddle mixer with tilting drum) is illustrated in Figure 5-13. Mixing time should be from three to five minutes after all ingredients have been added to the batch. A shorter mixing time may result in nonuniformity, poor workability, low water retention and less than designed air content. Too long a mixing time may adversely affect the air content of mortars made with air-entraining cement.

Fig. 5-13. Mortar mixer. (*Courtesy Portland Cement Assn.*)

The best method of charging will be learned only by experience, but good results have been obtained when about three-quarters of the required water, one-half of the sand and all of the cement are briefly premixed. The balance of the sand is then added, followed by the remainder of the water. Mixing is done most effectively if the mixer is loaded to its design capacity, which will normally range from 0.11 to 0.20 m^3 (4 to 7 ft^3).

ORGANIC MORTARS

A recent development in the masonry field has been the introduction of *organic mortar,* which is not a mortar in the conventional sense but a synthetic adhesive. It is applied to masonry units with a gun as a thin coating. Consequently units must be made practically full modular size and have smooth, straight meeting edges in order to provide close contact along the entire surface of the unit.

APPLICATION OF MORTAR

Mortars may be applied either by *hand* or by *machine.* Hand application involves the use of a *bricklayer's trowel* for the actual application (see Fig. 5-14), a *mortar board* to hold the freshly mixed mortar and a *hawk* (a flat plate with a handle on the bottom) to carry mortar for immediate use.

Two methods of hand application are used by masons on the job. In one, known as *threading,* relatively long beds of mortar are spread on the face shells of units already laid (see Fig. 5-14), and the ends of a number of units are buttered and laid in succession in that bed. The length of bed that may be threaded in this manner will depend on the temperature, relative humidity and the speed at which the mason works.

Fig. 5-14. Mortar "threading" on brick wall.

The other method is known as the *pick-and-dip* method, in which only enough bedding mortar is spread to accommodate one unit. Units to be placed are picked up, buttered (see Fig. 5-15) and placed, one by one.

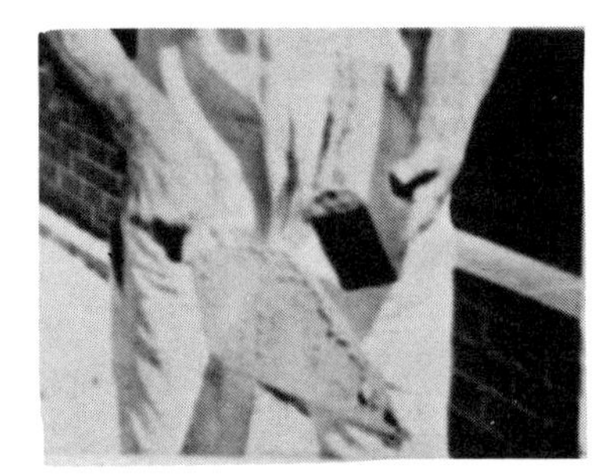

Fig. 5-15. "Pick-and-dip" mortar application.

Two methods of machine application are in use, depending on the type of mortar. Organic, threadline mortars are usually applied with a gun in much the same way as caulking or adhesives are applied. Conventional mortars are applied with a *mortar spreader,* similar to that illustrated in Figure 5-16. A mortar spreader is a self-propelled machine which spreads two face-shell beds at speeds of up to 8000 mm (25 ft) per minute. The machine is adjustable for various widths of wall and provides some consolidation of the mortar by vibration during the spreading process.

GROUT

Masonry grout is composed of a mixture of cement and aggregate, combined with enough water to produce a mix that will flow readily into the cores and cavities without segregation.

Cement

Portland cement, portland blast furnace slag cement, and portland-pozzolan cement are all used in making grout, the choice being determined largely by availability and the requirements of the particular project at hand.

Fig. 5-16. Mortar spreader in action. (*Courtesy Mortar Runner Inc.*)

Aggregate

The fineness or coarseness of the aggregate for a grout will be determined by the size of the space to be grouted, as well as the height of the lift. In low-lift grouting, fine aggregate only would be used where spaces were as small as 50 × 75 mm (2 × 3 in), or a cavity between wythes was as narrow as 20 mm ($\frac{3}{4}$ in).

In high-lift grouting, where the minimum horizontal dimension is 75 mm (3 in), a coarse aggregate of 12 mm ($\frac{1}{2}$ in) maximum size may be used. Under some specifications, 20 mm ($\frac{3}{4}$ in) maximum size coarse aggregate may be used when the grout space is 100 mm (4 in) or greater.

Mix Proportions

The amount of aggregate, fine and coarse, and lime, where required, to be used per unit of cement in making masonry grout, is indicated in Table 5-5.

Only enough water should be used to produce a fluid consistency that will allow the grout to be placed by chute or pump without segregation. This will normally require a slump—

Table 5-5. Mix Proportions for Masonry Grout

	Parts by Volume			
			Aggregate Measured in a Damp, Loose Condition	
Type	*Portland Cement, Portland Blast Furnace Cement, or Portland-Pozzolan Cement*	*Hydrated Lime or Lime Putty*	*Fine*	*Coarse*
Fine grout	1	0 to 1/10	2-1/4 to 3 times the total volume of cement and lime	
Coarse grout	1	0 to 1/10	2-1/4 to 3 times the total volume of cement and lime	1 to 2 times the total volume of cement and lime

(Courtesy Portland Cement Association)

measured according to the standard method for measuring concrete slump (ASTM C143)—of 200 mm (8 in) for units with low absorption and up to 250 mm (10 in) for units with high absorption. (See Fig. 5-17.)

Grout Admixtures

In cases where the masonry units involved have a high rate of absorption, it may be desirable to use a *grouting aid* admixture in the grout. It will have the effect of *reducing early water loss* to the masonry units, promoting the *bonding of the grout* to all the interior faces of the units and producing a *slight expansion of the grout* that will help to ensure complete filling of the cavities.

Grout Strength

The mix proportions indicated in Table 5-5 will, in most cases, produce grouts with a compressive strength of 4 to 17 MPa (580 to 2465 psi) in 28 days, depending on the amount of water used in the mix when tested by the standard laboratory methods. But the in-place compressive strength of the grout will often exceed 17 MPa because normally the units will absorb some of the mixing water from the grout in place, and this lowering of the water/cement ratio of the grout mix will increase the compressive strength. In addition, the presence of moisture in the units surrounding the grout will ensure favorable curing conditions, which is an aid to improved strength.

Mixing and Curing

Under normal conditions, the mixing of grout on the job site is not recommended. Whenever possible, grout should be batched, mixed

Fig. 5-17. Taking a slump test.

and delivered in the same way as ready-mixed concrete (see Fig. 5-5), in accordance with ASTM C94-1, Standard Specification for Ready-Mixed Concrete. Because of its high slump, ready-mixed grout should be continuously agitated after mixing until it is placed.

Normally, the high water content of the grout, the moist condition of the masonry units, and the shelter provided by them will ensure reasonably good curing conditions for grout. Grouts placed during cold weather do require special care because their high water content make them especially susceptible to frost action.

Glossary of Terms

Air-Entraining Agent A material that has the ability to introduce many small air bubbles into a mixture.

Cementatious Material A material that brings about a bonding action between two or more objects.

High-Early Strength Cement Portland cement with the ability to develop a higher rate of strength during its early life than normal portland.

Hydrated Lime The result of the combination of calcium oxide (quicklime) with water to produce calcium hydroxide (hydrated lime).

Hydration The process of combining with water to form a new product.

Initial Rate of Absorption The amount of water absorbed by the surface of a material on initial contact, measured within a given period of time.

Metallic Oxide The product of the combination of a metallic element with oxygen.

Nonhydraulic Lime Lime that does not have the capacity to harden and develop strength in a wet state.

Organic Derived from living organisms.

Permeability The rate at which a material will allow the passage of fluids.

Potential Strength The greatest strength possible under a given set of circumstances.

Synthetic Polymeric Emulsion A material produced artificially by the chemical combination of a number of relatively simple molecules and suspended in a liquid.

Wood Resin A material produced naturally by a growing tree.

Wythe A structure composed of a single course of masonry units.

Review Questions

1. Outline the basic differences between *mortar* and *grout.*
2. List five purposes for which masonry mortar is designed.
3. Define each of the following terms:
   (a) cementatious material
   (b) water retentivity
   (c) air-entraining agent
   (d) initial rate of absorption
   (e) mortar bond
   (f) high-lift grouting

4. Outline the specific purpose or purposes for which each of the following materials is added to basic masonry mortar or grout:
   (a) hydrated lime
   (b) a pozzolan
   (c) grouting aid admixture
   (d) cement dispersal agent

Selected Sources of Information

Canada Cement Company, Montreal, Quebec.
Dow Chemical Company, Sarnia, Ontario.
Medusa Cement Company, Cleveland Heights, Ohio.
Steel Bros. Ltd., Calgary, Alberta.

Chapter 6

MASONRY PROTECTIVE AND DECORATIVE COATINGS

Paints and other coatings that are designed for application to the surface of building materials are used for a variety of reasons, including *preservation, protection, sanitation, illumination, conservation of heat, improvement of working conditions, improved safety* and *economy.* Those that are used specifically for masonry materials have two basic functions, namely *protection* and *decoration,* though some of the other reasons listed above may also apply.

MASONRY PROTECTIVE COATINGS

Masonry in general requires protection against two potential sources of deterioration. First, masonry walls must be protected against the penetration of moisture or moisture vapor and against the passage of moisture from one wall face to the other. Second, masonry needs to be protected against the accumulation of dust, grease, oil, tar, chemical films, scuff marks or bacteria, that cannot readily be removed by washing with soap and water, solvent, chlorine bleach solution or disinfectant.

Moisture Penetration Protection

The passage of water through a masonry wall may occur in three ways. First, water may travel through the individual units by capillary action due to the porosity of the material. Second, water may pass through the mortar itself, again by capillary action. Finally, it may move through openings between the units and the mortar, in which case the water normally flows without hydrostatic pressure.

The steps that may be taken to prevent this transfer of moisture through masonry walls are known as *dampproofing* and *waterproofing.* Dampproofing refers to any method used to reduce the penetration of moisture by *capillary action,* while waterproofing refers to treatments intended to prevent the *flow* of water through a wall.

The most basic step that can be taken to waterproof a masonry wall is *proper construction.* The best mortar possible, the proper selection of masonry units based on density and initial rate of absorption, careful application of mortar and laying of the units and proper tooling of joints should result in the construction of walls that will resist moisture penetration for many years.

Further steps that may be taken to provide dampproofing and waterproofing to masonry walls involve the application of various *coatings* to the inside or outside of the walls. These coatings include: *parging* (a mortar coat),

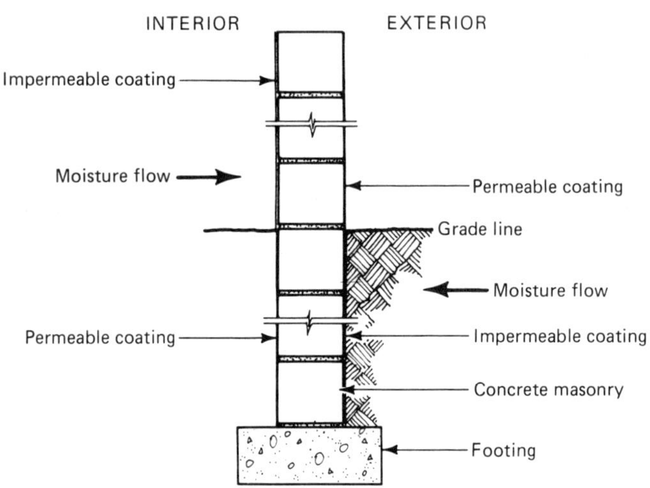

Fig. 6-1. Moisture flow through masonry walls.

bituminous coatings, latex paints, oil-based paints, portland cement paints, epoxy coatings, polyester-epoxy coatings, chlorinated rubber-based paints, silicone-based coatings, metallic oxide waterproofing compounds, elastomeric sheet coatings and various *ceramic glazes.*

Some of these coatings are *permeable* and some are *impermeable.* Permeable coatings are effective in preventing the flow of water, such as the infiltration of rain, but allow the transpiration of entrapped moisture or moisture vapor to the outside air. Impermeable coatings seal off the passage of water or water vapor completely.

Selection of Coating Type

The selection of the type of coating that will provide the best barrier against the movement of water and water vapor should be based on the *mechanics of moisture flow through masonry walls,* the *degree of alkalinity* of the wall, the amount of *soluble salts present* in the units and in the mortar, the *moisture content of the wall* to be treated and the *relative smoothness* of the units.

Mechanics of Moisture Flow. Above grade, moisture from within masonry structures will tend to migrate to the drier outside area. Below grade, it migrates in the opposite direction; it moves from the damp soil outside to the drier area inside the building. (See Fig. 6-1.) It is important, therefore, that an exterior barrier on walls above grade, designed to prevent the penetration of rain, should not interfere with the normal movement of moisture to the exterior of the wall, or most likely, trouble will be the result. An impermeable barrier applied to the exterior of a masonry wall above grade will trap moisture migrating outward. The freezing and thawing of the entrapped moisture may well result in the blistering and deterioration of the coating and crazing of the masonry.

Therefore, as illustrated in Figure 6-1, on walls above grade, any impermeable coating should be on the inside of the wall, while the outside should have a permeable barrier. Below grade, the reverse is true—the impermeable barrier should be on the exterior, the permeable one on the interior.

Alkalinity of Masonry. One of the chemical properties of masonry which affects its ability to hold a coating is its degree of alkalinity; that is, the amount of basic salts that it contains. Most stones, brick and tile are normally neutral, but the mortar in which they are set is quite basic. Most concrete masonry units are highly alkaline, unless they have been steam cured.

Alkalies react with the oils or oil-containing vehicles in some type of coatings in a process called *saponification.* As a result, the dry film becomes soft and tacky and, if the process is carried far enough, reverts to a liquid.

It is necessary, therefore, to guard against such a reaction. There are three possible steps which might be taken in this regard. One is to ensure that, if an appreciable amount of alkali is present, *coatings* will be used, at least for the prime coat, that *do not contain oils.* Another is to try to ensure that the affected surface remains so dry that the reaction cannot occur. With most exterior surfaces, this is practically impossible. The third alternative is to try to neutralize the alkali chemically. This is done by treating the surface with an acid solution. But this treatment only neutralizes a thin surface layer, and subsequent passage of water through the material can bring fresh alkali to the surface.

To test for alkalinity, dampen the surface to be tested and apply a small strip of *pink* litmus paper to the damp area. If the litmus paper turns *blue* or indigo, the surface is alkaline; if the color does not change, alkali is not present. A second method of testing may also be used. Mix 4 grams (0.14 oz) of phenolphthalein with 125 grams (4.4 oz) of pure grain alcohol. Add to this 1 liter (1 qt) of distilled water. Dip a glass rod into the solution and touch it to the surface to be tested. If a red color appears at the spot tested, alkali is present. The degree of alkalinity depends on the brightness of the color, which will range from light pink (satisfactory for painting) to bright red (unsatisfactory for oil-based paint).

Soluble Salts in Masonry. Another chemical characteristic of some masonry products and mortars is the presence of soluble salts such as sodium, potassium or magnesium sulfate. They may be present as a result of chemical reactions that take place during manufacture or curing; reactions resulting from soluble impurities in some of the ingredients or from materials used in mortar, particularly the cement.

As the water from the interior of masonry or mortar migrates to the surface and evaporates, it leaves behind a white, crystalline deposit known as *efflorescence* (see Fig. 6-2), that causes coating failure by mechanically destroying the film bond. Crystallization of the salts exerts a force sufficient to overcome the cohesive strength of the coating. It is not sufficient, therefore, that the masonry surface be free of efflorescence; whether or not there is any danger of these salts being brought to the surface in the future must be determined also. Further efflorescence indicates either that too much original water of construction is still present in the structure or that water is entering through some defect in the structure.

Fig. 6-2. (top) Efflorescence on concrete masonry. (*Courtesy Portland Cement Assn.*) (bottom) Efflorescence on brick.

The first step in the prevention of efflorescence is in the selection of materials. Brick and clay tile are available that do not contain soluble salts, and ASTM Specifications C358 and C91, for slag cement and masonry cement, contain requirements for cement specified to be nonstaining to limestone (i.e., "Nonstaining cement shall contain not more than 0.03 percent of water-soluble 'alkali'").

Moisture Content of Masonry. Water is the agent that has the greatest effect on the performance of masonry coatings. Water is necessary before either the process of saponification or efflorescence can proceed. In addition, oil- and solvent-based coatings cannot be successfully applied to a damp surface as proper adhesion cannot develop. Even though the surface may appear to be dry, water of construction may remain in the interior, and the application of a coating may trap this water, which, if it migrates to the coated surface, may cause a number of problems as previously mentioned.

A moisture meter is available for testing the moisture in masonry, and normally a reading of 12 percent or less is considered satisfactory for coating purposes.

Smoothness of Masonry Surfaces. The reative smoothness of the units to be coated will determine the kinds of coatings that may be applied. Smooth surfaces will usually permit the direct application of the finish coat, while rough or porous surfaces normally require a fill coat or primer before the finish coat is applied. (See Figs. 6-5 and 6-6.) The fill coat is used to fill up the open pockets in a rough surface and provide a more suitable base for the finish coat, while the primer is used to seal a porous surface to prevent too much surface absorption of the coating to be used.

Surface Preparation

Regardless of the type of coating selected for use, its success may depend largely on how well the surface is prepared. The masonry must be free from *dust, dirt, oil, grease, efflorescence, surface alkalinity* in some cases, or the *film left by strong cleaning fluids.*

Dust and dirt may be removed by washing down with a hose and scrubbing, brushing or blowing with compressed air, depending on the type of masonry, its location and the equipment available. Grease and oil may be washed off with solvent or a strong alkaline solution such as lye. The residue left by such treatments must then be removed by thorough rinsing with water. Most efflorescence can be removed by dry brushing or washing with a commercial masonry cleaner, followed by thorough rinsing with water. In cases where surface alkalinity must be reduced, it can be done by painting the surface, first with a three percent solution of *phosphoric acid,* followed by a two percent *zinc chloride* solution. Iron and other metallic objects embedded in the masonry surface should be covered with an anticorrosive primer before a water-based coating is applied.

COATING TYPES

Parging

Parging is simply a coating of mortar that may be applied to the exterior of masonry walls below grade. Where subsoil conditions do not cause a buildup of hydrostatic pressure, this may be all that is required to provide adequate waterproofing.

Type *M* mortar is recommended and is applied in two coats, each at least 6 mm (¼ in) in thickness. The first coat should be roughened when partially set, hardened for 24 hours and then moistened before the second coat is applied. (See Figure 6-3.) The second coat should be moist cured for at least 48 hours before backfilling.

Parging should be trowelled to a smooth, dense surface, extending from the footing to at least 150 mm (6 in) above the finished grade.

Fig. 6-3. Moistening first parging coat. (*Courtesy Portland Cement Assn.*)

It should be beveled at the top to form a wash and thickened at the bottom to form a cove between the base of the wall and the footing. (See Fig. 6-4.)

Bituminous Coatings

Bituminous coatings are materials that are produced from coal-tar or asphalt. They are furnished in *solid* form to be melted for hot application and in *liquid* form, either diluted (cut) with solvent or emulsified with water, for application at normal temperatures.

The coating is applied with broom, brush or spray, preferably over a bituminous primer coat. (See Fig. 6-5.)

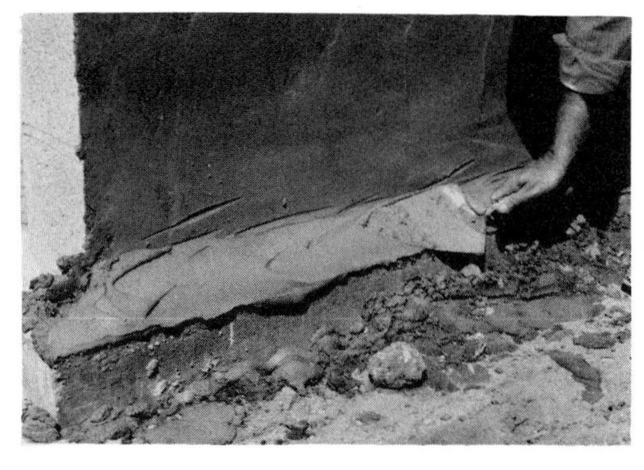

Fig. 6-4. Forming cove at footing with parging coat. (*Courtesy Portland Cement Assn.*)

Applications of bituminous coatings may be made in combination with some reinforcing material, such as bitumen-saturated felt paper, to form a built-up membrane. It is recommended that the membrane be made up entirely of coal-tar products or of asphaltic products but *not* a combination of both. *Cotton, felt, asbestos felt, woven glass fabric, glass fiber mat* and *woven burlap* are all available as bitumen-saturated ply materials.

The masonry surface should first be coated with a bituminous primer. The surface is then mopped with hot bitumen or cold bituminous emulsion, and a layer of fabric is rolled or broomed into the coating, overlapping about 300 mm (12 in) at the sides and ends.

The process is repeated until from two to six plies of fabric or felt have been applied, de-

Fig. 6-5. Bituminous coating applied by brush.

pending on the hydrostatic pressure to be resisted. A final coat of bitumen is applied, and some sort of rigid material, such as insulation board, is placed over the completed membrane to prevent it from being damaged during backfilling.

Membranes of this type have the advantage of maintaining the continuity of waterproofing over possible defects in the wall.

Bituminous coatings in general are impermeable and provide excellent resistance to the penetration of water. Their cost is low and their use recommended where appearance is not important.

Latex Paints

Latex paints are water emulsions of such resinous materials as *butadiene-styrene, polyvinyl acetate, epoxy resin* and *acrylic resin.* To these basic ingredients are added extenders, pigments, preservatives, defoaming agents, dispersing agents and flow agents to produce paints that dry throughout as soon as the water of emulsion has evaporated—usually in about one and one-half hours. They may be applied either to dry or damp surfaces, require no curing and provide a permeable barrier.

Polyvinyl acetate emulsions produce a much tougher film than the butadiene-styrene types and so can be used for exterior as well as interior coatings. One of its most important uses is for exterior finishes for masonry and stucco. Neither of the above types of latex paint can be successfully applied to a glossy surface and both must be protected from freezing before use.

Acrylic- and epoxy-resin-emulsion paints require no oxidation to form a film and remain flexible after drying. They have great resistance to weathering, tend not to lose adhesive qualities or color with age, contain no protein and so are not subject to deterioration. They are more expensive, however, than other latex paints.

Oil-Based Paints

Oil-based paints are manufactured from the resins of natural oils, such as *linseed oil, dehydrated castor oil, fish oil* or *soybean oil* or from *alkyd resins.* The latter are synthetic formulations made by combining a drying oil such as linseed oil, with glycerine and an acid component.

The oil-based paints designed for masonry use are usually reinforced with styrenated oils to improve their resistance to alkali, but to achieve success with them, surface alkalinity of the masonry must be reduced through aging of the masonry or by pretreating the surface. (See earlier discussion of alkalinity in masonry.)

They may be applied by brush, roller or spray to a dry surface. Since they form an impermeable coating, care must be taken that moisture from within the wall does not build up behind the film where it may cause the paint to fail by blistering and peeling.

Portland Cement Paints

Portland cement paints are produced in *standard* and *heavy-duty* types and are sold in powdered form in a variety of colors, to be mixed with water just before use. The standard-type paint contains a minimum of 65 percent portland cement by weight and is suitable for general use. The heavy-duty type contains 80 percent portland cement and is used where there is continuous and excessive contact with moisture, such as in a swimming pool. An additive, composed of fine siliceous sand, is available for use as a filler on porous or rough-textured surfaces.

Such paints are applied to a moist surface with a stiff brush and dampened by a fine water spray for from 48 to 72 hours, until the cement cures. They set by hydration of the cement, which bonds to the masonry surface.

Portland cement paints contain very little organic matter and are not subject to attack from alkali, which may be present on some masonry surfaces. (See previous discussion of alkalinity in masonry.) When properly applied and cured, they provide good protection against moisture penetration.

Epoxy Coatings

Epoxy coatings are based on epoxy or urethane resins and manufactured as two-component products, one of which is a catalyst that is added just prior to application.

These paints are highly resistant to alkali and form an impervious film but are relatively difficult to apply. Outdoor exposure of epoxy paints results in chalking, and this residue must be removed with soap and water to restore the original appearance.

Another coating involving epoxy consists of a two-component, *coal-tar-epoxy* product, that cures as a continuous film at room temperature. The film combines the characteristics of coal-tar pitch with the chemical resistance of epoxy resin to provide good immersion and environmental resistance to both fresh and salt water.

It is also highly resistant to many organic and inorganic acids, bases and salts, to most petroleum products, as well as to hydrogen sulfide liquors and sewage effluents. However, special equipment is required for application, and workers must be carefully protected against fumes, overspray or spills.

Epoxy coatings are relatively expensive and are intended for specialized requirements rather than for general use.

Polyester-Epoxy Coatings

The demand for heavier-bodied coating materials for use on masonry and concrete walls has led to the development of polyester-epoxy coatings, which contain a much higher percentage of solids than most other paints. They are two-part materials, a polyester component and an epoxidized oil component, that are

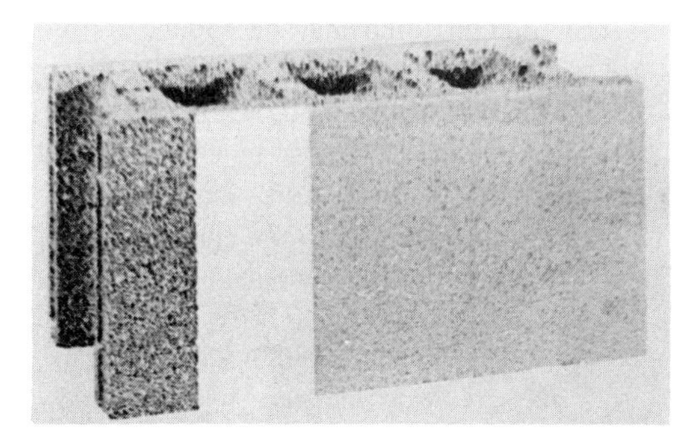

Fig. 6-6. Filler and finish coats on coarse-textured concrete masonry. (*Courtesy Canadian Pittsburgh Industries*)

mixed on the job at the time of application. Color is added to the polyester component before mixing.

Also available is a high-solids, vinyl filler material used in conjunction with the finish coating. It is applied directly over a coarse or porous masonry surface as a base for the finish coat. (See Fig. 6-6.) Where the surface has a fine, close texture, the finish coat may be applied directly to the masonry surface. (See Fig. 6-7.)

Fig. 6-7. Clear polyester-epoxy coating on brick: (left) before and (right) after. (*Courtesy Canadian Pittsburgh Industries*)

The filler material may be applied by brush, roller, or spray, to a thickness that will give approximately 406 μm (0.012 in) of dry film. The top coating, available in either semi-gloss or gloss finish, will add another 152 μm (0.0046 in) of dry film to the coating.

These coatings should not be used on exterior surfaces but on interiors. Here they provide a tough, long-lasting finish that is highly resistant to water, grease and many chemicals and can be cleaned with strong caustics. They are sometimes specified instead of ceramic tile because, with their use, the mortar joints, which would be exposed in a ceramic tile application, are also covered with an impervious coating.

A similar top-coating material is available for a clear finish, either in semigloss or gloss. It is to be used over previously painted surfaces or over natural surfaces, such as brick or stone, to preserve the natural appearance. (See Fig. 6-8.) One or two coats may be required, depending on the porosity of the surface and the degree of gloss required.

Both filler and top coat require overnight drying time before a second coat may be applied and approximately two weeks for complete cure.

Chlorinated Rubber-Based Paints

These paints are composed of chlorinated natural rubber blended with resins and pigments. They are applied as a three-coat system, preferably by roller, over a dry or slightly damp surface and produce a film about 205 μm (0.006 in) in thickness.

Because of its heavy consistency, such a paint easily fills the voids on porous surfaces and provides an impermeable barrier. It possesses good resistance to alkali and is useful in conditions of high humidity, in cold storage plants and to control corrosion in such industrial situations as pulp plants, sewage treatment plants and so forth. It should not be used in situations in which the coating will be in contact with animal or vegetables fats and oils.

Fig. 6-8. Polyester-epoxy coating on concrete masonry unit. (*Courtesy Pittsburgh Paints*)

Silicone-Based Coatings

Silicone is a colorless resinous material produced synthetically from *silicon dioxide* which, when applied to masonry surfaces, causes no change in color or texture.

The material does not actually seal openings but it does retard water absorption by changing the contact angle between water and the walls of the capillaries in the masonry. It will not bridge large openings, and when masonry with a coarse, open texture is involved, a fill coat should be applied under the silicone.

Application of silicone-based coatings is commonly accomplished by flooding the surface with a low pressure spray.

Metallic Oxide Waterproofing Compounds

Metallic oxide waterproofing compounds are used to waterproof the interior of masonry walls by filling the cracks and sealing the pores and voids in the masonry.

The compound consists of an oxide made from *powdered cast iron,* treated with a chemical *oxidizing agent* such as sodium peroxide, and mixed with *portland cement, sand* and *water.* A premixed, factory product is available, but the ingredients may also be mixed on the job.

A stiff mixture of the compound is first packed into any existing cracks and then a parge coat, fluid enough to be applied by trowel, is applied to the entire wall. This coat is cured for a minimum of 24 hours and then an additional thin coat may be trowelled on to provide a smooth finish.

Elastomeric Sheet Coatings

Instead of using a built-up bituminous membrane to waterproof a wall, an elastomeric sheet membrane may be used in the same way.

Elastomeric materials include *butyl, neoprene* and *ethylene-propylene-diene-monomers,* manufactured in sheets of up to 6 m × 30 m (20 ft × 100 ft) in size. The largest sized sheet that can conveniently be handled should be used, in order to minimize the number of joints necessary.

A cement that is recommended by the sheet manufacturer is used to secure the sheets to the masonry surface. Sheets are overlapped and spliced at joints with splicing tape.

Ceramic Glazes

Ceramic glazes are materials that may be applied to brick, tile or concrete masonry (see Fig. 2-16) during the manufacturing process, to give them a hard, durable, impermeable, often colored surface. Thus the units are laid with their protective coating already in place so that only the mortar joints have to be treated to render them as impervious as possible.

Comparison of Coatings

All of the materials described above have a place as protective coatings for masonry. Their suitability for a particular location and comparisons of the various coatings are outlined in Table 6-1.

MASONRY DECORATIVE COATINGS

In many cases it is desirable to add color to the exposed surface of some masonry walls, particularly those constructed of concrete masonry. In others it may be desirable to maintain or enhance the existing natural colors, such as those of brick or tile. In both situations, this can be done by the application of a decorative coating. Usually the same material that is applied for protective purposes is used also for decorative purposes.

Any of the coatings described in the foregoing discussion, to which colors may be added, can be used for decorative purposes. They include *portland cement paints, oil-based paints, latex paints, epoxy coatings, polyester-epoxy coatings* and *chlorinated rubber-based coatings.* Some are suitable for either interior or exterior work, while others are confined to interior surfaces. (See Table 6-1.) The same precautions and preparations observed for the utilization of these materials for protective purposes also apply to their use as decorative coatings.

Those materials that are available in a colorless state can be used as a coating to protect and enhance natural color in masonry products. They include *silicone, epoxy* and *polyester-epoxy coatings* (see Fig. 6-7); only the silicone is recommended for exterior use.

Glossary of Terms

Basic salts Salts that are alkaline in reaction.

Capillary A minute, thin-walled tube.

Chlorinated Combined with chlorine.

Table 6-1. Masonry Protective Coatings

Type of Coating	Suitability for Masonry Walls: Above Grade		Below Grade		Comparative Coating Factors									
	Int. Surface	*Ext. Surface*	*Int. Surface*	*Ext. Surface*	*Permeable Film*	*Alkali Resistant*	*Surface Condition Required*	*Usual Application*	*Moist Curing Required*	*Type of Thinner Used*	*Type of Finish*	*Coats Generally Required*	*Resultant Coating*	*Expected Service Life (Years)*
Parging			x	x	Yes	Yes	Moist	Trowel	Yes	Water	Flat	2	Opaque	Indefinite
Bituminous coatings				x	No	Yes	Dry or moist	Brush or spray	No	Solvent or water	Flat	2	Opaque	Indefinite
Portland cement paints	x	x	x		Yes	Yes	Moist	Brush	Yes	Water	Flat	2	Opaque	5–8
Latex paints	x	x	x		Yes	Yes	Dry or moist	Brush or roller	No	Water	Flat	2–3	Opaque or clear	4
Oil-based paints	x				No	No	Dry	Brush or roller	No	Solvent	Flat or gloss	2	Opaque	4
Epoxy coatings	x				No	Yes	Dry	Spray	No	None	Flat or gloss	2	Opaque or clear	4–6

Table 6-1. Continued

Type of Coating	Suitability for Masonry Walls				Comparative Coating Factors									
	Above Grade		*Below Grade*											
	Int. Surface	*Ext. Surface*	*Int. Surface*	*Ext. Surface*	*Permeable Film*	*Alkali Resistant*	*Surface Condition Required*	*Usual Application*	*Moist Curing Required*	*Type of Thinner Used*	*Type of Finish*	*Coats Generally Required*	*Resultant Coating*	*Expected Service Life (Years)*
Polyester-epoxy coatings	x				No	Yes	Dry	Brush roller, or spray	No	None	Semi-gloss or gloss	2–3	Opaque or clear	8–10
Chlorinated rubber-based paints	x				No	Yes	Dry or moist	Brush or roller	No	Solvent	Semi-gloss	2	Opaque	5
Silicone-based coatings		x			Yes	Yes	Dry	Spray flood	No	Water or solvent	Flat	2	Clear	5–8
Metallic oxide waterproofing compounds			x		No	Yes	Moist	Trowel	Yes	Water	Flat	2	Opaque	Indefinite
Elastomeric sheet coatings				x	No	Yes	Dry	Brush	No	None	Flat	2–3	Opaque	Indefinite

Epoxy A class of synthetics derived from certain special types of organic chemicals.

Hydrostatic Pressure Pressure exerted by water.

Litmus Paper A paper containing a dye that reacts to acids or bases.

Monomer A simple, unpolymerized form of a compound.

Oxidizing Agent A material used to supply oxygen for the oxidation process.

Phenolphthalein A white, crystalline compound, whose solution has a color reaction with acids or bases.

Polyester A combined form of an ester.

Saponification The process of conversion into soap.

Review Questions

1. Explain how a coating applied to a building material might affect each of the following:
   (a) sanitation
   (b) illumination
   (c) improvement of working conditions
2. Outline the methods by which water may penetrate into or pass through a masonry wall.
3. Explain in some detail the steps to be taken during construction of a masonry wall to try to ensure that it is waterproof.
4. Explain what is meant by a *permeable* coating and an *impermeable* coating and where each would be utilized in masonry construction.
5. Explain clearly:
   (a) the meaning of efflorescence
   (b) the cause of efflorescence
   (c) the effect of efflorescence on masonry coatings
   (d) how efflorescence may be controlled

Selected Sources of Information

Brick Institute of America, McLean, Virginia.

Canadian Building Digests, N.R.C., Ottawa, Ontario.

Canadian Pittsburgh Industries Ltd., Toronto, Ontario.

National Concrete Masonry Association, McLean, Virginia.

Chapter 7

PROPERTIES OF MASONRY WALLS

Masonry walls serve many purposes in a building. They serve as enclosures and protect against the elements. Load-bearing walls also carry loads from above such as floors and roofs. Masonry walls must not only have *strength* (compressive, flexural and shear), but must also provide a shield against *sound, fire, heat transfer* and *moisture.*

Masonry walls will behave in different ways depending on the forces to which they are exposed. An investigation will be made of masonry details showing some of the failure modes and how they are overcome. These areas of investigation will include *compression, thermal movement, unit shrinkage and growth, moisture infiltration and exfiltration.*

STRENGTH

Most building codes in masonry tend to deal almost totally with compression, particularly from the engineering aspect. The key tests on masonry also tend to be in the area of compression. There are a number of factors that affect the compressive strength of masonry walls; *compressive strength of units, eccentricity of load, slenderness of the wall, mortar bedding, workmanship, mortar strength* and *reinforcing.*

Testing seems to indicate that the compressive strength of the unit is the most important variable for a given situation. When using a specific unit, the type of loading and slenderness have more effect on the strength of the wall than does the mortar strength. The type of loading as illustrated in Figure 7-1 will determine the stress diagrams for the wall. Axial loading will result in a rectangular stress diagram, while eccentric loading will cause excessive compression at one face and very little compression or maybe even tension at the other face. A wall will support a greater axial than eccentric load. The strength of a wall is not affected by the slenderness providing it does not pass certain limits (20 for unreinforced walls and 30 for reinforced walls).

There is an increase in wall strength when hollow walls are fully bedded over face-shell bedding (see Fig. 14-47), but it is less than the increase in the net surface area. Full mortar bedding will increase the wall strength by 10 percent to 20 percent over face-shell bedding as illustrated in Figure 7-2. The mortar strength has very little effect on the compressive strength of the wall, especially in the stronger mortars (M, S and N) as shown in Figure 7-3.

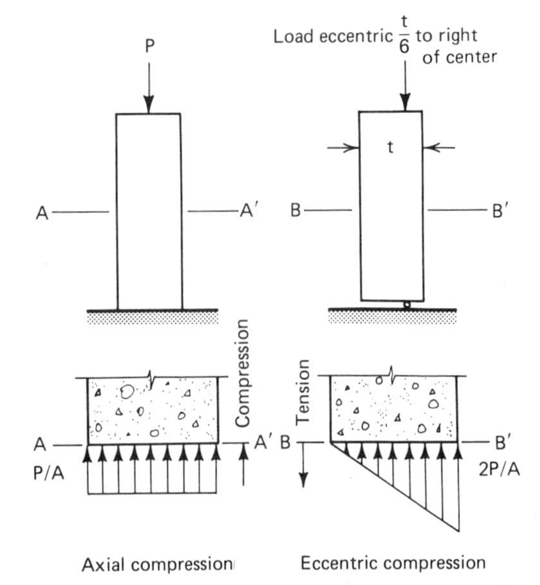

Fig. 7-1. Unit stress diagrams for solid masonry walls subjected to vertical loading. (*Courtesy National Concrete Masonry Assn.*)

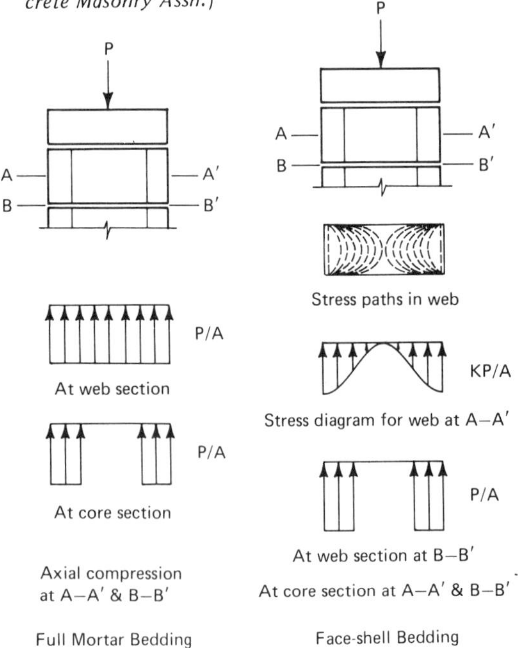

Fig. 7-2. Unit stress diagrams for block walls with full and face-shell bedding. (*Courtesy National Concrete Masonry Assn.*)

Workmanship has a significant effect on wall strength, especially where there has been poor attention paid to things like an incorrect adjustment of I.R.A. (initial rate of absorption) which will result in poor bond and shear. Mortar joints that are not fully filled or joints that are excessive in size affect the strength of the wall. Plumbness and/or poor curing conditions will also reduce wall strength.

When a wall has been grouted, fully or partially, the amount of reinforcing in these grouted spaces will show up in increased wall strength. The grout strength generally has little effect on the wall strength since the grout will separate from the wall with the shrinkage that occurs. This effect may be overcome by using expansive cements.

Masonry walls in high-rise buildings are built to resist lateral as well as axial loads. Vertical reinforcing can resist this racking, and for this reason, in the upper floors of a tall building, only vertical reinforcing is used; most of the horizontal reinforcing is placed at the floor level. Shear walls are generally placed in buildings to resist lateral forces and to transfer these loads to the foundation. (See Fig. 7-4.)

SOUND

Masonry walls are well suited to resist the passage of sound or noise from one side to the other. It is an important function of buildings to hold noise to a tolerable level. Noise is commonly defined as sound that is not wanted and this will vary depending on the individual's level of tolerance and on the activity he or she is engaged in. For noise that cannot be eliminated or reduced, some effort is made to absorb it or to reduce its transmission. Absorption will reduce the amount of sound generated within an area. The reduction of transmission affects the amount of sound that will be transferred from one area to another.

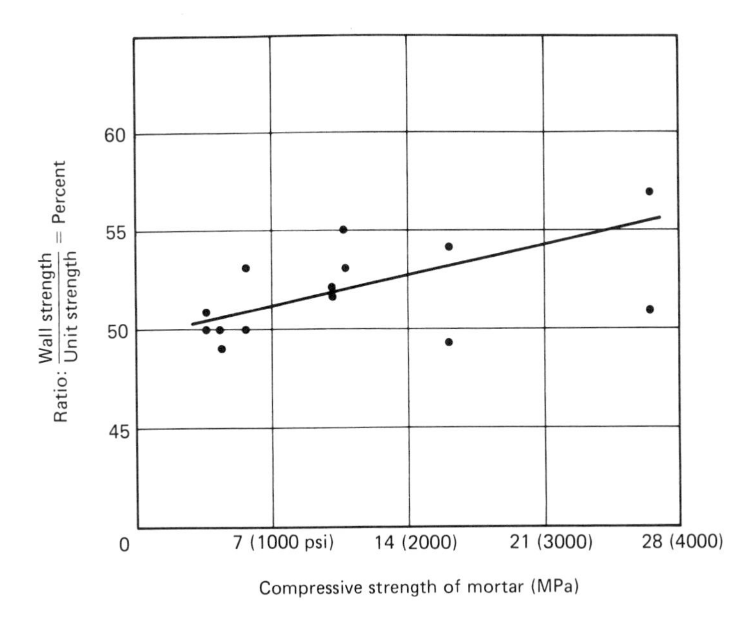

Fig. 7-3. Effect of mortar strength on compressive strength of hollow block prisms axially loaded. (*Courtesy National Concrete Masonry Assn.*)

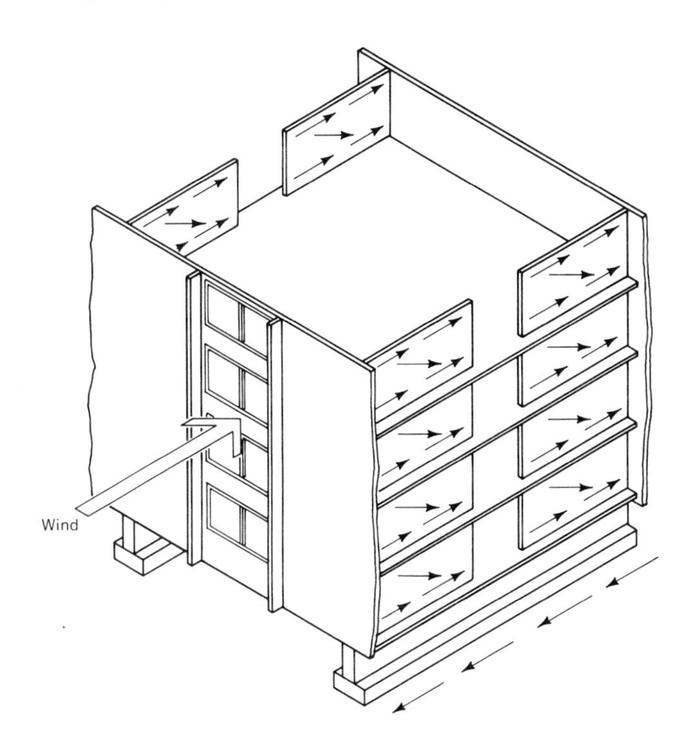

Fig. 7-4. Shear walls load transfer. (*Courtesy National Concrete Masonry Assn.*)

Sound Absorption

Sound is absorbed by using special absorbent materials on the surfaces to keep the sound from bouncing around within a space. If the surfaces are capable of absorbing 50 percent of the sound within an area, they would have a sound absorption coefficient of 0.50. In order to find the sound absorption of a room it is necessary to consider the sound absorption of a material and the area that is covered by the material in comparison to the total area. Materials that absorb sound will have a porous surface and will not tend to reflect sound. Carpets, furniture, and draperies effectively resist the flow of sound. Lightweight concretes, similar to the types used in concrete block, absorb sound well. (See Table 7-1.)

The frequency and the intensity of the sound will have an effect on the overall noise level. The frequency depends on the number of vibrations per second. One vibration per second is called a hertz. The intensity of the sound is measured in decibels which is the smallest change in sound that the human ear can detect. This range of sound can vary from the threshold of audibility to a level of intensity that will cause hearing loss. (See Table 7-2.)

Sound Transmission

Sound transmission deals with the problem of sound travelling through barriers from one space to another. Sound was once limited by isolating the source, but the trend now is to specify a sound transmission class for the wall and the impact noise resistance of the floor and design to them.

To deal with sound transmission, the wall must form an effective barrier to stop the passage of sound. Sound waves are made up of particles of energy that must be converted to

Table 7-1. Noise Reduction Coefficients (Absorption)

Material	Surface Texture or Finish	Approximate NRC
Brick unpainted	All	0.05
Concrete floor, bare	All	.02
Wood floor	All	.03
Glass	All	.02
Plaster	Rough	.05
Plaster	Smooth	.04
Wood panel	All	.06
Acoustical tile	All	.55
Carpet on concrete	Heavy	.45
Clay brick, unpainted	Smooth	.04
Lightweight aggregate block, unpainted	Medium	.45
Normal weight aggregate block, unpainted	Medium	.27
Clay brick, painted	Smooth	.02
Lightweight aggregate block, painted	Medium	.20
Normal weight aggregate block, painted	Medium	.12

(Courtesy National Concrete Masonry Association and Brick Institute of America)

Table 7-2. Sound Intensity Levels

	Decibels	Source	Comment
DEAFENING	150		Short exposure can cause hearing loss
	140	Jet plane takeoff	
	130	Artillery fire Machine gun Riveting	
	120	Siren at 30 m Jet plane (passenger ramp) Thunder–Sonic boom	Threshold of pain
	110	Woodworking shop Accelerating motorcycle Hard rock band	Threshold of discomfort
VERY LOUD	100	Subway (steel wheels) Loud street noise Power lawnmower Outboard motor	
	90	Truck unmuffled Train whistle Kitchen blender Jackhammer	
LOUD	80	Printing press Subway (rubber wheels) Noisy office Average factory	Intolerable for phone use
	70	Average street noise Quiet typewriter Train at 30 m Average radio	
MODERATE	60	Noisy home Average office Normal conversation	
	50	General office Quiet radio or television Average home Quiet street	
FAINT	40	Private office Quiet home	
	30	Quiet conversation Broadcast studio	
VERY FAINT	20	Empty auditorium Whisper	
	10	Rustling leaves Sound proof room Human breathing	
	0		Threshold of audibility

(Courtesy of the Brick Institute of America)

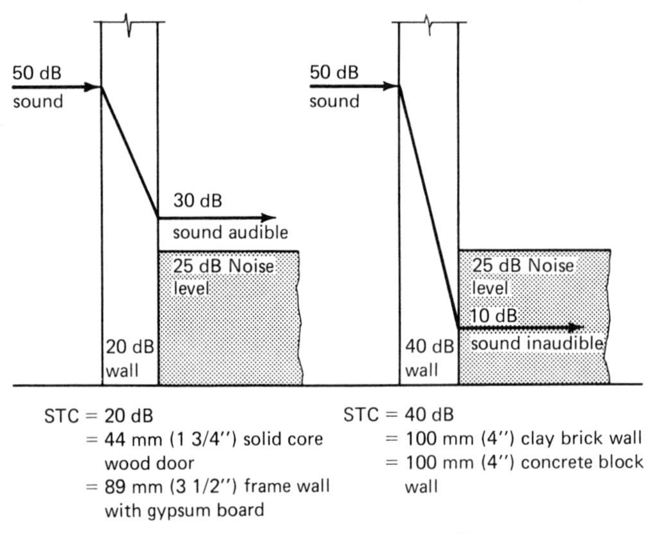

Fig. 7-5. Sound transmission of walls.

heat to stop the sound. This is accomplished by having a wall of sufficient mass (see Fig. 7-5) so that particles of energy will not penetrate to interfere with the sound level on the other side of the wall. This wall must be constructed so that there are no leaks, the results of which are illustrated in Figure 7-6.

Transmission loss will occur *through walls, around doors,* and *along plumbing pipes.* A single masonry wall having a mass of 390 kg/m^2 (80 lb/ft^2) will provide a barrier that will reflect a transmission loss of 50 dB as will a masonry cavity wall with two wythes spaced 50 mm (2 in) apart, each wythe having a mass

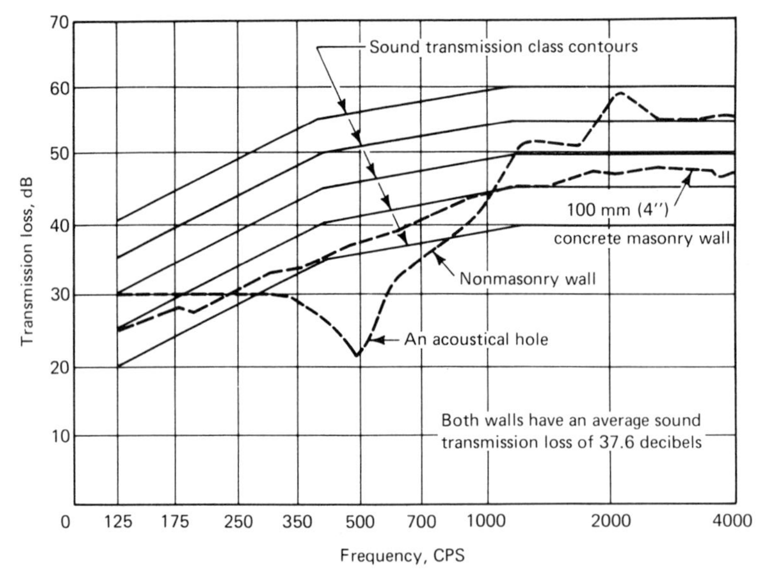

Fig. 7-6. Sound transmission loss. (*Courtesy National Concrete Masonry Association*)

of 98 kg/m² (20 lb/ft²). Cavity walls as shown in Figure 7-7 will provide a good barrier. Sealing a wall to make it airtight will decrease the sound transmission but will also decrease the sound absorption.

Doors should be avoided in sound barrier partitions, but if they must be used, they should be of solid wood and have gaskets on the top and on the sides. A sill or a strip should be provided at the bottom to make a complete seal.

Plumbing can be a good pipeline for sound transmission; quiet fixtures should be used in critical areas, and the pipes should be resiliently attached as illustrated in Figure 7-8. Complete soundproofing is not practical so some effort must be made to avoid placing quiet areas near noisy regions. Table 7-3 illustrates some sound transmission class limitations and will indicate which walls require extra soundproofing.

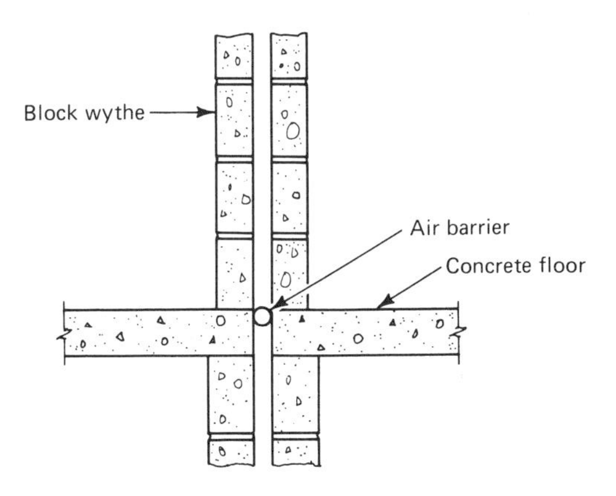

Fig. 7-7. Separating cavity walls.

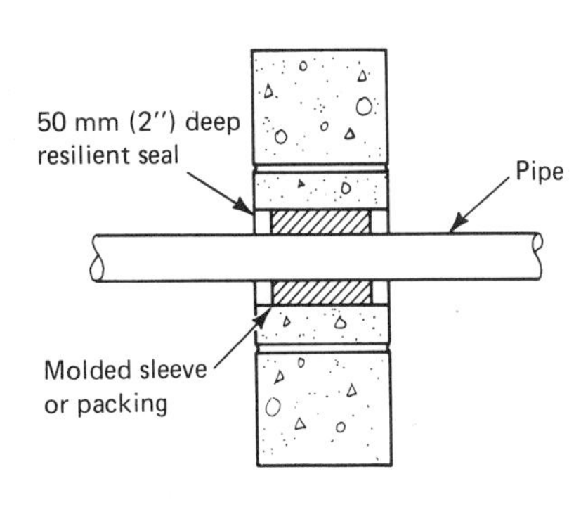

Fig. 7-8. Isolation of pipe through a masonry wall. (*Courtesy National Concrete Masonry Assn.*)

Table 7-3. Sound Transmission Class Limitations (dB)

	Low Background Noise		High Background Noise	
Location of Partition	*Bedroom Adjacent to Partition*	*Other Rooms Adjacent to Partition*	*Bedroom Adjacent to Partition*	*Other Rooms Adjacent to Partition*
Living unit to living unit	50	45	45	40
Living unit to corridor	45	40	40	40
Living unit to public space (average noise)	50	50	45	45
Living unit to public space and service areas (high noise)	55	55	50	50
Bedrooms to other rooms within same living unit	45	NA	40	NA

(Courtesy of National Concrete Masonry Association)

FIRE

Fire is a major hazard to human life and to the loss of property and therefore constitutes a major consideration in building bylaws. As codes are minimum regulations only, the careful designer should not be satisfied with these regulations and should design to achieve fire safety. Masonry walls have the ability to resist the spread of fire. They will not twist and buckle, dropping beams, floors and roofs into the interior of the building. Masonry does not contribute to the combustibility of the building, does not depend on sprinkler systems and in many cases can be repaired rather than replaced after a fire.

When designing buildings, the *combustibility, flammability* and *fire endurance* of building materials are vital factors. Although combustibility is important, it is the flammability of lining materials that is critical. It influences the speed with which a fire develops; a corridor lined with a flammable material can be rendered useless as an avenue of escape. There are several tests available to determine the degree of flammability of materials but the most common is ASTM Standard E84. The flammability should be kept to a minimum to provide time for occupants to escape and for fire fighting. (See Table 7-4.)

In tall buildings it is less possible to provide for rapid means of escape, so compartmentation is a logical consideration for limiting the spread of fire. In smaller buildings the fire walls will extend through the building, for height and width and sometimes beyond the face of the building, as in balconies. Generally the fire resistance of the enclosing elements will determine their effectiveness, whether in separation of occupied areas, either vertical or horizontal, or the enclosure of stairways.

Consideration should also be given to doors at exits. The doors should be of noncombustible materials and should be self-closing since an open door provides a very easy pathway for fire.

Table 7-4. Flammability of Conventional Lining Materials

Material	Minimum Thickness mm (in.)	Unfinished	Pigmented Paint Gloss Finish	Pigmented Paint Flat Latex
Masonry Sheet metal Asbestos board	7 (1/4)	25	25	25
Gypsum plaster and wallboard	9.5 (3/8)	25	75	25
Dressed wood	17 (11/16)	150	150	150
Exterior grade plywood	7 (1/4)	150	X	X
Hardboard	7 (1/4)	150	150	150
Fiberboard	11 (7/16)	X	150	150
Hardwood plywood	4 (1/8)	X	–	–
Particleboard	7 (1/4)	X	–	–

(Courtesy of National Building Code of Canada)

X = Rating of over 150
– = No test information available

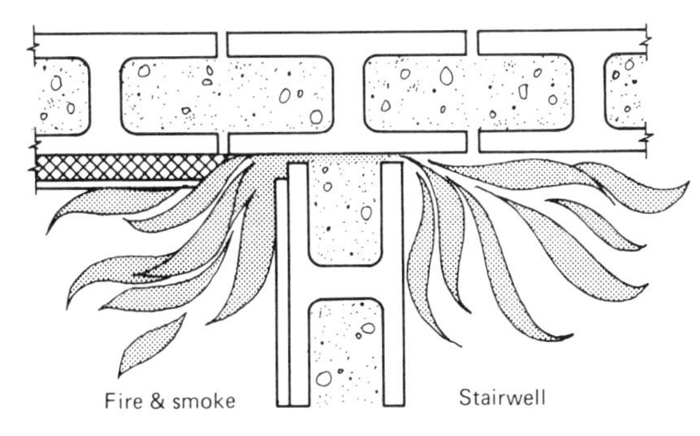

Fig. 7-9. Fire and smoke spread into stairwells. (*Courtesy Alberta Masonry Institute*)

The tightness of wall and floor junctions helps to stop the passage of flames into stairwells and should be considered. A small angle in the corner will inhibit the spread of fire (see Figs. 7-9 and 7-10).

Factors that influence the spread of flames:

1. Smooth hard surfaces normally will not spread flame as fast as soft or fuzzy surfaces.

2. Increasing the thickness of surfacing material will tend to retard the spread of flames.

3. The moisture content of a material will affect the rate of flame spread.

4. In general, materials such as stone, glass, metals, masonry products, ceramic tile, plaster, asbestos products and stucco will exhibit a low flame spread.

Design should endeavor to keep the flame spread to a minimum by considering flammability ratings of materials.

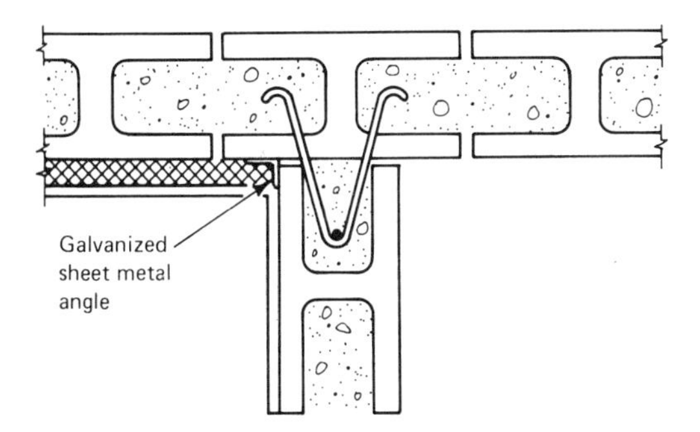

Fig. 7-10. Stopping fire and smoke spread into stairwells. (*Courtesy Alberta Masonry Institute*)

In order to assess the performance of walls, columns, floors, and other building members, standard fire endurance tests have been developed. The most common of these tests are ASTM E119, E152 and E163. (See Fig. 7-11.) To pass the test, the component must prevent the passage of fire, must not collapse or develop fissures, and must not transmit enough heat to the unexposed side to endanger combustibles stored there. The temperature on the exposed side should conform to the standard time temperature curve shown in Figure 7-12.

When dealing with hollow masonry, the fire endurance is based on the amount of solid material in the wall thickness as illustrated in Figure 7-13, although there is some fallacy to this statement as shown in Figure 7-14. The fire endurance of a wall made up of a number of layers, such as a cavity wall, is greater than the fire endurance of each layer. If the material is concrete masonry, the aggregate used will

Fig. 7-11. Masonry wall undergoing fire test. (*Courtesy Portland Cement Assn.*)

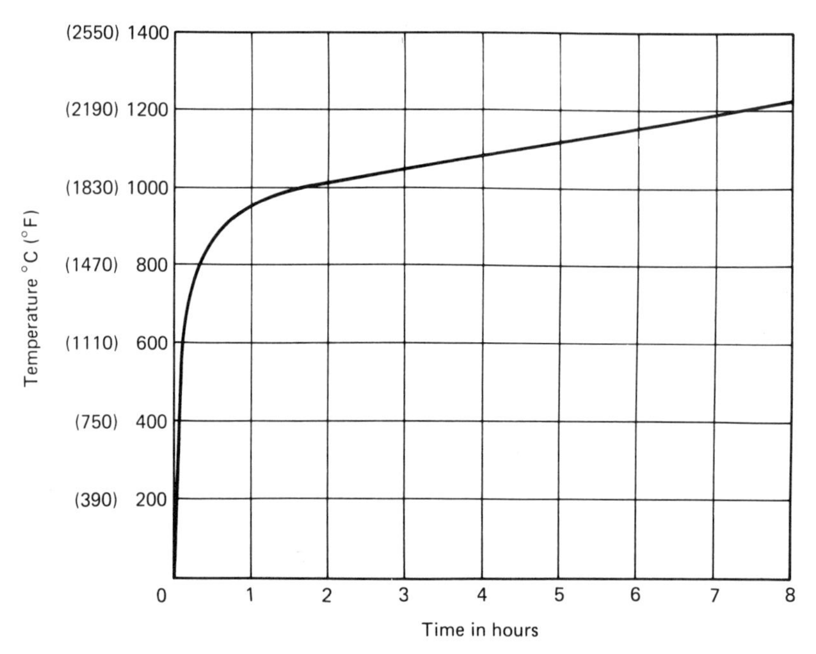

Fig. 7-12. Standard time-temperature curve for fire tests. (*Courtesy Portland Cement Assn.*)

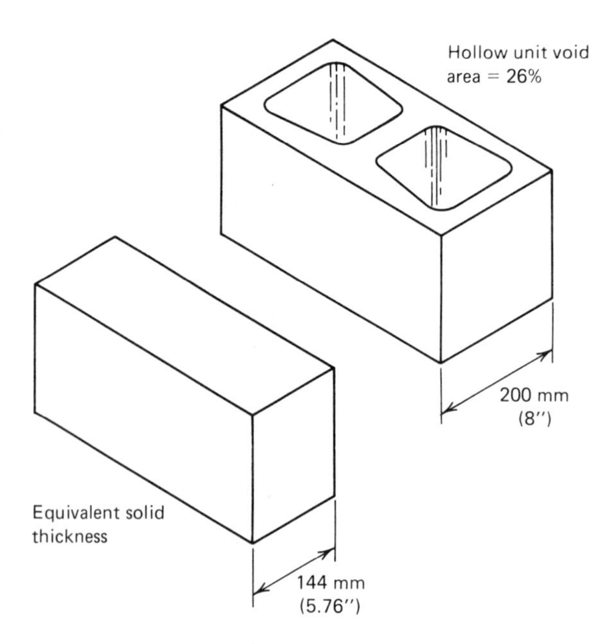

Equivalent solid thickness = % solid material × thickness
100 − 26 = 72 × 200 = 144 mm. (5.76")

Fig. 7-13. Equivalent solid thickness.

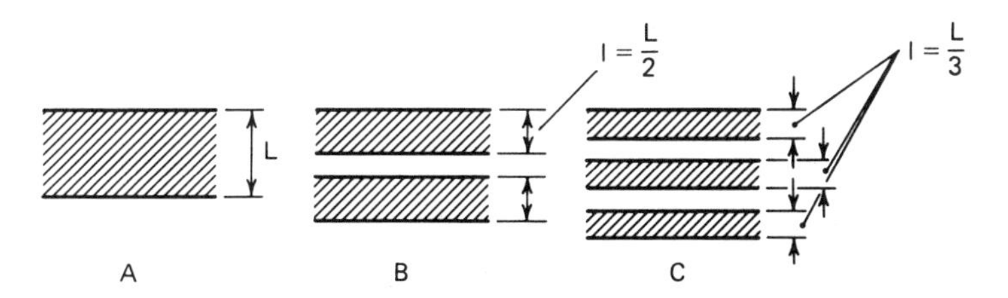

Fig. 7-14. Fallacy of equivalent thickness concept. (*Courtesy Canadian Building Digest #140*)

have an effect on the endurance as shown in Figure 7-15. The fire endurance is also increased by plastering the face of the masonry wall with at least 13 mm (½ in) of gypsum sand plaster. The endurance of hollow units can also be increased by filling the voids with dry granular material such as vermiculite or sand.

Smoke is more dangerous than fire; more people are injured by smoke than by the actual fire. Some effort should be made to control the smoke within a building. The smoke should be isolated to the floor or story where the fire has broken out; this is sometimes difficult as there is a natural tendency for smoke to move upwards. There are elements within a building that make natural smokestacks such as elevator shafts, stairways and ventilation and air conditioning systems. An opening at the top of a shaft will cause all smoke to flow into the shaft; to reverse the flow the shaft must be

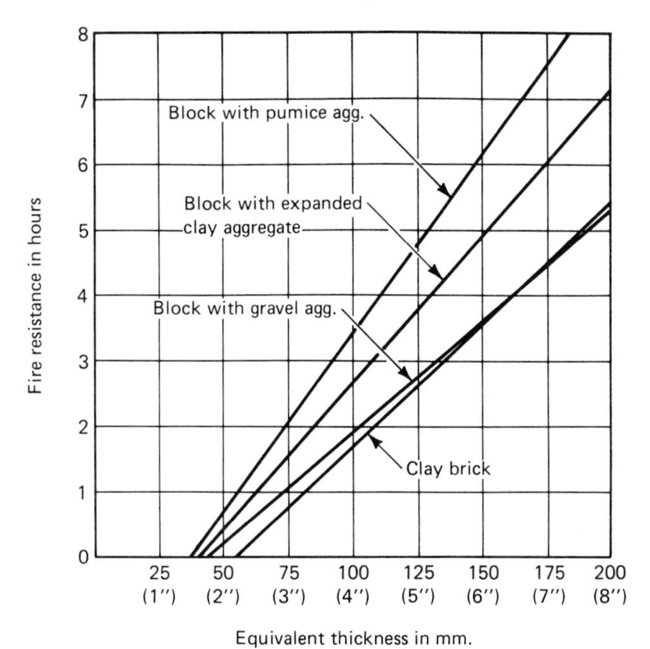

Fig. 7-15. Equivalent thickness vs. fire resistance. (*Courtesy National Research Council and Brick Institute of America*)

closed at the top and be pressurized. Ventilation and air conditioning systems should be shut off at the outbreak of a fire to cut down spread. Doors should also be closed at the outbreak to close off the compartments and to localize the smoke.

HEAT TRANSFER

The primary purpose of thermal insulation in exterior walls is to reduce the heat flow and to maintain desired air temperatures in winter and summer. *Interior air temperature* is not the only factor that affects comfort. *Humidity* and *air movement* are also important considerations. As the relative humidity is reduced, the temperature must be increased to keep the same comfort level. In winter conditions for example, the comfort level at 20°C (68°F) and 100 percent relative humidity is about equivalent to 26°C (79°F) and 10 percent humidity.

Air movement has significant effect on comfort. An increased air flow across the body can increase the convection losses, and when perspiration is occurring, heat is lost due to evaporation. This movement will increase the comfort in high temperatures but can have the reverse effect in low temperatures.

The rate of heat flow through walls, floors, ceilings, windows and doors depends on the difference in inside and outside temperatures and the resistance to heat flow of the materials used in the element. This heat flow, from a region of high temperature to one of low temperature, occurs by either *conduction, convection, radiation* or a combination of these. Conduction is usually given the most attention when insulating an element.

Other factors affecting heat flow are as follows:

1. The presence of moisture in the units raises the *U* factors and increases heat flow.

2. Surface color of the wall determines the amount of heat absorbed and radiated.

3. Thermal bridges such as ties tend to increase heat flow.

Coefficient of heat flow is the amount of heat that is transmitted through an element from one side to the other in a unit of time through a unit of area per unit of temperature difference. Any change in the properties of the materials forming the element that will cause a different quantity of heat to flow will reflect in a fluctuation in the *U* value. (See Table 7-5.) The placement of the elements will also affect the performance of the wall as illustrated in Figure 7-16.

Thermal resistance is the reciprocal of conductance and is the insulating value of a construction material. The coefficient of heat transmission (*U*) is the reciprocal of the total resistance of a construction assembly and is measured in watts per square meter, per one degree Celcius. (See Fig. 7-17.) An ability to calculate the thermal gradient will permit the forecasting of the magnitude of movement, the location of condensation and freezing planes. To illustrate the procedure of calculating the thermal gradient, an arithmetic or graphical method can be used as in the following example.

All components of the wall are listed in a tabular form, including air films with their thermal resistances listed opposite them. The total temperature drop across the wall is 50°C (90°F) and can be distributed among the individual components in proportion to their resistances. The interface temperatures then can be calculated by multiplying the resistance of the component by the temperature difference and dividing by the total resistance of the wall. This will give the temperature drop across the component. This can then be plotted on a graph as shown in Table 7-6. Consider a dewpoint of 7°C (45°F), which corresponds to a relative humidity of 35 percent in a room with a temperature of 24°C (75°F). There is sufficient insulation in the wall so that condensation will not occur inside the vapor barrier where the temperature is 11.8°C (53°F).

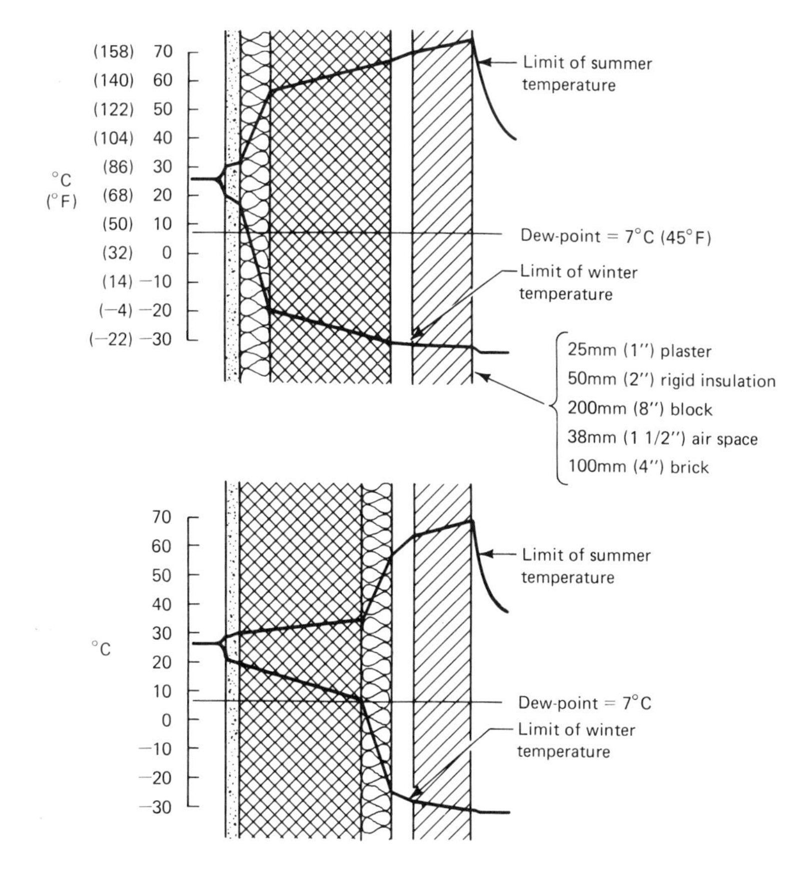

Fig. 7-16. Effect of placement of elements in a wall.

Table 7-5. *U* Value Fluctuations. Construction: 350 mm (14 in) Cavity Wall: 100 mm (4 in) face brick outer wythe, 50 mm (2 in) air space, 200 mm (8 in) hollow unit inner wythe (L.W. Agg.)

Wall Details	*U* Value
No insulation	1.24 (.22)
No insulation, 12.7 mm (1/2) gypsum board on strapping	.95 (.18)
Loose-fill insulation in hollow unit	.58 (.10)
Loose-fill insulation in hollow unit and gypsum board on strapping	.51 (.09)
25 mm (1) rigid foam in cavity	.70 (.12)
25 mm (1) rigid foam in cavity and gypsum board on strapping	.60 (.11)
Loose-fill insulation in hollow unit and 25 mm (1) rigid foam in cavity	.43 (.08)
Loose-fill insulation in hollow unit and 25 mm (1) rigid foam in cavity and gypsum board on strapping	.39 (.07)
Loose-fill insulation in hollow unit and 50 mm (2) rigid foam in cavity and gypsum board on strapping	.33 (.06)

Note: *U* Values are in watts/m^2/°C;
( ) Values are in Btu/ft^2/°F.

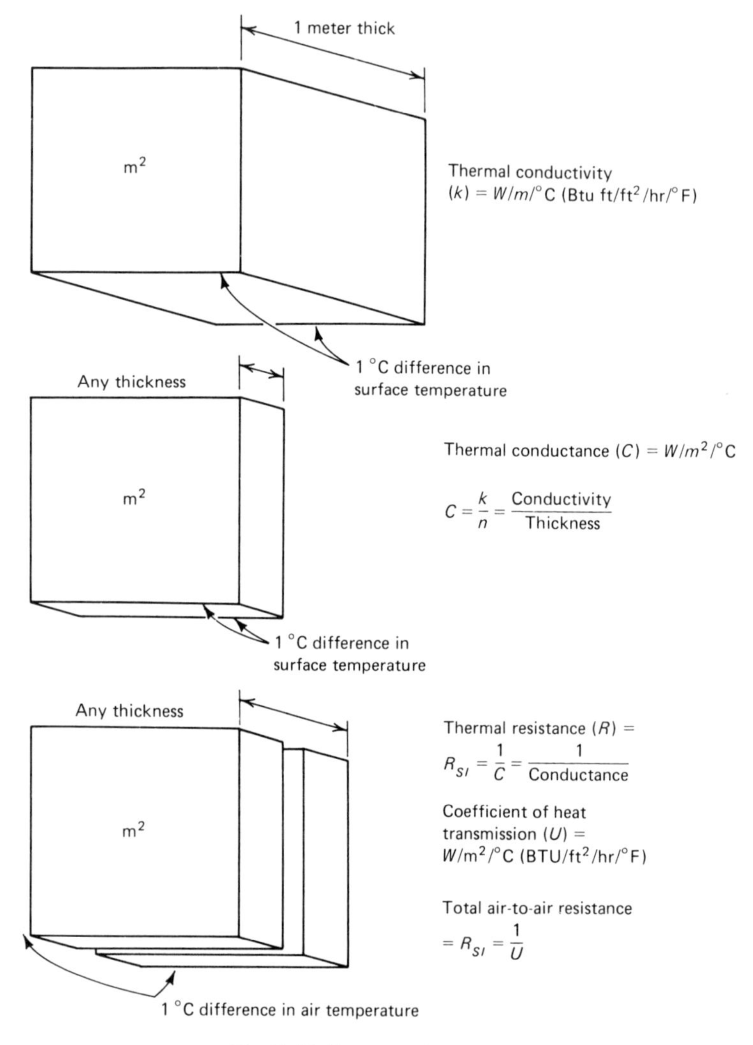

Fig. 7-17. Terms used.

MOISTURE

Most buildings are exposed to water in some way so consideration must be given to providing weathertight walls. In order for walls to be weathertight they must be *properly designed,* built of *quality materials* and show evidence of *good workmanship.*

Rain is the primary source of water that will penetrate masonry walls. Rain penetration results from a combination of water on the wall, openings to permit passage and forces to drive or draw it inward. It can be prevented by eliminating any one of these conditions. As it is almost impossible to design an exterior wall that will not get wet, and as openings are bound to occur in the form of pores, cracks and poorly bonded joints between elements or

Table 7-6. Arithmetic Determination of Temperature Gradient

Component	Thickness n, Meters	Conductivity k	Conductance C = k/n	Resistance $R_{SI} = 1/C$	Temperature Drop °C	Interface Temperature °C
						24
Internal air			8.29	0.12	3.0	
Film (still air)						21
Plaster finish	0.019	0.98	51.58	0.02	0.5	20.5
Concrete block	0.200	0.57	2.85	0.35	8.7	11.8
Vapor barrier		—	—	—	—	11.8
Wall insulation	0.050	0.04	0.80	1.25	30.9	-19.1
Air space	0.025	0.15	6.00	0.17	4.2	-23.3
Face brick	0.100	1.28	12.80	0.08	2.0	-25.3
External air			34.07	0.03	0.7	
Film (25 kmph)						-26
Total				2.02	50.0	

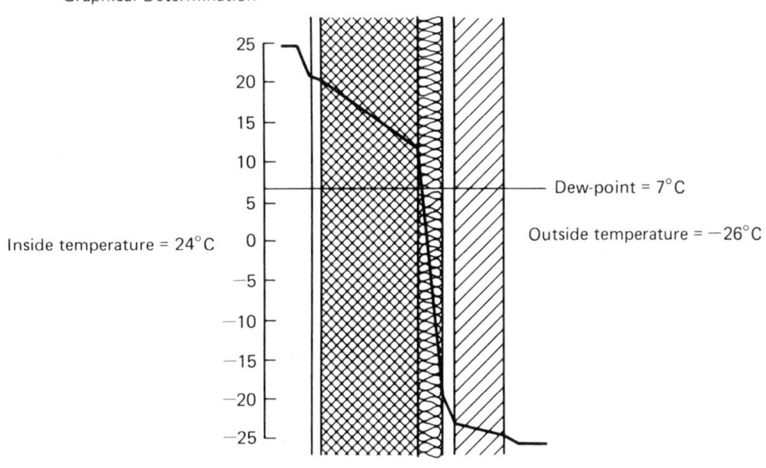

materials, consideration must be given to the force that will cause penetration. The forces contributing to rain penetration are *kinetic energy* of the rain drop, *capillary suction, gravity* and *air pressure differences.* They can be controlled by introducing an air space and an air barrier into the wall. (See Fig. 7-18.)

Design

Solution to moisture penetration must start with the designer. If quality materials and good workmanship are assumed, the design must be checked to see if water can enter and where it goes once it has entered. The three most common types of masonry walls (single-wythe, multiwythe and cavity) have differing degrees of moisture penetration. (See Figs. 12-1, 12-2 and 12-4.) The cavity wall provides the best resistance to water penetration since the vented space or cavity will eliminate the pressure difference and also provide a place for moisture that does penetrate. The bottom of this space can easily be flashed to the outside. The single-wythe wall, on the other hand, may require a coating. Consideration should also be given to the style of mortar joint (see Fig. 14-5) as the

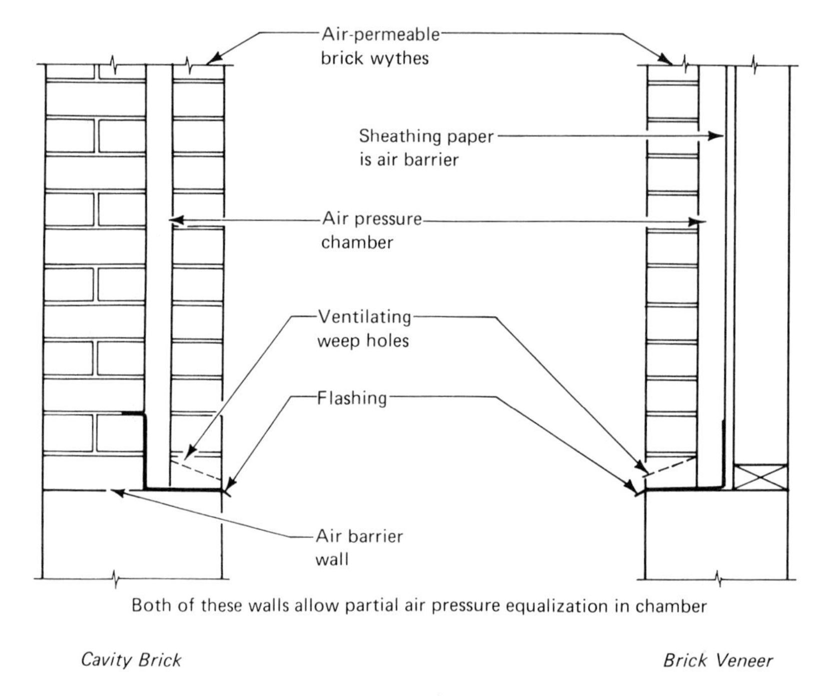

Fig. 7-18. Walls that resist rain penetration. (*Courtesy Canadian Building Digest #40*)

tooled ones are more weathertight. Parapets should be provided with a coping that slopes and extends over the edges complete with drips. (See Fig. 7-19.) Flashing is the most important consideration in the design and provides a method of diverting the moisture to the outside at openings, beams, parapets, roofs and floors. Workmanship, design and performance characteristics of masonry materials have been further researched and are given in ASTM and CSA specifications.

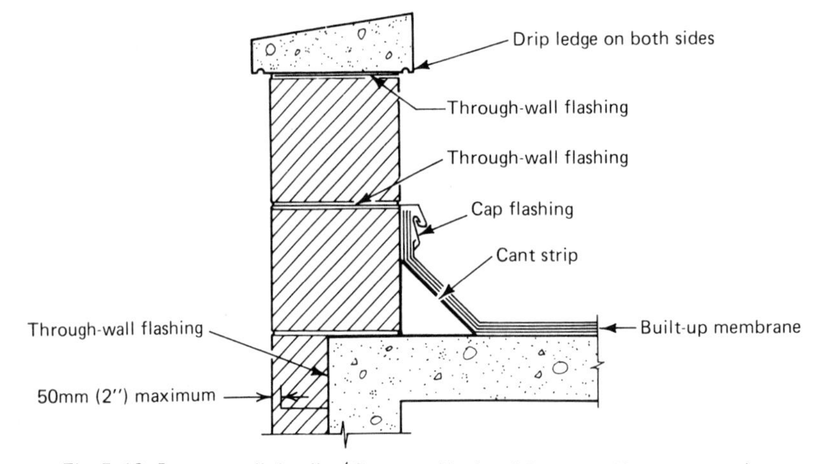

Fig. 7-19. Parapet wall details. (*Courtesy National Concrete Masonry Assn.*)

Workmanship

Workmanship applies to the care and technique that is used in the placement of the units and to the elements of design that depend on it for execution. Workmanship is probably the single most important factor that affects weathertightness. This involves placement of mortar, setting of units, installation of flashing and application of wall coatings.

Most of the discussion centers around water penetration from the outside to the inside, but movement in the other direction in the form of vapor should also be given serious consideration. The proper placement of vapor barriers can solve the condensation problem. This prevents the water vapor from reaching the condensation plane. If condensation does occur where wall surface temperatures are substantially below air temperatures, it may be eliminated by reducing the humidity of the air, by increasing air movement at wall surfaces or increasing the heat resistance of the wall.

MASONRY MODES OF FAILURE

Construction of masonry walls and details from the design intent into a finished structure is a difficult procedure unless precautions are taken. Even with the best intentions, failures can occur. Failure is usually considered as total collapse due to compressive or lateral loads or a combination of these. Total collapse without warning in a building is rare, but it has taken place. Failures usually show in the form of cracks in the wall that will let in moisture and air. Failures in a wall are usually attributed to *compressive and lateral loads, thermal movement, unit shrinkage* and *water and vapor movement.* Most failures are caused by poor design, poor workmanship and poor or no inspection.

Compressive and Lateral Loads

Failure due to axial loads normally shows up in the form of vertical splitting due to the horizontal tension in the masonry units. (See Fig. 7-20.) The reason for this type of failure is mainly the action of the mortar joint and the unit. The mortar is usually less rigid than the unit and, under load, it has a tendency to spread laterally. This spreading puts the mortar in triaxial compression and the unit in biaxial tension. ~~(See Fig. 8-33.)~~ Failure in the masonry occurs when the tensile stress in the unit reaches the ultimate tensile strength. It is obvious that the elastic properties of both the unit and the mortar influence the ultimate strength of the masonry. There is a tendency to make the mortar as strong as the unit; this would make the walls very brittle. The reason that masonry walls have stood for centuries is the flexibility of the joint, and for that reason you should keep a softer joint.

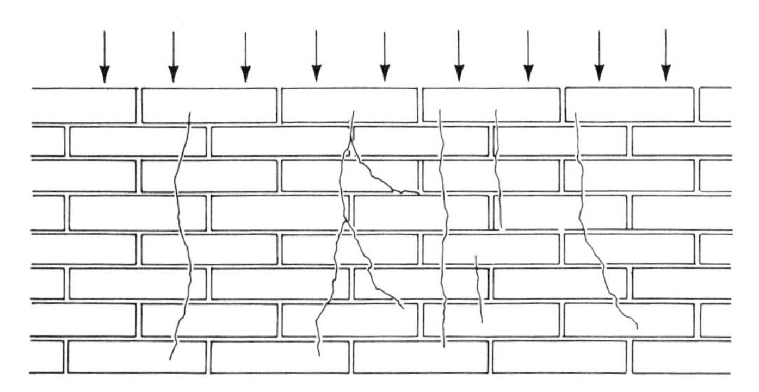

Fig. 7-20. Axial loads. (*Courtesy Alberta Masonry Institute*)

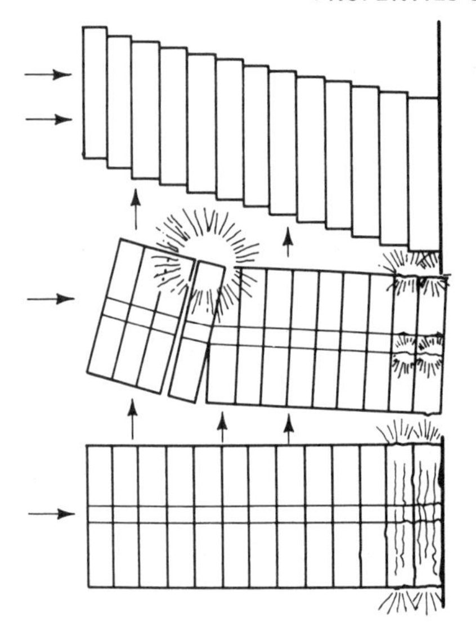

Fig. 7-21. Compression and shear. (*Courtesy Alberta Masonry Institute*)

Axial loads cause the classical failure of bottom floors that are overloaded and that start to fail as in Figure 7-21(a). When axial and wind loads act on a wall the failure is similar to Figure 7-21(b) where tension occurs on the windy side of the building and compression occurs on the opposite side. When the wind is the stronger of the two forces there is a tendency for the floor to slide as in Figure 7-21(c) though this is not too common. If this does occur, it will happen on the lower floors. These types of failure can be avoided by using stronger units, larger grout spaces and vertical reinforcing.

In a reinforced concrete frame building constructed with block infill panels and brick facing, failure can occur; the frame will shorten due to *elastic, moisture, thermal* and *creep* strains and the brick cladding will expand. (See Fig. 7-22.) Using soft mortar and leaving a space at each shelf angle will help alleviate the problem.

Failures in grout-filled block walls can also occur due to compressive loads and due to cores that are not completely filled as shown in Figure 7-23. This usually occurs when low-

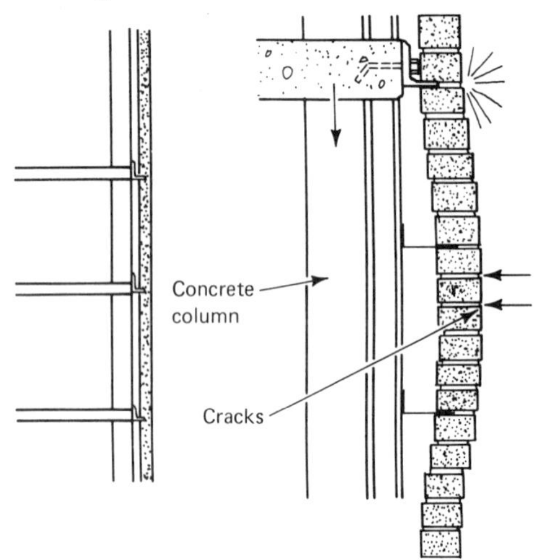

Fig. 7-22. Frame shortening. (*Courtesy Alberta Masonry Institute*)

Fig. 7-23. Failure of improperly filled void. (*Courtesy Alberta Masonry Institute*)

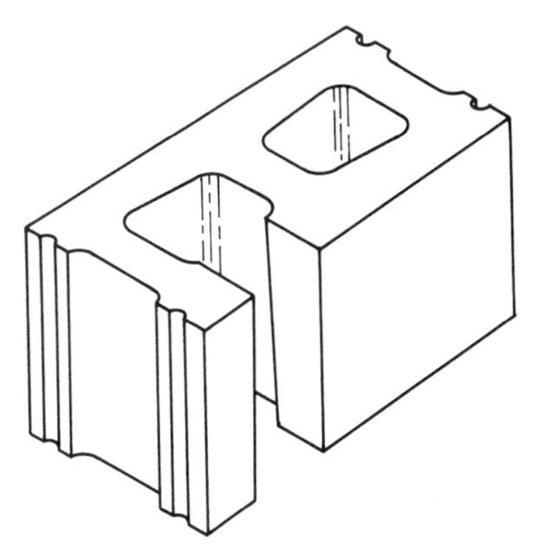

Fig. 7-24. Cleanout detail.

slump grout is used; even a vibrator will not cause it to settle because of the suction to the block. The cure is to use grout that has a slump of at least 200 mm (8 in) and to check the cavity at the bottom by tapping to see if it is filled. Cleanouts in the bottom row will ensure that the mortar drippings can be cleaned out. (See Fig. 7-24.)

Thermal Movement

Changes in temperature will cause movement in masonry walls. Concrete can withstand a certain amount of expansion and contraction, but masonry being more brittle often cannot.

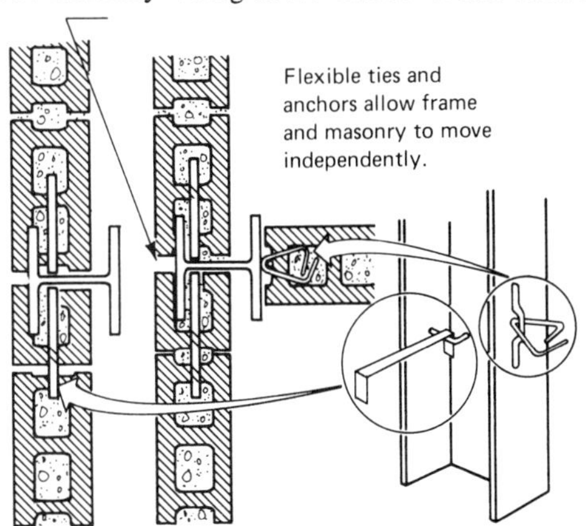

Fig. 7-25. Flexible ties. (*Courtesy Alberta Masonry Institute*)

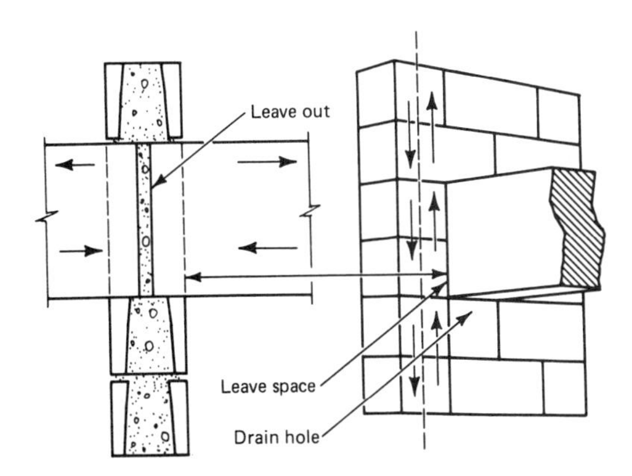

Fig. 7-26. Beam elongation. (*Courtesy Alberta Masonry Institute*)

When these cracks appear, water can enter and cause problems. Allowance should be made where differential movement due to temperature changes is likely to occur as in Figure 7-25.

Figures 7-26 to 7-29 illustrate some problem cases and possible solutions. Figure 7-26 shows the case of a center wall with a concrete beam coming into it. Failure can occur when the ends of the beams push against one another, but this can be solved by leaving a space between them

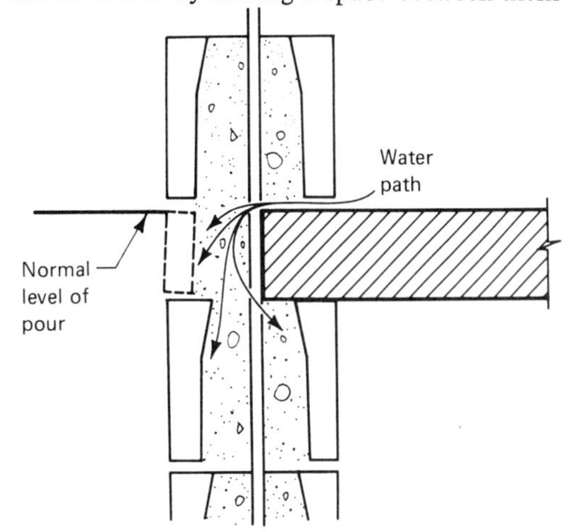

Fig. 7-27. "Slip" at floor level. (*Courtesy Alberta Masonry Institute*)

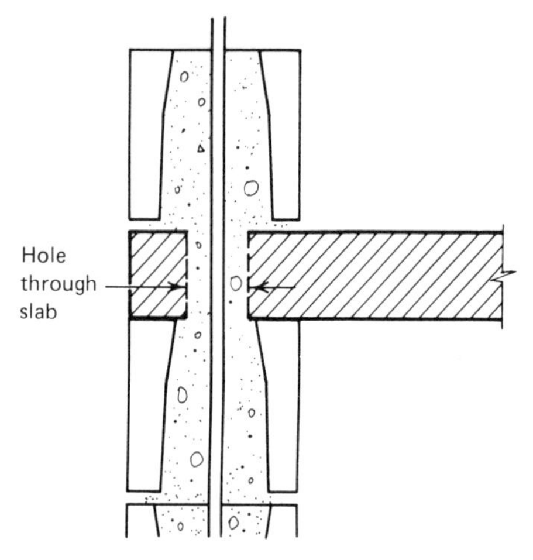

Fig. 7-28. Floor slab exposed at exterior wall. (*Courtesy Alberta Masonry Institute*)

and a slip surface underneath. This leaves a path for water to enter but if a drain hole is left the moisture can drain out. A common error in design is illustrated in Figure 7-27 where the floor level meets the exterior wall and is covered with a slip. Water has a tendency to work down behind the facing, and when freezing occurs, the facing pops. A better design is shown in Figure 7-28 where the floor slab extends to the outside perimeter of the building, with holes left for the wall reinforcing to pass through. Figure 7-29 indicates

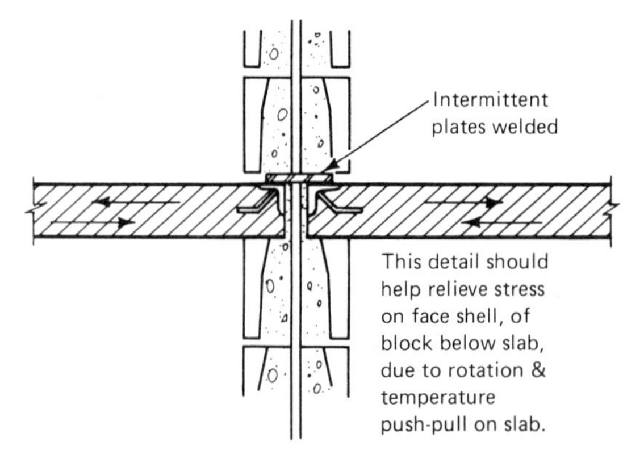

Fig. 7-29. Face-shell stress relief. (*Courtesy Alberta Masonry Institute*)

a method that will reduce the chance of the face shell popping off an interior wall due to stress placed on it by slab rotation and expansion and contraction.

Unit Shrinkage

Shrinkage has been one of the largest problems that faces designers of walls containing concrete masonry units. Although there are many factors that affect the shrinkage of concrete masonry, the major factors are the types of aggregates and the method of curing. Units made from sand gravel aggregates were found to have less shrinkage than units made of lightweight aggregates. When high pressure steam is used in curing, there is a reduction of up to 30 percent over units cured by other means. Clay units on the other hand have a tendency to grow upon contact with moisture. This cracking that occurs between the units and the mortar due to shrinkage can be minimized by letting units dry before laying and by using reinforcing in the joints.

Water and Vapor Movement

Water and vapor movement is generally broken into two parts; *infiltration* from the outside in the form of rain, snow and vapor; and *exfiltration* from the inside in the form of warm air and vapor.

Moisture infiltration can cause problems when water goes through the facing, runs down the back of the veneer, collects at the angle level as illustrated in Figure 7-30, and freezes. This can be remedied by installing a flashing at the angle level along with weep holes. Consideration should be given in the design to avoid water running continuously on a wall which will cause stains.

In high-rise buildings there is a tendency to arrive at situations where there is a negative pressure on one side of the building due to air movement and wind. (See Fig. 7-31.) This

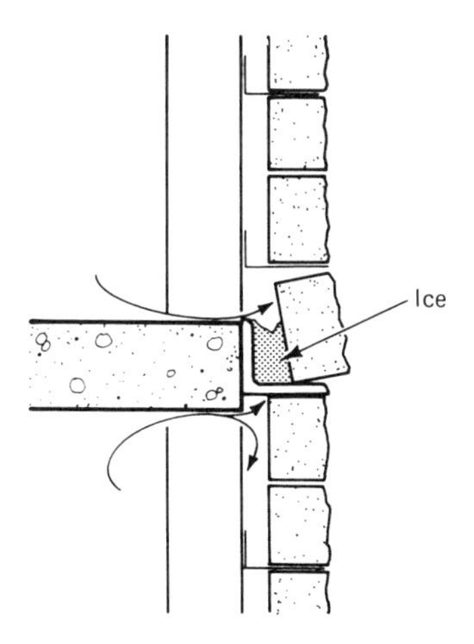

Fig. 7-30. Shelf-angle detail. (*Courtesy Alberta Masonry Institute*)

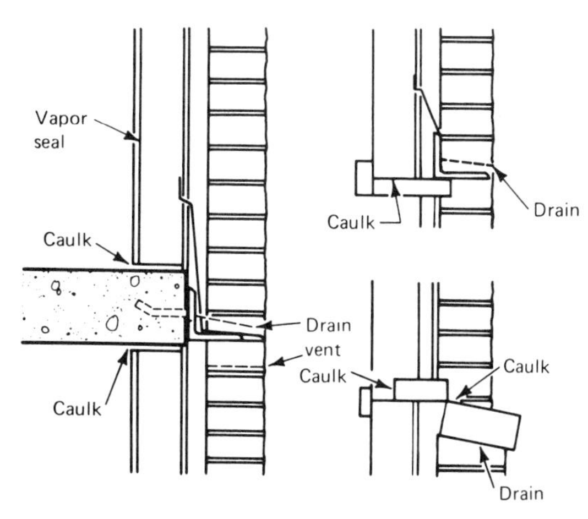

Fig. 7-32. Vapor seal. (*Courtesy Alberta Masonry Institute*)

pressure difference can cause vapor exfiltration. In many regions there is an additional vapor pressure buildup due to inside warm moist air moving towards the cooler drier air outside. The solution to this problem is to provide a complete seal on the inside as in Figure 7-32. This complete inner seal will also improve the moisture penetration resistance of a thick single-wythe wall as in Figure 7-33, where the movement of the moisture from the outside is impeded by the air within the material being driven in the other direction by penetrating moisture.

Other designs that cause failure are shown in Figure 7-34 where a precast facing is used at the floor level. This type of detail can cause construction sequence difficulties:

1. The fastening method of the precast; it must be installed ahead of the block, thus

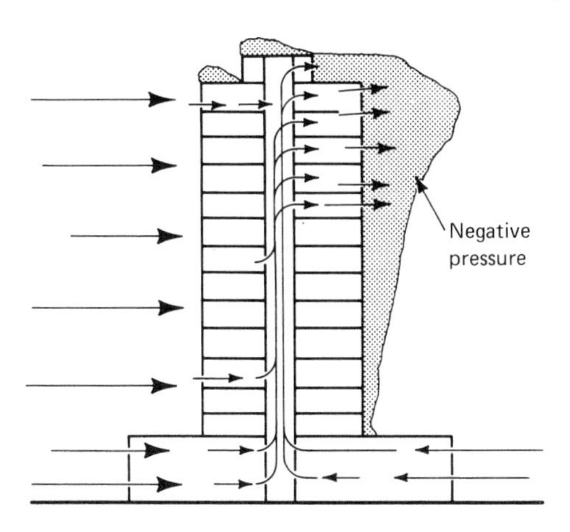

Fig. 7-31. Negative pressure. (*Courtesy Alberta Masonry Institute*)

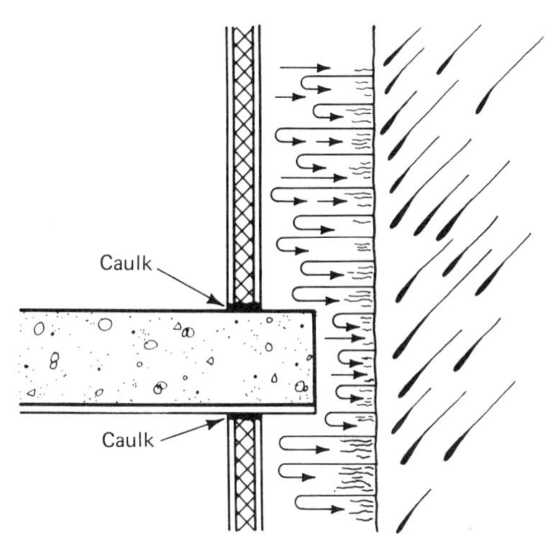

Fig. 7-33. Moisture penetration resistance of a thick single-wythe wall. (*Courtesy Alberta Masonry Institute*)

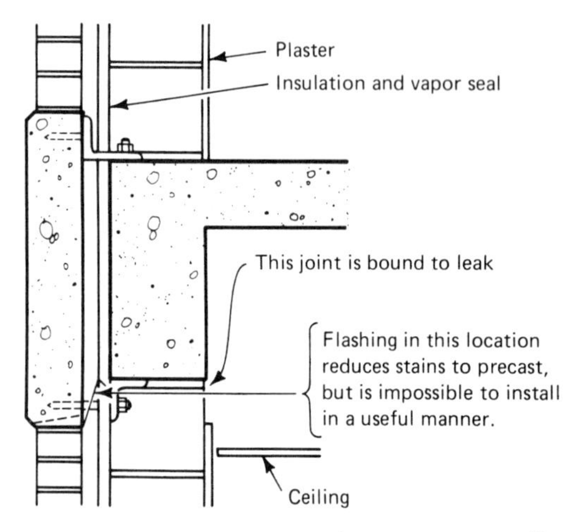

Fig. 7-34. Precast facing detail (*Courtesy Alberta Masonry Institute*)

making it almost impossible to properly place the flashing and insulation.

2. The block infill is placed next from the inside, ruling out tight joints in the last two rows, where the block also has to be broken around the fastener.

3. Any movement will cause a crack between the beam and the top row of block.

4. Insulation must be pushed behind precast where extruded joints rule out a tight fit.

5. Water vapor is now allowed to escape into the cavity, causing problems.

A better design, as far as potential failure is concerned, is to avoid the horizontal precast as shown in Figure 7-35.

Probably the classic example of a design that can only cause problems is shown in Figure 7-36 where block backup and brick facing are used in conjunction with a structural steel frame. This detail easily encounters problems such as:

1. Air leakage.

2. Difficulty in placing veneer and in maintaining a cavity.

3. Difficulty in anchoring block and in maintaining flashing around the corner.

The obvious solution to this detail is to eliminate the steel and reinforce the block so that it has load carrying capacity.

Inaccuracies in workmanship will cause many problems as Figure 7-37 and 7-38 show, but good inspection should alleviate this problem.

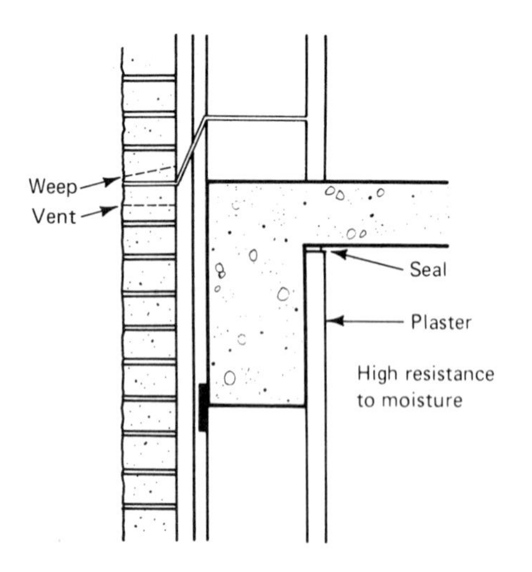

Fig. 7-35. Exterior wall-floor level detail. (*Courtesy Alberta Masonry Institute*)

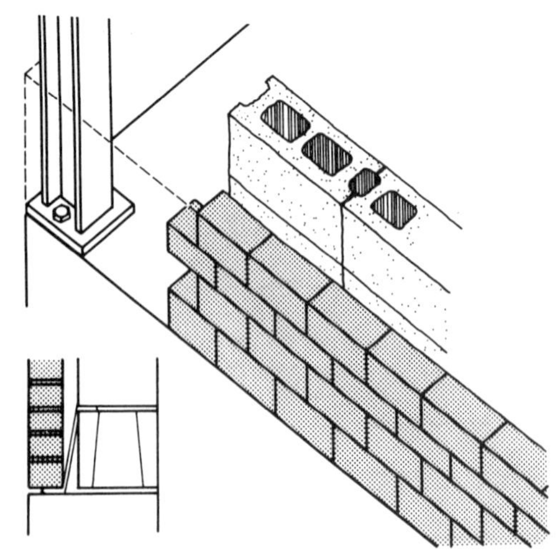

Fig. 7-36. Structural steel corner detail. (*Courtesy Alberta Masonry Institute*)

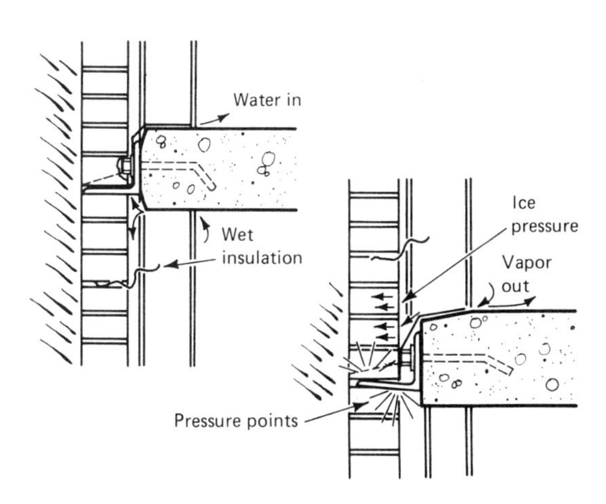

Fig. 7-37. Inaccuracies in walls. (*Courtesy Alberta Masonry Institute*)

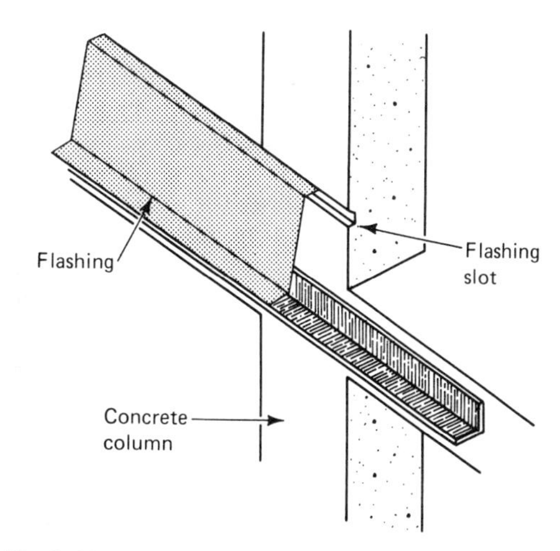

Fig. 7-38. Flashing detail at column. (*Courtesy Alberta Masonry Institute*)

Glossary of Terms

Axial Load A vertical load that bears vertically on top of a member.

Combustibility Capable of being burnt.

Compartmentation Dividing an area into smaller rooms to confine a fire.

Conduction Transfer of heat by molecular disturbance within a substance.

Convection Transfer of heat from one place to another by actual motion of the hot material.

Creep Strain Strain caused by creep within the material.

Flammability Ease with which a material burns.

I.R.A. Initial rate of absorption.

Moisture infiltration The passage of moisture through small openings from outside to the inside.

Radiation Thermal energy is transformed into radiant energy and may travel long distances before being changed back into heat.

U Value Coefficient of heat flow.

Vapor Exfiltration The passage of vapor through materials from inside to outside.

Review Questions

1. What element of a wall has the most effect on the compressive strength of the wall?
2. When is the strength of a wall not affected by the slenderness of the wall?
3. List three things that affect the amount of sound transmission through a wall.
4. What effect will sealing a wall have on the absorption and transmission of sound?

5. Why is the flammability of lining materials a vital consideration in the designing of buildings?
6. How can the stack effect of smoke in elevator shafts be reduced?
7. When a wall is built to the following specifications:

Exterior wall surface air	R_{SI} = 0.03 (R = .17)
100 mm (4 in) Limestone facing	R_{SI} = 0.06 (R = .33)
25 mm (1 in) Air space	R_{SI} = 0.17 (R = .94)
38 mm (1.5 in) Rigid insulation	R_{SI} = 1.06 (R = 6.0)
200 mm (8 in) Lightweight block	R_{SI} = 0.35 (R = 1.94)
19 mm (.75 in) Plaster finish	R_{SI} = 0.02 (R = .11)
Inside wall surface air	R_{SI} = 0.12 (R = .67)

would the wall:
(a) Have a *U* value of over 0.50 $W/m^2/°C$ (.088 $Btu/ft^2/°F$)?
(b) Create a condensation problem if the outside temperature was –20°C (–4°F), the inside temperature was 21°C (70°F) and the dew-point occurred at 8°C (46°F).
8. What affects the heat flow through a masonry wall?
9. How are the forces that contribute to rain penetration through a masonry wall controlled?
10. What effect does workmanship have on the performance of a masonry wall?

Selected Sources of Information

Alberta Masonry Institute, Calgary, Alberta.
Brick Development Assn., Lancaster, England.
Brick Institute of America, McLean, Virginia.
National Concrete Masonry Assn., McLean, Virginia.
National Research Council Canada, Ottawa, Ontario.
Portland Cement Assn., Skokie, Illinois.

Chapter 8

TESTING

This chapter will examine some of the standard tests used in the testing and evaluation of masonry materials and assemblages. Included in these tests will be investigations of *aggregates, mortar* (fresh and hardened), *grout, masonry units* and *assemblages.* Some of these tests are ideal for a laboratory setting while others are more suited to a field setting.

AGGREGATES FOR MASONRY MORTAR

The aggregates used in masonry mortars must consist of natural or manufactured sand. The material should be sound, free of deleterious substances and organic impurities, and should fall within an acceptable gradation limit. The most common tests deal with *sampling, sieve analysis, bulking, silt, color, deleterious substances* and *compression.*

Sampling

The important thing in sampling is to obtain a sample that is representative of the true nature and conditions of the material it represents. The size of the sample is usually determined by the size of the aggregate, but with fine aggregate used in masonry mortar, 10 kg (22 lb.) is satisfactory. This sample is then reduced to the necessary size by the use of a sand splitter.

Sieve Analysis Test

Aggregate used for masonry mortar should fall within the limits specified in Figure 8-1. If the fineness modulus varies by more than 0.20 from the value assumed, the sample should be rejected unless suitable adjustments are made. The test sample should weigh, after drying, 100 g for a sample of which at least 95 percent will pass through a No. 2.5 mm (0.1 in) sieve. Pass the sample through a standard set of sieves containing the sizes ranging from No. 5 mm (0.2 in) to No. 75 μm (0.003 in) as illustrated in Figure 8-1. The amount of aggregate retained on each sieve and the percentage retained is determined. A cumulative percentage is then determined, added and divided by 100 to obtain the fineness modulus as illustrated in Table 8-1.

When choosing sand for mortar, the finer side of the specification, as shown in Table 8-1, is recommended. Care must be taken not to use too fine a sand as shrinkage will increase, water retention will be low and more cement will be needed.

Bulking Test

This test is used to determine the amount a fine aggregate will bulk at varying degrees of moistness. Fill a 15L (0.5 cubic foot) "unit measure" with dry sand rodded in three lifts and weigh. Add water equal to 5 percent of the

Fig. 8-1. Sieve analysis.

mass of the aggregate and mix. Fill the unit mass measure with the damp, loose sand and weigh. Compute the bulking factor using this formula:

$$\text{Bulking factor} = \frac{A}{B-C}$$

where A = Unit mass of dry, rodded aggregate.

B = Unit mass of damp, loose aggregate.

C = Mass of water added to the dry aggregate.

This operation is repeated two more times, each time adding water equal to 5 percent of the original aggregate mass to the aggregate. Chart the results. The increase in volume will

Table 8-1. Sieve Analysis

Sieve Size	Percent Retained	Cumulative Percent
No. 5 mm (0.2 in)	0.0	0.0
No. 2.5 mm (0.1 in)	2.7	2.7
No. 1.25 mm (0.05 in)	14.7	17.4
No. 600 µm (0.02 in)	20.0	37.4
No. 300 µm (0.01 in)	36.7	74.1
No. 150 µm (0.005 in)	17.1	91.2
No. 75 µm (0.001 in)	4.6	222.8
Pass No. 75 µm (0.001 in)	4.2	

Fineness Modulus of the sand is 222.8/100 = 2.23   F.M. Range 2.6–2.9 Medium sand
2.2–2.6 Fine sand   2.9–3.2 Coarse sand

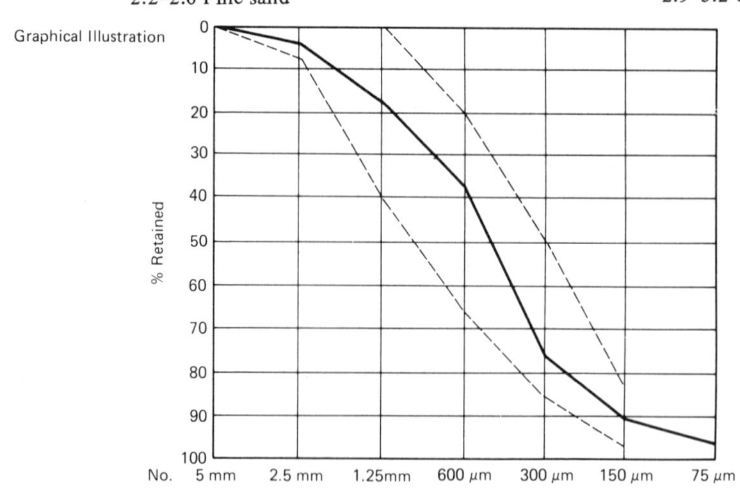

Fig. 8-2. Bulking test: before (top) and after (bottom).

depend on moisture content as illustrated in Figures 5-12 and 8-2.

Silt Test

An excessive amount of silt (clay, silt) in the sand used for mortar will affect the bond to the sand particles. The strength and workability of the mortar will also be affected and corrective measures may have to be taken. To perform the test, use a one liter jar and fill to 50 mm (2 in) with the sand sample; then fill the jar three-fourths full of water. Shake the container vigorously and let set for 24 hours. Measure the thickness of the layer of silt on top of the sand. (See Fig. 8-3.) If the layer is over 3 mm (1/8 in), the sand should not be used for masonry mortar without washing.

Color Test

This test is used to determine the amount of injurious organic material in sand. Fill a 350 ml bottle to the 130 ml level with the sand sample. Fill the bottle to the 200 ml level with a 3 percent solution of NaOH; shake the solution and let it stand for 24 hours. The color of the solution is compared with the standard color solution. (See Fig. 8-4.) If it is darker, it should not be used without further investigation of corrective measures.

Deleterious Substances

The amount of deleterious substances must not exceed the amounts shown in Table 8-2.

Compressive Strength

Aggregate used in mortar should develop a compressive strength of not less than 95 percent of strength developed when substituting Ottawa sand (fineness modulus of 2.40) in place of the aggregate.

Fig. 8-3. Silt test.

Fig. 8-4. Colorimetric test.

Table 8-2. Deleterious Substances

	Maximum Permissible Percent by Weight
Coal and lignite	0.25
Clay lumps	1
Shale, alkali, coated grains, soft or flaky particles	1
Other deleterious substances	as specified
Total deleterious substances	3

(Courtesy American Society for Testing and Materials)

MORTAR

Mortar is a main element in a masonry wall as it must perform a number of tasks. To meet performance needs, it is tested to meet certain specifications. Common tests include *flow after suction and water retention, compressive strength, proportion of cementious material, consistency* and *air content.*

Flow After Suction and Water Retention

To determine the flow after suction, the mortar is placed in a truncated cone (see Fig. 8-5), levelled off, and the cone is removed. The table is then dropped 25 times in a period of 15 seconds, and the diameter of the mortar is measured. (See Fig. 8-6.) The diameter of the cone at the base is 100 mm (4″). Should the mortar flow out to a size of 240 mm, (9.6″) the increase in diameter is 140 mm and the flow is 140/100 = 140 percent. The mortar is placed into the suction apparatus for one minute (see Fig. 8-6) and returned to the flow table. The flow is measured and, for example, it is 219 mm; (8.76″) the flow after suction would be 119/100 = 119 percent. The water retention is the ratio of the flow after suction to the flow before suction and would equal 119/140 = 85 percent. This is well within the 70 percent allowed and is very good for mortar.

Fig. 8-5. Flow test.

Construction mortars generally require more flow and water retention than laboratory mortars. To increase water retentivity the use of well-graded sand, highly plastic lime or increased air content are methods that are recommended.

Compressive Strength of Mortar

Twelve 50 mm (2 in) cube specimens are made to determine the compressive strength of mortar. Six of these cubes are tested at 7 days and the remainder at 28 days. The average strength of laboratory and job-prepared mortar mixed to a flow of 100-115 percent must not be less than the strengths shown in Table 5-2. If the mortar does not meet the 7-day strength

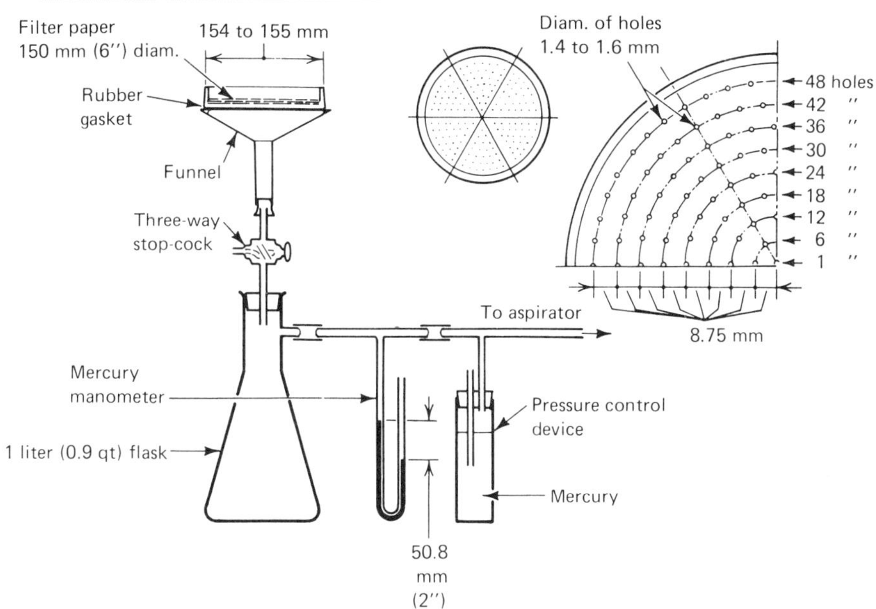

Fig. 8-6. Measuring flow after suction/water retention test. (*Courtesy Portland Cement Assn.*)

but meets the 28-day strength it is acceptable. If the mortar fails to meet the 7-day strength but is two-thirds of it, the contractor may continue at his own risk awaiting 28-day results.

The 50 mm (2 in) cubes should be filled after the mortar mix has been allowed to stand for 90 seconds. The cubes are filled in two layers; each layer is tamped 8 times in such a pattern that the entire surface is tamped 4 times. (See Figs. 8-7 and 8-8.) The cubes should be

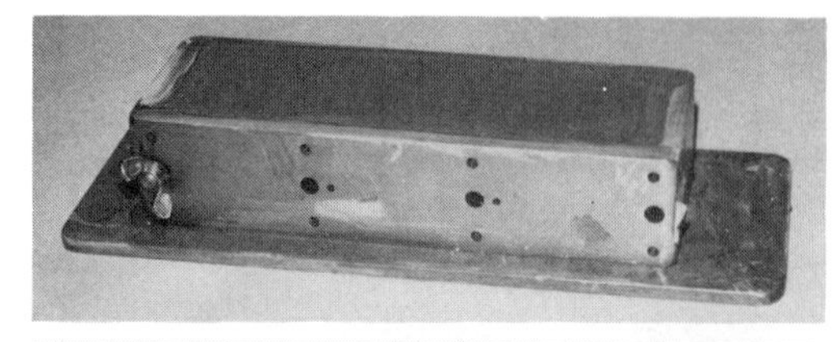

Fig. 8-7. Compressive strength of mortar. (*Courtesy Portland Cement Assn.*)

kept at the jobsite, covered in polyethylene and free of vibration, for 24 hours. The cubes are then taken to a curing room and stripped in 48 to 52 hours depending on the type of mortar. The cubes are stored in a curing room until it is time to test the specimens in a compression machine. (See Fig. 8-9.) The

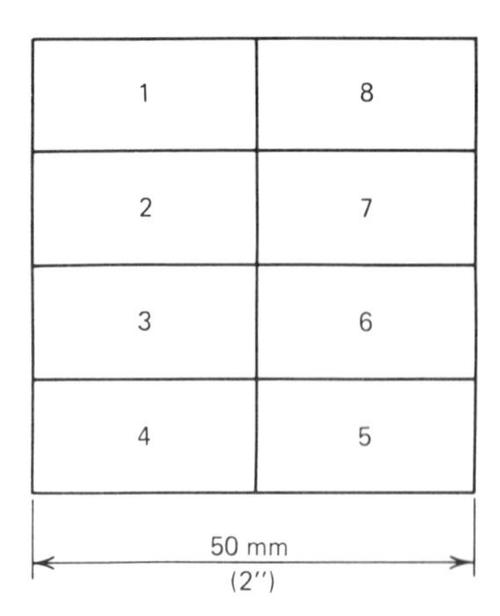

Rotate a quarter of a turn after tamping and repeat three times.

Fig. 8-8. Cube test tamp pattern.

Fig. 8-9. Compression machine.

compressive strength is calculated by the following formula:

$$\text{Compressive strength, } C = W/A$$

where C = Compressive strength in MPa (psi).

W = Load in N (lb).

A = Area in m^2 (in^2).

Test for Determining Proportion of Cementious Material in Fresh Mortar

This method is based on the wet screening of fresh mortar through a No. 150 μm (0.005 in) sieve and then adjusting for the fines in the aggregate. Obtain two 500 to 700 g samples of mortar to be tested and place each in a one-liter jar which contains 250 ml of methyl alcohol. Shake well. (See Fig. 8-10.) Determine the mass of the mortar by weighing jars before and after placing mortar into them. Obtain a sample of the sand used in the mortar, dry the sand, and then wet wash it through a No. 150 μm (0.005 in) sieve. The amount of sand left on the sieve is then dried and weighed. One of the samples of mortar is then wet washed through a No. 150 m (0.005 in) sieve and the portion retained is also dried and weighed. The alcohol is burned off the other sample of mortar and it is oven dried. (See Fig. 8-11.)

Fig. 8-10. Cementious material test.

The moisture content is determined by the following calculations:

$$\text{Mortar moisture content} = \frac{b - a - c}{c} \times 100$$
$$= \text{percent}$$

where a = Mass of container plus alcohol in grams.

b = Mass of container plus alcohol plus mortar in grams.

c = Mass of oven dried mortar sample in grams.

Fig. 8-11. Oven for drying samples.

The proportion of cementious material in the mortar is determined by the following calculation:

$$\text{Mass of wet mortar} = j = i - h$$

$$\text{Mass of dry mortar} = k = \frac{100j}{1+g}$$

Mass of + 150 μm mortar, dry, corrected

$$= q = \frac{yxr}{w}$$

Mass of - 150 μm mortar, dry, corrected

$$= p = k - q$$

$$\text{Proportion of cementious material} = 1 : \frac{q}{p}$$

$$\text{Percent by mass} = \frac{p}{p + q} \times 100$$

where h = Mass of container plus alcohol.

i = Mass of container plus alcohol plus mortar.

g = Moisture content of mortar.

y = Mass of mortar of fraction +150 μm sieve, dry.

r = Mass of sand dry.

w = Mass of sand of fraction -150 μm seive, dry.

Consistency of Mortar

Consistency of mortar can be measured by using a flow table or by the cone penetration method. The flow table method was discussed previously. The cone penetration test method determines the consistency by measuring the penetration of a conical plunger into a mortar sample. (See Fig. 8-12.) A cylindrical measure, having an inside diameter of 76 mm (3 in) and

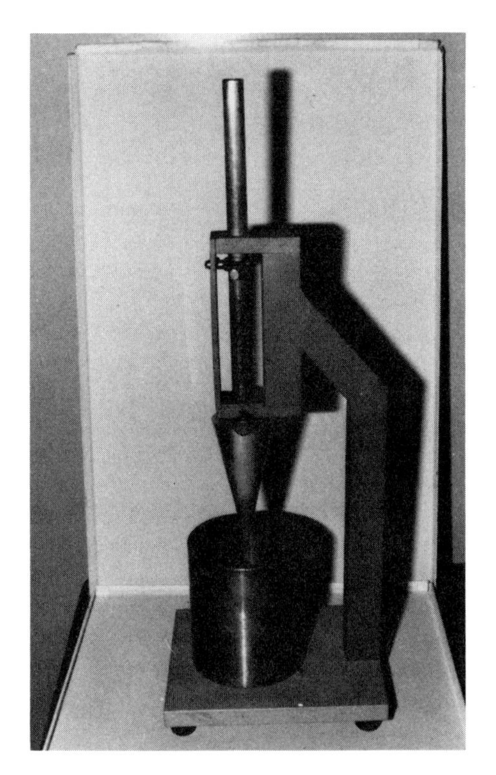

Fig. 8-12. Cone penetrometer.

a depth of 88 mm (3.5 in), is filled with mortar in three equal layers. Each layer is spaded 20 times with a metal spatula. The top is levelled and a cone 41.3 mm (1⅝ in) in diameter and 92.08 mm (3⅝ in) long is released into the mortar. The depth of penetration is measured in millimeters (or inches).

Air Content of Mortar

The air content of mortar can be measured by the *pressure meter method* or by the *volumetric method.* An indication of the amount of air in mortar can also be obtained by using the *air indicator.*

Pressure Method. A representative sample of mortar is placed in the bowl of the air meter in three layers. Each layer is tamped 25 times. The side of the bowl is then tapped a few times, the top is levelled, the top rim is cleaned and the top of the meter is fastened on. Water is put into the stopcocks to fill the air cavity above the mortar. Then the stopcocks are closed. Pressure is built up in the upper chamber to the calibrated level of the meter. This pressure is then released into the main bowl, and a reading of the air content is taken from the gauge as shown in Figure 8-13.

Volumetric Method. A sample of mortar is placed in the bowl of the meter in three equal layers. Each layer is tamped 25 times. The side of the bowl is then tapped a few times, and the top is levelled. The funnel-shaped top of the meter is fastened to the bowl, and water is added to the zero mark.

Fig. 8-13. Air content—pressure method.

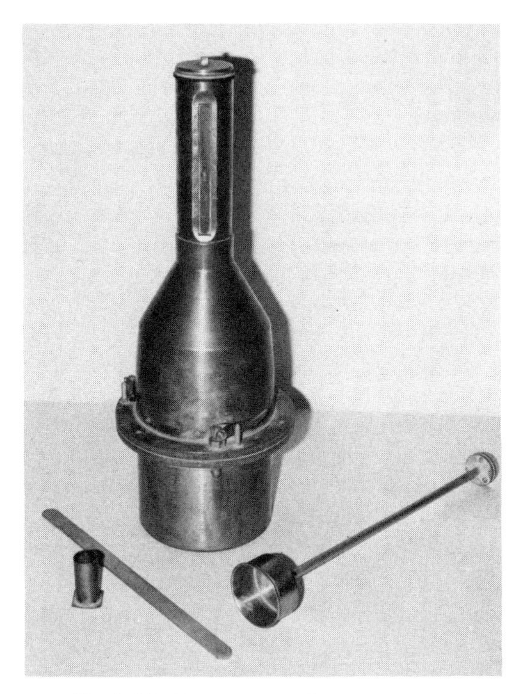

Fig. 8-14. Air content—volume method.

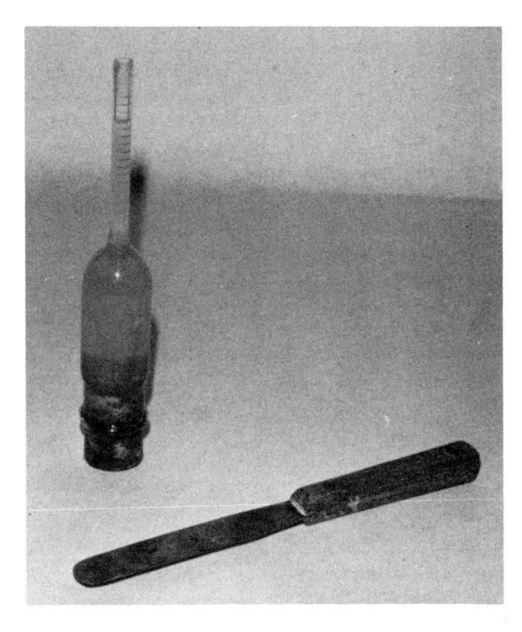

Fig. 8-15. Air content—air indicator.

The cap is screwed on and the meter is agitated to remove the air from the mortar. The meter is then placed in a vertical position. The drop in the water column will indicate the air content. If there is an excessive amount of foam on top of the water, alcohol should be added in measured increments to disperse the foam. The air content is determined by taking the reading off the column and by adding the amount of alcohol used to get the total reading as shown in Figure 8-14.

Air Indicator. To perform this test place a small sample of mortar in the cup of the indicator and tamp with a paper clip. Then pour the liquid in the funnel holding your thumb over the bottom. Invert the cup and place in the top of the funnel and shake the indicator. The drop in the level of the liquid will indicate the amount of air in the sample. (See Fig. 8-15.)

GROUT

Grout is used to fill voids in hollow units and the cavity between two wythes of masonry. This grout is often placed with reinforcing in a load-bearing wall to increase the strength of the wall. The grout should have a specific *consistency* and *compressive strength* to meet desired specifications.

Consistency or Fluidity Test of Grout

Two methods can be used for determining the consistency or fluidity of grout; the *slump test* and the *spread test.* (See Figs. 8-16 and 8-17.) The slump test is more common as it requires less equipment and is the same test commonly used for concrete. A truncated cone with a top diameter of 100 mm (4 in), a bottom diameter of 200 mm (8 in), and a height of 300 mm (12 in) is used in both tests. In the slump test, the cone is used with the narrow end at the

Fig. 8-16. Slump cone.

top. The mixture is placed in the cone, one-third of the volume at a time; then each layer is rodded 25 times. The cone is removed and the distance the grout settles is the slump of the mixture. An acceptable slump for grout is 250 mm (10 in) or over as a quantity of water is absorbed by the masonry units, thus reducing the amount of water in the grout. To perform the spread test, the same cone is used but with the large end up. Once the cone is filled, it is raised 50 mm (2 in) and the diameter of the grout is measured. The spread usually measures from 380 mm to 530 mm (15 to 21 in).

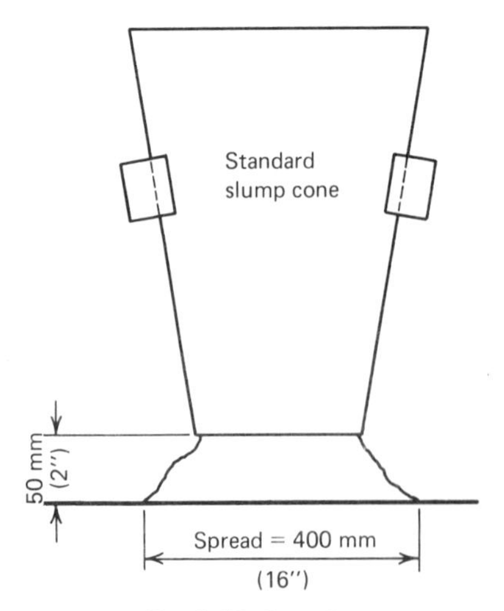

Fig. 8-17. Spread test.

Grout Compressive Strength (Field Test)

To perform this test, set up brick or block units as shown in Figure 8-18 in a flat, undisturbed location. Place a piece of wood 40 mm ($1\frac{5}{8}$ in) thick and 75 mm (3 in) square on a flat surface and arrange the units around it. The inside of the units are covered with a layer of absorbent paper towelling. Pour the grout into the 75 mm by 75 mm by 150 mm (3 in by 3 in by 6 in) mold in two layers, rodding each layer 25 times with a small stick. Level the top and cover to prevent evaporation. After 48 hours, place the prism into a curing room (100 percent relative humidity) for a 7- to 28-day curing period. The prism is then capped with a sulfur or equivalent cap and compressed in a testing machine. (See Fig. 8-19.) The compressive strength is then calculated. This is accomplished by measuring the area of the end of the prism and dividing it into the total load.

Fig. 8-18. Grout compressive strength test.

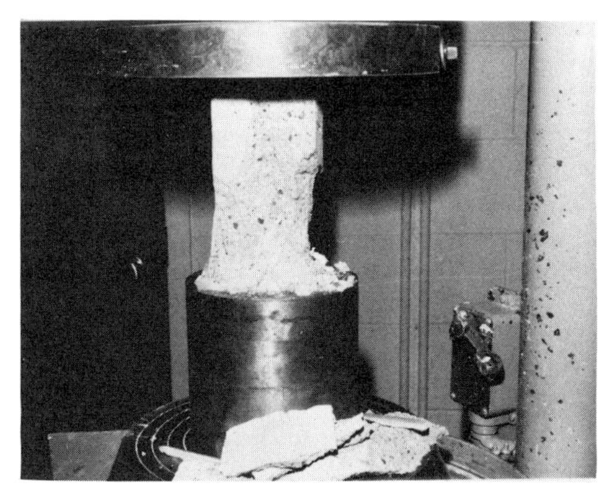

Fig. 8-19. Testing grout prisms.

Compressive strength = $C = W/A$

where C = Compressive strength in MPa (psi).

W = Load in N (lb).

A = Area in m^2 (in^2).

SAMPLING AND TESTING BRICK

This section deals with the procedures used for *sampling* and testing brick for *size* and *warpage, flexure, compression, absorption* and *suction.*

Sampling

For the purpose of tests, bricks are selected to represent the entire shipment. At least ten bricks should be selected per 50,000 and should be tagged for identification.

Size

Ten dry, full-sized bricks should be measured with calipers or gauge graduated in a minimum of 1 mm ($\frac{1}{32}$ in) graduations. The length, width and depth should be measured along both beds and along both faces at the midpoint and then averaged. These averages should be calculated to the nearest 0.5 mm ($\frac{1}{64}$ in). (See Fig. 8-20.)

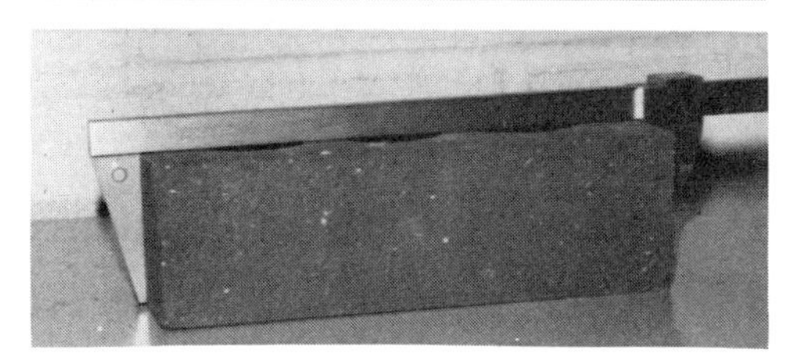

Fig. 8-20. Measuring brick size.

Warpage

Ten bricks should be alternately placed on a ground metal or glass surface. Where a concave surface is evident, a straight edge is laid lengthwise or diagonally, whichever has the greatest warpage, and the distance from the straight edge to the brick is measured. When a convex surface is encountered, the brick is laid with the corners approximately equidistant from the plane. An average of the distances at each corner is recorded to the nearest 0.5 mm ($\frac{1}{64}$ in). (See Fig. 8-21.)

Flexure

Five dry test specimens are supported on the flat on a span of 175 mm (7 in) and loaded at mid-span by means of a metal-bearing plate

Fig. 8-21. Brick warpage.

6 mm (1/4 in) thick and 38 mm (1½ in) in width. The supports for the specimen should be free to rotate in a longitudinal and transverse direction. (See Fig. 8-22.) The rate of loading should not exceed 9 kN/min (2000 lb/min). The flexure-stress is calculated as follows:

$$S = 3Wl/2bd^2$$

Where S = Stress in specimen in MPa (psi).

W = Load indicated by testing machine in Newtons (lb).

l = Distance between supports in m (in).

b = Width of specimen in m (in).

d = Depth of unit in m (in).

Compression

The five test specimens consist of half bricks with the length equal to the width, or if the bricks have cells, the entire unit can be tested.

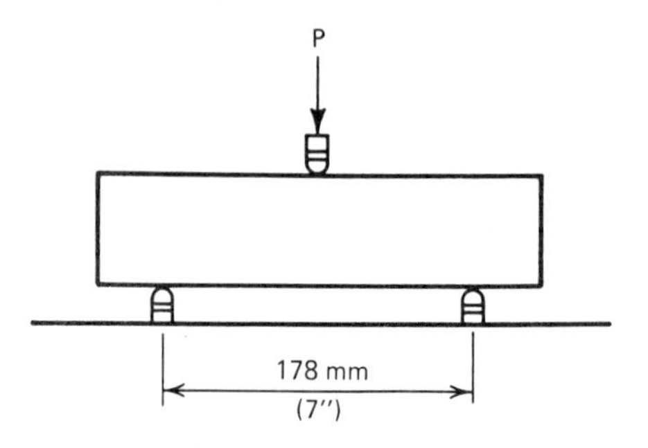

Fig. 8-22. Brick flexural test.

The units are capped with sulfur, cement-gypsum or gypsum plaster to give two flat parallel surfaces. The specimens must be compressed flatwise with an acceptable testing machine at any rate up to half the load. Finish the loading in one to two minutes. The compressive strength is calculated as follows:

$$\text{Compressive strength, } C = W/A$$

where C = Compressive strength in MPa (psi).

W = Load in N (lb).

A = Area in m^2 (in^2).

The average compressive strength of all the specimens shall be considered as the strength of the lot.

Absorption

Five test specimens (half bricks) must be tested. The test specimens are dried at 110°C to 115°C, (230°F to 240°F), cooled, weighed, and then submerged in distilled water (15.5°C to 30°C or 60°F to 85°F) for 24 hours. The surface water is wiped off and then the specimens are weighed. The absorption of each piece is calculated as follows:

$$\text{Absorption, percent} = \frac{100\,(W_2 - W_1)}{W_1}$$

where W_1 = Dry mass of specimen.

W_2 = Saturated mass of specimen.

An alternate method is also possible. After the specimen is submerged for 24 hours, it is returned to the bath and the water is heated to boiling. It is boiled for five hours, then cooled for 16 to 18 hours and the calculation is the same as before.

Initial Rate of Absorption (Suction)

The specimens are set in a shallow tray on two brass supports so that they will hold the unit 6.35 mm (0.25") up from the bottom.

Water is added and controlled at a level of 3 mm (0.12 in) ± 2.5 mm (0.1 in) above the lower face of the brick. Five weighed specimens are measured and then set in the water on the supports for one minute. They are then removed, wiped and weighed. The initial rate of absorption is calculated as follows:

$$X = 200\ W/LB$$

where X = Gain in mass on basis of 200 cm^2 (31 in^2).

W = Actual gain in mass of the specimen in g (oz)

L = Length of specimen in cm (in)

B = Width of specimen in cm (in)

The corrected gain in weight of the specimen is the initial rate of absorption in one minute. If specimens are cored, the net area should be used. For maximum bond strength, brick should have an initial rate of absorption of 21 grams/minute/200 cm^2 (31 in^2). (See Fig. 8-23.)

Field Test for I.R.A. A rough but effective test for I.R.A. (initial rate of absorption) is to place a twenty-five cent coin on the brick surface and draw a line around it with a pencil. Then trace around it with a wax crayon. Using a medicine dropper, quickly drop water within the circle until 20 drops have been placed, taking care not to spill over the circle. Note the time for the water to be absorbed into the brick, beginning with the time the circle is first filled. If the time exceeds one and a half minutes, the brick need not be wetted, but if it takes less than one and a half minutes, then the brick must be prewetted. The wetting should be done several hours before laying so that the water has time to penetrate and so that the surface is almost dry to permit bonding. (See Fig. 8-24.)

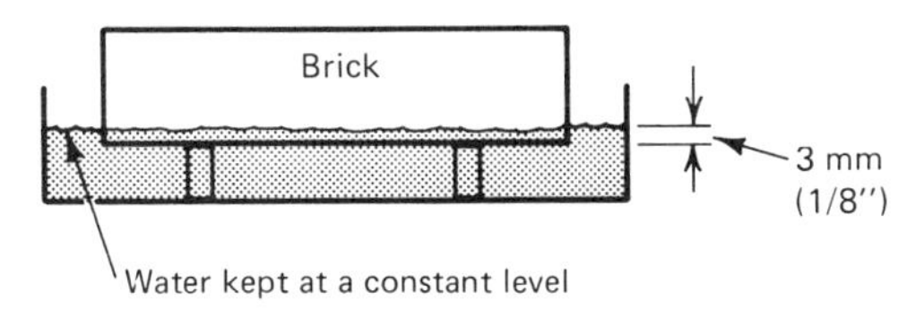

Fig. 8-23. I.R.A. test.

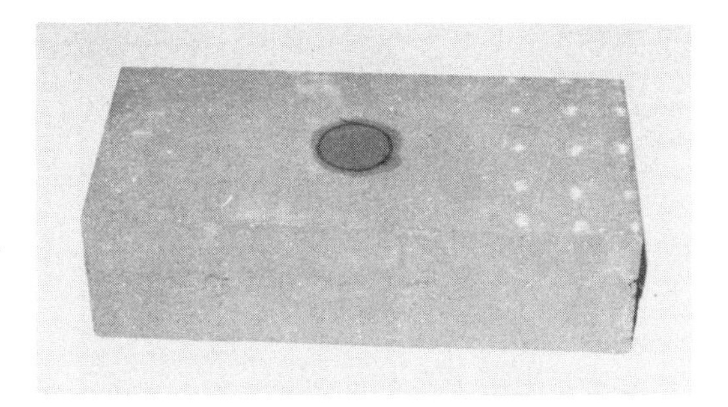

Fig. 8-24. I.R.A. field test.

METHODS OF SAMPLING AND TESTING CONCRETE MASONRY UNITS

These methods cover the *sampling* and testing of masonry units for *compressive strength, absorption, moisture content, measurement of dimensions* and *drying shrinkage* of concrete block.

Sampling

Ten full size specimens are selected from each lot of 10,000 units that will represent the entire lot. The samples are marked prior to testing.

Compressive Strength

Five full size specimens are capped with sulfur and granular materials or gypsum plaster capping. Each specimen is set in a standard testing machine with the block set in the normal service position. The load is applied up to half the expected load at any speed and the remainder is loaded in one to two minutes. The compressive strength of the unit is the total load in newtons divided by the gross cross-

Fig. 8-25. Compressive strength of block.

sectional area. The result should be reported to the nearest 50 kPa (5 psi). (See Fig. 8-25.)

Absorption

Five test specimens are immersed in water at room temperature for 24 hours. The specimens are weighed while suspended in water, removed from the water and drained for one minute, whipped and weighed. They are then dried in an oven at 100°C to 115°C (212°F to 240°F) for 24 hours. The percent absorption is calculated by:

$$\text{Absorption percent} = [(A - B)/B] \times 100$$

$$\text{Absorption} = [(A - B)/(A - C)] \times 100$$

where A = Wet mass of unit (g)

B = Dry mass of unit (g)

C = Suspended mass of unit (g)

The moisture content as sampled is calculated as follows:

$$\text{Moisture content, percent} = [(A - B)/(C - B)] \times 100$$

where A = Sampled mass of unit (g)

B = Dry mass of unit (g)

C = Wet mass of unit (g)

Measurement of Dimensions

Five full size units are measured for length, width, and height to the nearest millimeter. The measurements are made at the midpoints of both faces for the length and the height, while the width is measured at the midpoint of top- and bottom-bearing surface. The face-shell and web thicknesses are measurements at the thinnest point of each element at a point 13 mm (½ in) above the mortar bed to the nearest 0.5 mm (1/64 in). The report should show the average measurements for each dimension checked. (See Fig. 8-26.)

Drying Shrinkage of Concrete Block

Three whole units, free of cracks and structural defects, are immersed in 23°C (73°F) water for 48 hours. The initial length of each unit is measured while the unit is partially immersed in the water. Weigh each unit both while saturated and while surface dry by draining and then wiping it before taking a reading. The units are placed in a drying oven at 122°C

Fig. 8-26. Measuring block size.

(250°F) for 5 days. Cool to 23°C (73°F) and obtain the lengths and masses of the units.

Return specimen for an additional drying period of 48 hours; cool, weigh and measure. This 48-hour period should be repeated until the average change in length is 0.002 percent over a period of 6 days of drying, and the loss of mass is 0.2 percent or less over the previous mass. The drying shrinkage percent is calculated as follows:

$$S = (\Delta L/G) \times 100$$

where S = Linear drying shrinkage percent.

ΔL = Change in length.

G = Initial length.

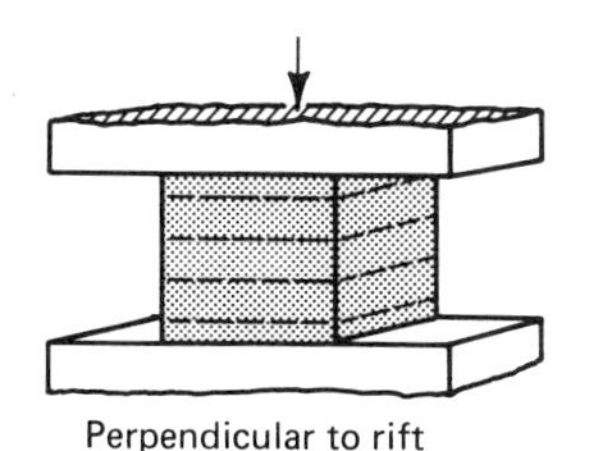

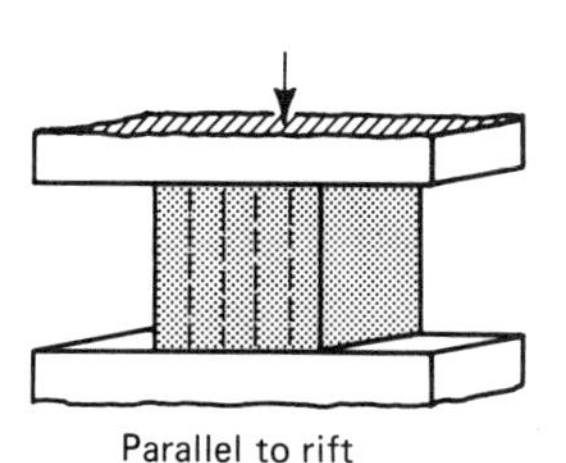

Fig. 8-27. Compressive strength of stone.

STONE TESTING

The methods of testing stone do not vary much from those of other masonry products as illustrated in the following tests.

Compressive Test

Samples are cut from stone with minimum dimensions of 50 mm (2 in) and a height-to-width ratio of at least 1 : 1. The samples may be made with the height parallel or perpendicular to the rift. (See Fig. 8-27.) The pressure surfaces are ground and loaded similar to mortar cubes. The calculations are the same as those used with cubes or prisms for grout, but variation must be made in the calculation if the height-to-width ratio is other than 1 : 1.

Flexure

Samples are cut to the size of brick, and the test is conducted in the same way as the flexure test for brick. The modulus of rupture is determined by the same formula that was used previously for brick.

Absorption

The samples are cut into cube prisms and cylinders which are dried, then immersed as in the absorption test for brick. The formula used to determine absorption percent is the same one used in the brick calculation.

MASONRY ASSEMBLAGES

When a rational design is used for masonry, it is essential to establish allowable design stresses early in the design process. These design stresses (*compression, flexure, shear* and *bond*) are established by testing such masonry assemblages as prisms and walls.

Compressive Strength and Behavior

In order to design a building with load-bearing walls, the allowable design stresses of the wall must be determined. The design stresses of concrete are based on f'_c (compressive strength of concrete). The strength is determined by compressing a concrete cylinder, where the height of the cylinder is twice the diameter. The compressive strength will vary

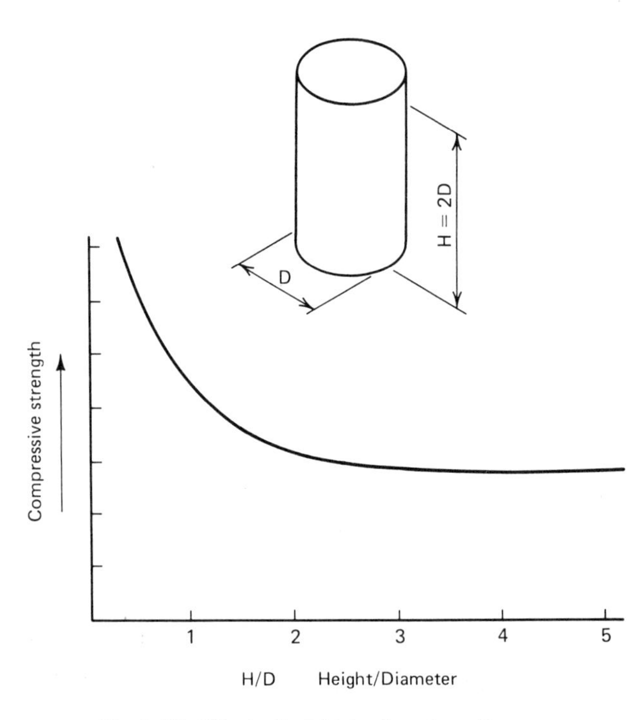

Fig. 8-28. Effect of height-to-diameter ratio.

with this height-to-diameter ratio as illustrated in Figure 8-28, where the compressive strength is very high for very low h/d ratios and levels off at a ratio of two or more. If you use a cylinder with an h/d ratio of less than two, the top of the cylinder tends to expand laterally making the test inaccurate. Similarly, the design compressive strength of masonry is based on f'_m (compressive strength of masonry). These stresses are based on the 28-day strength of the masonry and are determined in three ways:

1. Unit strength.
2. Prism strength.
3. Compressive strength of full size walls.

Table 8-3. Compressive Strength of Masonry
(Gross Area for Solid Units and Net Area for Hollow Units)

Compressive Strength of the Units		Assumed Compressive Strength of the Masonry (f'_m) MPa (PSI)	
MPa	*(PSI)*	*Type M&S Mortar*	*Type N Mortar*
10.00	(1450)	7.7 (1116)	5.4 (783)
15.00	(2175)	10.0 (1450)	7.1 (1029)
20.00	(2900)	12.0 (1740)	8.4 (1218)
30.00	(4350)	15.0 (2175)	8.6 (1247)
40.00 and over	(5800)	16.0 (2320)	8.6 (1247)

Unit Strength. The rational method that assumes that the wall strength is related as a percentage of the unit strength is shown in Table 8-3.

Prism Strength. Prism tests are conducted using the same materials and workmanship as used in the structure. A prism is a small assembly of individual units that may be used to determine the compressive strength of masonry. (See Fig. 8-29.) The compressive strength of masonry is affected by masonry units, mortar and grout. The height-to-minimum-thickness ratio should be considered to avoid the platen effect in the prism. (See Fig. 8-30.) Testing has determined that the h/t ratio of prisms should be from two to five. Correction factors should be used as the ratio varies according to Table 8-4.

Fig. 8-29. Masonry prisms. (*Courtesy Portland Cement Assn.*)

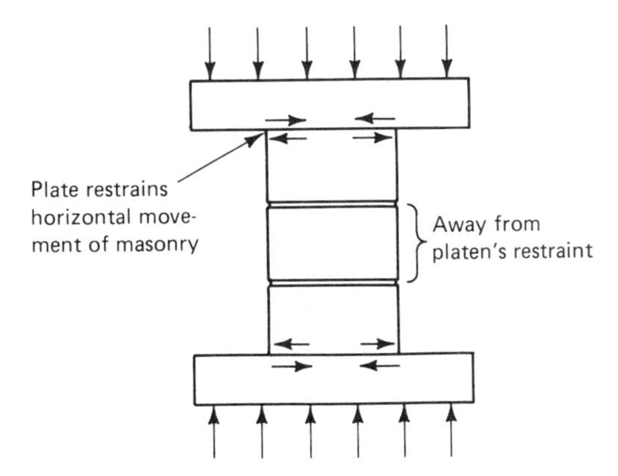

Fig. 8-30. Platen's restraint.

Table 8-4. Height-to-Thickness Ratio

h/t	Brick	Block
1	–	0.9
2	0.7	1.0
3	0.8	1.1
4	0.9	1.2
5	1.0	1.3
6	1.1	1.4
7	1.2	1.5

(Courtesy Canadian Standards Association)

Laboratory tests have indicated that the f'_m of masonry is determined without a correction factor at an h/t ratio of five for brick and of two for block. The compressive strength is determined by taking the ultimate load and dividing it by the cross-sectional area. The gross cross-sectional area is used for solid units and the net-cross-sectional area is used for hollow units. If the results do not vary more than 10 percent, the average of the test specimens can be used for f'_m. If the test results vary more than 10 percent, the average compressive strength must be modified by the following formula:

$$f'_m = \left(1 - 1.5\left(\frac{\nu}{100} - 0.10\right)\right)\bar{X}$$

where f'_m = Ultimate compressive strength (MPa)

ν = Coefficient of variations of specimens tested, percent = $100 \frac{S}{\bar{X}}$

$\bar{X}$ = Average compressive strength of the specimens.

s = Standard deviation in MPa = $\frac{\sqrt{\Sigma(X - \bar{X})^2}}{n - 1}$

X = Compressive strength of individual specimen.

The number of units seems more critical than the height, as illustrated in Table 8-4.

The test specimens will fail by tensile splitting when not restrained by platen effect. The failure occurs in this method since the mortar is compressed and, tending to expand, puts the bricks into tension as shown in Figure 8-31. The bricks split as they are very weak in tension. The mortar is in triaxial compression (confined) and can carry a greater load than when in an unconfined situation. If masonry units with a strength of 35 MPa (5250 psi) were made into a prism with mortar with a strength of 10.5 MPa (1522 psi), the prism would probably fail at a load of 17.5 MPa (2537 psi) (approximate). If the mortar joint thickness in a wall is reduced, the mortar confinement and the strength are increased, but if joint thickness is increased, the strength decreases.

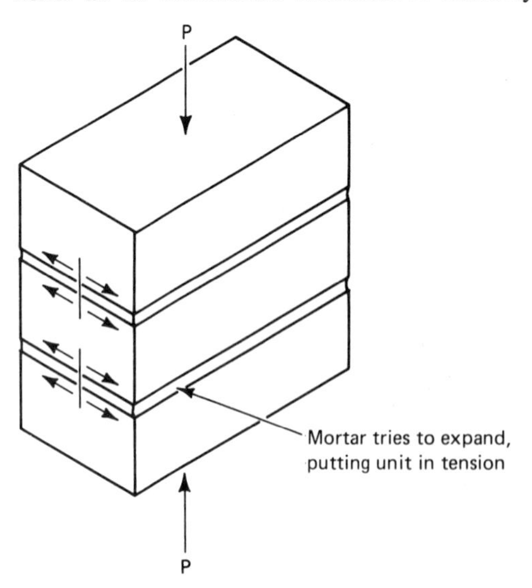

Fig. 8-31. Triaxial compression.

The prism specimens are built of similar materials used in the wall under similar conditions. They are usually made about five units high for standard brick and two units high for concrete block. Care should be taken, where there is a slight variation in the size of the units, to lay the units plumb and straight on one side. The straight side will allow easier placement of a cap perpendicular to the specimen.

The prisms should be cured under jobsite conditions for 48 hours in a place where they are protected from damage. They should be protected top and bottom with a piece of plywood banded to them for transportation to the laboratory. The specimens are then cured for an additional 26 days at a temperature of 21°C (70°F) and a relative humidity of 30 percent to 70 percent. They are then capped with a sulfur cap or equivalent and compressed in a compression machine. The ultimate load is then divided by a cross-sectional area of the prism in m^2 to find the compressive strength (MPa). The cross-sectional area is found by using the average of the dimensions at quarter points and midpoints of heights. (See Fig. 8-32.) Correction should be made if the h/t ratio is other than that specified in Table 8-4.

$$\text{Compressive Strength} = C = W/A$$

where C = Compressive strength in MPa (psi).

W = Load in N (lb).

A = Load in m^2 (in^2).

Compressive Load Testing of Full Scale Walls. Where the capacity of test equipment

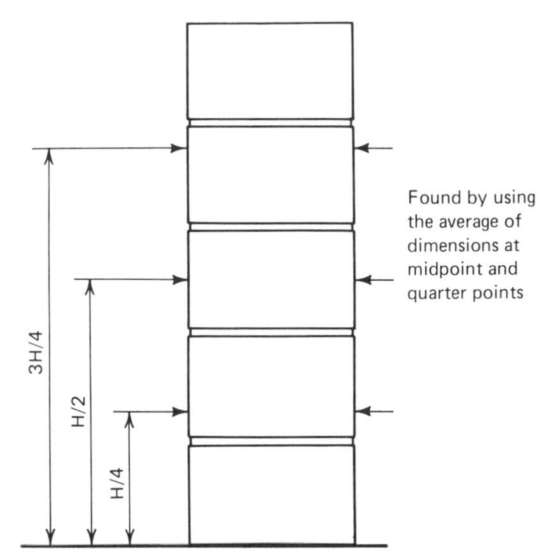

Fig. 8-32. Prism cross-sectional area.

merits, it is possible to test full size wall sections. This method is more representative of the actual load-carrying capacity of a masonry wall. These include the compressive strength of the units, eccentricity of the load, slenderness of the wall, mortar bedding, workmanship, mortar strength and amount of reinforcing. Figure 8-33 illustrates installation of wall section into a test frame. Figure 8-34 shows a sample being tested with zero eccentricity, whereas Figure 8-35 illustrates a case involving eccentricity. Figure 8-36 illustrates a failure where there is eccentricity at the top where Fig. 8-37 has eccentricity top and bottom. Figs. 8-34 and 8-35 are grouted and reinforced in three cells

Fig. 8-33. Placing wall panel in test frame. (*Courtesy University of Alberta*)

Fig. 8-34. Failure—grouted and reinforced. (*Courtesy University of Alberta*)

Fig. 8-35. Wall eccentrically loaded—single curvature. (*Courtesy University of Alberta*)

Fig. 8-36. Wall failure—eccentrically loaded at top. (*Courtesy University of Alberta*)

Fig. 8-37. Wall failure—hollow masonry wall. (*Courtesy University of Alberta*)

and are 16 units high. Figs. 8-36 and 8-37 are hollow masonry walls 22 units high.

Flexural Strength

Flexural tests are conducted on prisms and full size wall sections in order to obtain allowable design stresses. The flexural test conducted on prisms is primarily a method used as a quality control check of comparative *flexural bond strengths* developed between different types of masonry units and mortar. This test is not intended to determine flexural design stresses. In order to determine the design *flexural tensile strengths* for masonry, full size wall sections are subjected to transverse loads.

Flexural Bond Strength. This procedure requires the building of five test specimens in stack bond prisms at least 460 mm (18 in) high. The number of courses should be ade-

Fig. 8-38. Flexural strength—third-point loading. (*Courtesy American Society for Testing and Materials*)

quate to enable third point loading so that the points occur midway between joints as illustrated in Figure 8-38. The samples are cured for 28 days and are tested horizontally as a simply supported beam. The samples are tested in third point loading or uniform loading as in Figure 8-39. This test is generally used only for high bond mortars, as values for standard mortars would be so low it would be difficult to compare two different types of mortars.

Fig. 8-39. Flexural strength—uniform loading. (*Courtesy American Society for Testing and Materials*)

Flexural Tensile Strength. Flexural wall strengths are a measure of the wall's tensile strength capacity. This is the only method used at present to determine the design flexural strength of masonry. These panels are tested in horizontal or vertical positions and should have a minimum size of 1200 mm by 2400 mm (48 in by 96 in). The panels are subjected to concentrated loads as shown in Figure 8-40. The deflection readings are taken at the center of the span. These specimens should be cured for 28 days prior to the testing date.

Shear Strength

Shear strength is the ability of a wall to resist the lateral shearing forces occurring within the plane of the member. In order to obtain allow-

Fig. 8-40. Flexural wall test. (*Courtesy Portland Cement Assn.*)

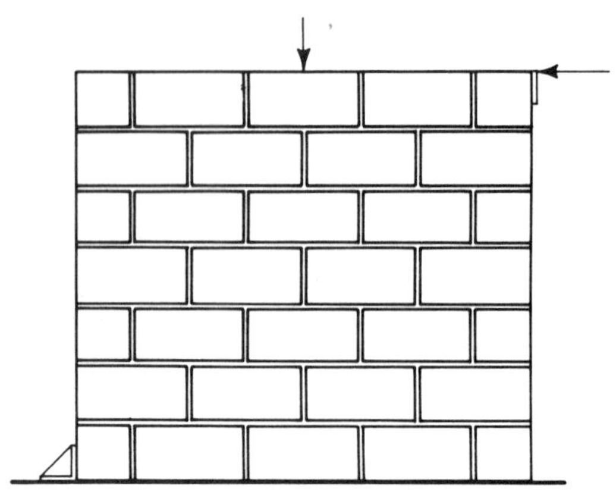

Fig. 8-41. Racking test.

able design shear stresses, two tests can be used, the *racking test* and the *diagonal shear test.*

In the racking test, the wall section is placed in a frame and held at the bottom; then a horizontal load is applied at the top as shown in Figure 8-41. This method requires the specimen to be held down so that it does not overturn. These hold down rods produce an indeterminate load on the wall section so that it is difficult to determine the stresses in the wall.

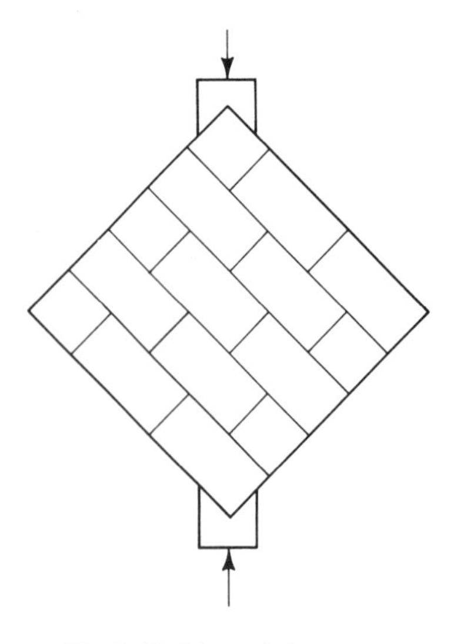

Fig. 8-42. Diagonal shear test.

The diagonal shear test does not have an overturning problem as the wall section is set in the frame in a diagonal direction as in Figure 8-42. The load is then applied along this diagonal and will produce a condition of pure diagonal tension, thus measuring the shear capacity. This test is of special concern in seismic areas and only applies to masonry. The specimen used in the diagonal shear test is about one-quarter the size used in the racking test.

Bond Strength

With concrete, the water-cement ratio is important, but with mortar, the aggregate-cement ratio is more important. Mortar strength may not be as critical as consistency and bond. Tests for consistency were dealt with in the mortar tests earlier in this chapter. Bond strength in buildings is of critical importance, at least of

Fig. 8-43. Bond couplet test. *(Courtesy American Society for Testing and Materials)*

equal importance as compressive strength. No codes or standards require a bond strength evaluation because there is no test at present that can be performed that will give adequate results. Some work has been done with simple cross-brick couplets (see Fig. 8-43), with block bond tensile tests (see Fig. 8-44), and with bond shear tests (see Fig. 8-45). Probably the best way to test the bond between masonry units and mortar is to physically lift the unit off the mortar bed a few moments after it has been laid. This is not a very practical test however, as there is no effective means of illustrating comparative results.

Fig. 8-44. Bond tensile test. (*Courtesy Portland Cement Assn.*)

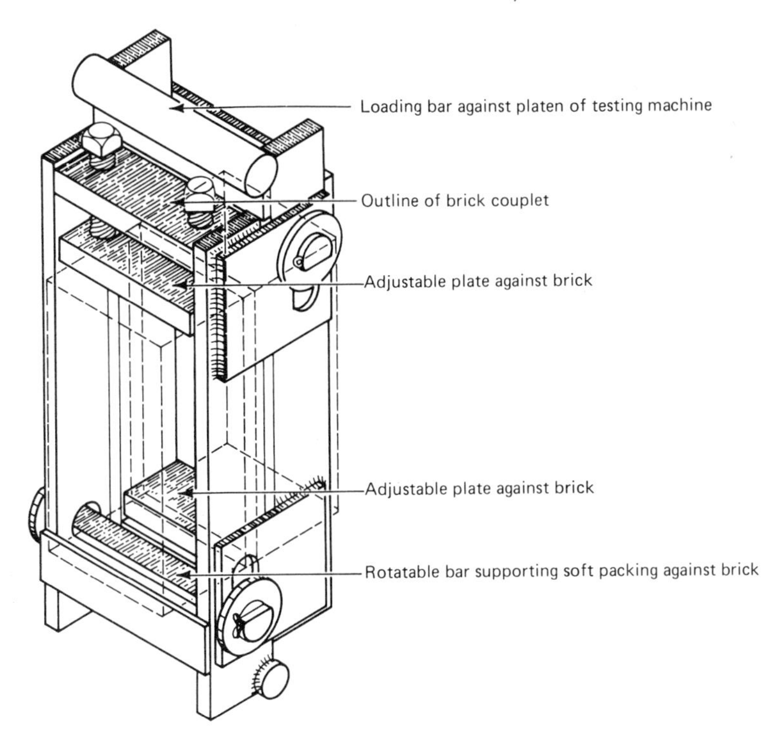

Fig. 8-45. Bond shear test. (*Courtesy American Society for Testing and Materials*)

Glossary of Terms

Absorption The weight of water a masonry unit absorbs when immersed in water, expressed as a percentage of the weight of the dry unit.

Air Content The amount of air bubbles entrapped in mortar, grout, or concrete, expressed as a percentage of the total volume.

Compressive Strength The measured amount of load a masonry material or assemblage will resist before it fails.

Consistency A method of comparing the wetness or fluidity of different batches of mortar, grout, and concrete.

Eccentricity The distance between the centroidal axis of the member and the axis through which the load is applied.

Efflorescence A powder or stain usually found on the surface of a masonry wall, resulting from soluble salts in the masonry and the mortar that are drawn to the surface by moisture.

Fineness modulus The relative fineness or coarseness of aggregate used in mortar, grout, and concrete.

Flextural Strength The measured amount of lateral force a masonry unit or assemblage will resist before failing.

NaOH Sodium hydroxide.

Ottawa Sand A clean white sand used for testing purposes.

Platen Effect The resistance to horizontal movement on the test specimen, caused by the head or platen of a testing machine. (Compression.)

Shear Strength The resistance of a wall or assemblage to lateral shearing forces occurring within a plane in the member.

Slenderness Ratio The ratio of the effective height to the effective thickness of a member.

Rift The indication on the surface of the stone as to the plane along which it will split.

Ultimate Load The maximum load that is required to cause a member to fail.

Water Retention The property of the mortar that will prevent the rapid loss of water to masonry units.

Review Questions

1. What is the purpose of doing a sieve analysis on a sample of sand?
2. Why is it critical to know the amount a sample of sand will bulk?
3. What corrective measure should be taken with a sand sample if the silt test and the color test indicate an excessive amount of impurities?
4. Explain the importance of matching the water retentivity of mortar to the absorption of a masonry unit.
5. Does the compressive strength of the mortar in a wall have a more significant effect on the compressive strength of the wall than does the unit?
6. What is the purpose of determining the amount of cementious material in mortar?
7. Compare the methods used to determine the consistency of mortar.

8. Why is it necessary to find the amount of air in a sample of mortar?
9. What effect does eccentricity have on the strength of a masonry wall?
10. What property of mortar is more critical than strength?

Selected Sources of Information

Alberta Masonry Institute, Calgary, Alberta.
American Society for Testing and Materials, Washington, D.C.
Brick Institute of America, McLean, Virginia.
Canadian Standards Association, Rexdale, Ontario.
National Concrete Masonry Association, McLean, Virginia.
Portland Cement Association of America, Skokie, Illinois.
R.M. Hardy and Associates, Calgary, Alberta.
University of Alberta, Edmonton, Alberta.

Chapter 9

DESIGN METHODS

The preceding chapters have discussed, in some detail, the different materials used in the manufacture of the various masonry units and the mortars that bind them together. Standard tests outlined in Chapter 8 describe methods that enable the laboratory technician to determine the performance of these materials under laboratory conditions.

Masonry includes a number of basic materials that produce sound, durable, relatively inexpensive and aesthetic structures when proper design methods are applied. The responsibility of the designer, using appropriate design methods, is to transform all available data into a building that has the foregoing qualities.

Two design methods have evolved: the *empirical design method* and the *rational design method.*

THE EMPIRICAL DESIGN METHOD

The empirical design method is the more classic of the two methods. It originated when man began to arrange blocks of stone and mud in some orderly fashion. Through centuries of experience and observation, masons developed "rules of thumb," which enabled them to build with relative confidence. These rules are now being standardized and printed as design and construction codes.

In Canada, the Canadian Standards Association has published CSA Standard Can-S304-M78, Masonry Design and Construction for Buildings, which contains the basic guidelines for masonry design as dictated by Canadian conditions. This, hopefully, will encourage the development of standard construction practices across the country. The ultimate result should be better buildings, both in terms of initial cost and long-term maintenance. The United States has several codes, both national and state, that deal with masonry design. The Uniform Building Code, the Southern Standard Building Code, the Basic Building Code and the American Standard Building Code are some of the major codes that deal with design criteria for masonry.

In this text, The Alberta Building Code 1978 will be used as the reference standard, and portions are reprinted for convenience in the Appendix. All references to tables in the design section are to tables in the Appendix. Subsection 4.4.4 of this standard deals with the empirical design method. This clause deals primarily with wall design and two basic types are described: *load-bearing* and *non-load-bearing* walls.

Load-bearing Walls

Load-bearing walls are walls that will subsequently support vertical roof or floor loads. To ensure adequate performance of a load-bearing wall, the following two conditions must be examined: (a) wall compressive load capacity and (b) wall stability. The allowable load capacity of a wall depends upon its cross-sectional area and a maximum allowable compressive stress. These values are given in Table 10 of the code. (See Appendix.) Clause 4.4.4.7 page 337, deals with the conditions under which the values in Table 10 apply. The application of these clauses is better illustrated with an example.

Example 9–1

Calculate the maximum allowable vertical force per meter on a 250 mm (10 in) solid wall using solid concrete block units. The concrete block ultimate strength is 20 MPa (2900 psi) and the mortar is type N.

Solution

Sentence 4.4.4.7(1) states that the allowable force on a masonry wall shall be the product of its maximum allowable stress and its cross-sectional area. Written in equation form,

$$F = A \times f_c \qquad \text{Eq. (9-1)}$$

where F = Force in newtons (lb).

A = Actual cross-sectional area in m^2 (in^2).

f_c = Maximum allowable compressive unit stress in kPa (psi).

For Equation (9-1) to be applicable, the resultant of the load must fall within the middle third of the wall. This being the case, the maximum allowable compressive stress for the solid concrete block as selected from Table 10 is 1.0 kPa. Substituting into equation (9-1):

$$\text{Force/meter} = \frac{250 \times 1 \times 1.0}{1000}$$

$$= 0.25 \text{ MN}$$

$$= 250 \text{ kN (56 kips)}$$

Had the wall height been given, a check of the stability conditions would ensure that the wall would perform satisfactorily. Clause 4.4.48 of the code deals with these requirements which are illustrated in Example 9-2.

Example 9–2

What minimum wall thickness will be required for the load-bearing walls in a four-story building with a 3 m (9.8 ft) between floors? The walls are solid wall construction using hollow concrete blocks.

Solution

First, calculate the total wall height: 4 × 3 = 12 m (39.4 ft) for wall total height not including the basement walls. Referring to clause 4.4.4.11, hollow block walls to 11 meters (36 ft) must be 290 mm (11.4 in) in thickness. For walls above 11 meters (36 ft) in height, sentence 4.4.4.11.(2) states that the wall shall be increased 100 mm (4 in) in thickness for each increment of 11 meters (36 ft) in height measured from the top down.

The bottom meter of the first floor wall must be increased to 290 + 100 = 390 mm (15.4 in).

Furthermore sentence 4.4.4.11.(6) states that should a change in thickness occur between floor levels, the greater thickness shall be taken to the next higher floor level. This makes the first floor level 390 mm (15.4 in) thick with the remainder of the wall being 290 mm (11.4 in) thick. To ensure that the change in wall thickness does not produce an adverse effect on the wall construction, sentence 4.4.4.11(7) requires the top 190 mm (7.5 in) of the thicker wall to be of solid units or hollow units filled with grout.

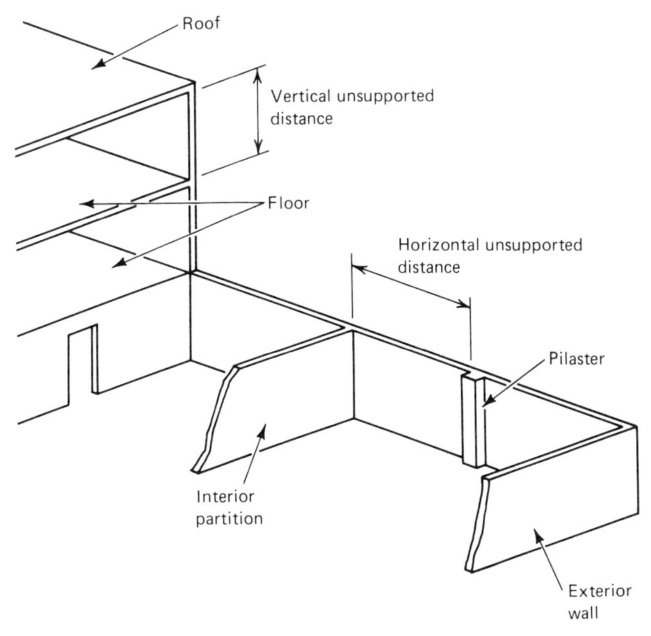

Fig. 9-1. Building elements providing lateral supports to walls.

Additional stability is required in wall design in the form of lateral supports located at specified distances in either the vertical or horizontal directions. Sentence 4.4.4.8.(1) deals with these requirements. (See Fig. 9-1.) For example, hollow units require lateral supports at a distance of eighteen times the thickness when used in solid wall construction. For the wall in Example 9-2 the maximum distance between lateral supports for the 290 mm (12 in) portion must be

$$18 \times 290 = 5220 \text{ mm}$$
$$= 5.22 \text{ m (17 ft)}$$

The distance between floors is 3 m (9.8 ft), which is less than the 5.2 (17 ft) meters allowed.

Non-Load-bearing Walls

Non-load-bearing walls may be either exterior walls or interior partitions. Conditions for exterior walls, panel walls, and curtain walls are given in clauses 4.4.4.12 to 4.4.4.18 inclusive. Clause 4.4.4.17 deals with the height and thickness requirements for partition walls.

Example 9-3

Calculate the required thickness of a partition wall between columns spaced at 10 meters (32.8 ft) o.c. and having a height of 4 meters (13 ft).

Solution

Clause 4.4.4.17 of the code requires that the spacing between supports shall be not more than 36 times the partition thickness and the height of a partition shall not exceed 72 times its thickness.

Minimum thickness for the horizontal conditions:

$$\frac{10 \times 1000}{36} = 277 \text{ mm (10.9 in)}$$

Minimum thickness for the vertical conditions:

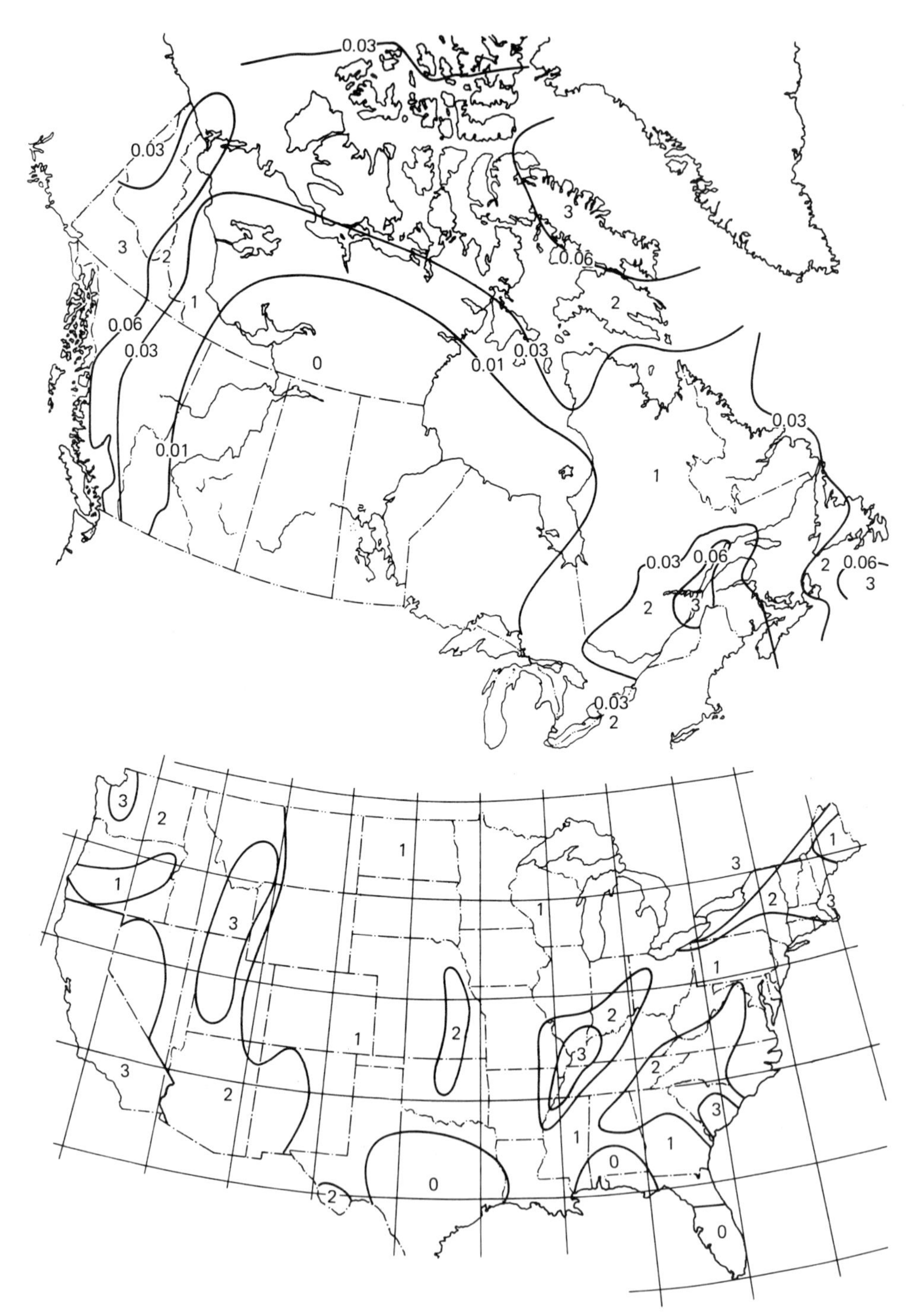

Fig. 9-2. Seismic zones in (a) Canada per The National Building Code of Canada 1975 and (b) the United States per the 1973 Uniform Building Code.

$$\frac{4 \times 1000}{72} = 55 \text{ mm } (2.16 \text{ in})$$

The larger of the two values governs the minimum thickness of the wall, and in this case the minimum unit size required would be 290 mm (12 in).

From the preceding examples it can be seen that the empirical design method is limited in scope. One further limitation should be brought forward at this point. The empirical design method is restricted to zone 1 seismic areas and areas with no seismic activity. (See Fig. 9-2.)

THE RATIONAL DESIGN METHOD

Designers are familiar with the ultimate strength design method used in reinforced concrete design and the limit states design method used in structural steel design. In each of these methods, the structural components must be able to withstand some factored load above the maximum working load expected before failure occurs.

In the design of masonry components a maximum allowable stress is determined by engineering analysis. The design method then attempts to ensure that all stresses due to actual working loads will not exceed the established maximum allowable stresses. This method is known as the *working stress design method.* The success of this method depends largely on the following assumptions.

1. The materials used act elastically in a linear sense. That is, the unit stress is directly proportional to the unit strain under load, and the strain disappears when the load is removed.

2. The materials are homogeneous throughout. Of the materials involved, brick, concrete block, mortar and steel, only the steel can be considered truly elastic. To ensure that the assumptions made are relevant, the stress levels are kept relatively low in the nonelastic materials.

Application of the above assumptions to plain masonry produces a somewhat conservative design. This is to ensure that tensile stresses occurring under working loads are kept to an absolute minimum to avoid cracking in the mortar joints. Introduction of steel reinforcing to absorb tensile stresses allows for better use of the relatively high compressive strength of the masonry units. Thus, the working stress design method may be applied to both *plain* masonry and *reinforced* masonry with reliable results.

Plain Masonry

Because plain masonry has very little tensile strength, the design of its components within the code is limited to walls and columns. Tables 4 and 5 of the code contain the allowable stresses that are to be applied to the various masonry units. It should be noted that all the allowable stresses are based on the value f'_m, the ultimate compressive strength of masonry at twenty-eight days. Consider a simple example of a short masonry column as illustrated in Figure 9-3 loaded with an axial compressive load, P. The axial stress due to this load can be calculated by using the basic formula:

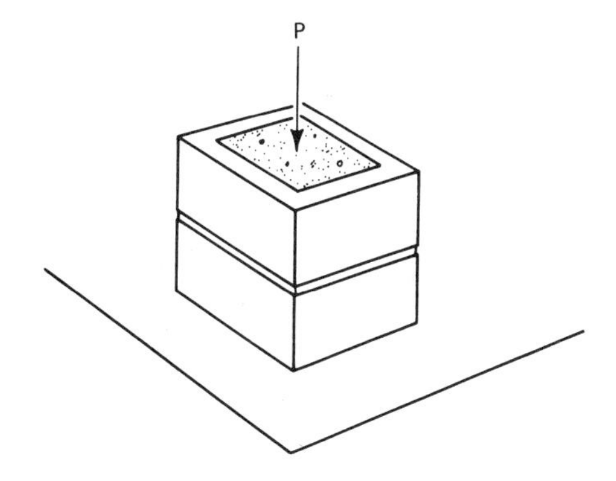

Fig. 9-3. Short masonry column with axial compressive load *P*.

$$f_c = \frac{P}{A_n} \qquad \text{Eq. (9-2)}$$

where f_c = The axial compressive unit stress.
P = The applied load.
A_n = The net cross-sectional area of the masonry unit.

Should the wall be subjected to a load as shown in Figure 9-4, producing bending in the wall, the tension in the wall is calculated by the formula:

$$f_t = \frac{M}{S} \qquad \text{Eq. (9-3)}$$

where f_t = The tension stress due to flexure.
M = The bending moment due to the applied load.
S = The elastic section modulus of the masonry wall with respect to the plane of bending.

In the preceding examples, only stresses due to the applied axial and horizontal loads were considered. When designing walls and columns by the rational design method, as with the empirical design method, stability must be considered as well as the stresses due to the applied loads.

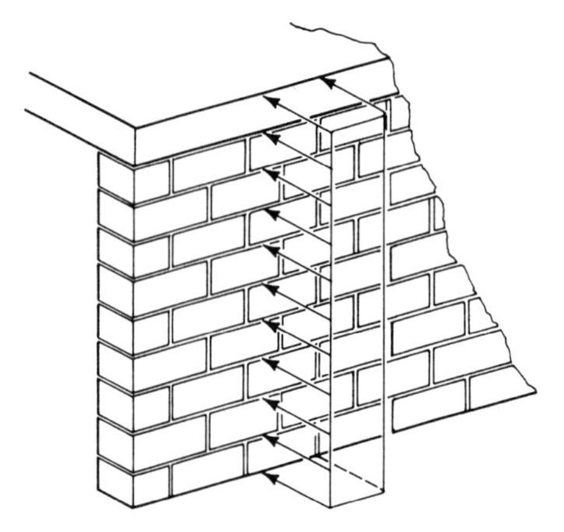

Fig. 9-4. Wall bending due to horizontal load.

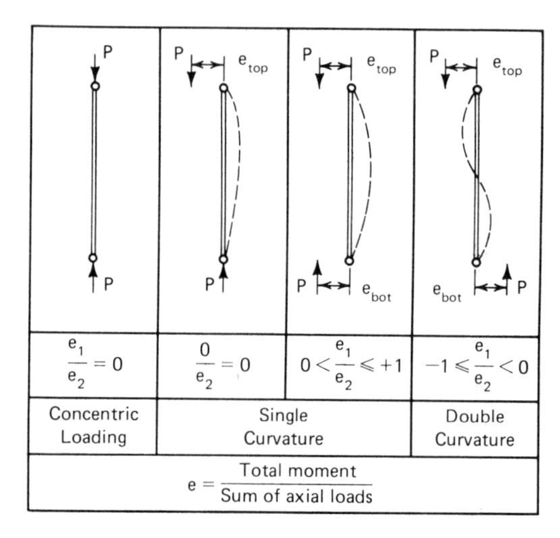

Fig. 9-5. Curvature effects on the value of e_1/e_2.

The stability of a wall depends on its slenderness ratio. Clause 4.4.3.18 of the building code (see Appendix), defines slenderness ratio as the ratio of the wall effective height, h, to the effective thickness, t. For walls, clause 4.4.3.18 requires that the slenderness ratio, h/t is less than $10(3 - e_1/e_2)$ where e_1 and e_2 are the virtual eccentricities produced by the loading conditions at the top and bottom of the wall. (See Fig. 9-5 and Fig. 8-37.) The ratio e_1/e_2 is obtained by dividing the smaller eccentricity by the larger. This produces an absolute value of unity or less. Where the wall is bent in single curvature, the ratio is positive. Where the wall is in double curvature, the ratio is negative. Figure 9-5 illustrates various loading conditions and the resulting curvatures that they produce.

When considering stability in a wall, buckling (see Fig. 8-37) will occur perpendicular to its plane. The effective height of the wall is determined by the degree of end restraint with respect to its thickness. Figure 9-6 illustrates the various conditions of end restraint and their effects on the effective wall height. The less restraint that is applied to the wall, the greater the effective height becomes. When dealing with columns, both axes of the column must be

Degree of end restraint of wall or column	Minimum effective height h	Symbol
Effectively held in position and restrained against rotation at both ends.	0.65 H	H
Effectively held in position at both ends, restrained against rotation at one end.	0.80 H	H
Effectively held in position at both ends, but not restrained against rotation.	1.00 H	H
Effectively held in position and restrained against rotation at one end; restrained against rotation but not held in position at other end.	1.20 H	H
Effectively held in position and restrained against rotation at one end, and at the other partially restrained against rotation but not held in position.	1.50 H	H
Effectively held in position at one end but not restrained against rotation, and at the other end restrained against rotation but not held in position.	2.00 H	H
Effectively held in position and restrained against rotation at one end but not held in position nor restrained against rotation at the other end.	2.00 H	H

Fig. 9-6. End restraint on walls or columns and corresponding effective heights.

considered, as the end restraint may differ about each of the two major axes. Consider the column shown in Figure 9-7. The effective height with respect to the Y - Y axis is equal to 2H. Due to the lateral stability provided by the roof beam, the effective height of the column with respect to the X - X axis remains at the original height H.

The slenderness ratio conditions being satisfied, the allowable vertical load on the wall may now be determined. Sentence 4.4.3.28.(1b) states that provided the maximum virtual eccentricity does not exceed one-third of the wall thickness, the following expression may be applied:

$$P = C_e \times C_s \times f_m \times A_n$$

where P = Allowable load.

C_e = Eccentricity coefficient.

C_s = Slenderness coefficient.

f_m = Allowable compressive stress.

A_n = Net cross-sectional area.

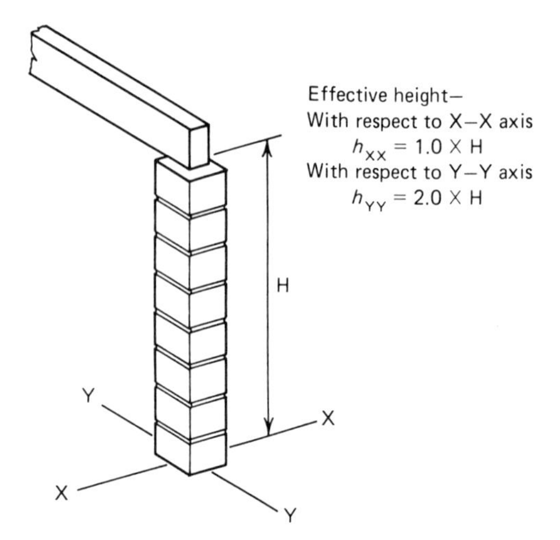

Fig. 9-7. Column stability about two axes.

In the event that the virtual eccentricity exceeds one-third of the wall thickness, the maximum compressive stress is substituted by the allowable flexural tensile stress normal to the bed joints.

To avoid confusion the preceding discussion has been aimed primarily at walls. The design of plain masonry columns is very similar, as becomes evident when referring to the code.

Reinforced Masonry

The addition of reinforcing steel to masonry adds greatly to the versatility of masonry design. To this point, only walls and columns have been discussed under somewhat limited conditions. This is due to the fact that plain masonry has very little strength in tension. The addition of reinforcing steel introduces ductility, allowing masonry components to withstand tensile stresses with no material failure.

To illustrate the application of the working stress design method to reinforced masonry, consider a typical reinforced flexural section as shown in Figure 9-8. To comply with the theory on which the working stress design method is based, the following assumptions must be made:

1. A section that is plane before bending remains plane during bending.

2. The tensile stresses are resisted only by the reinforcing steel.

3. The moduli of elasticity of the masonry and the reinforcing steel remain constant.

4. The reinforcing steel must be completely surrounded by and bonded to the masonry.

In the analysis of the section, the area of steel must be transformed to an equivalent area of masonry. This is accomplished by multiplying the area of steel by the modular ratio (n) where,

$$n = \frac{E_s}{E_c} \qquad \text{Eq. (9-4)}$$

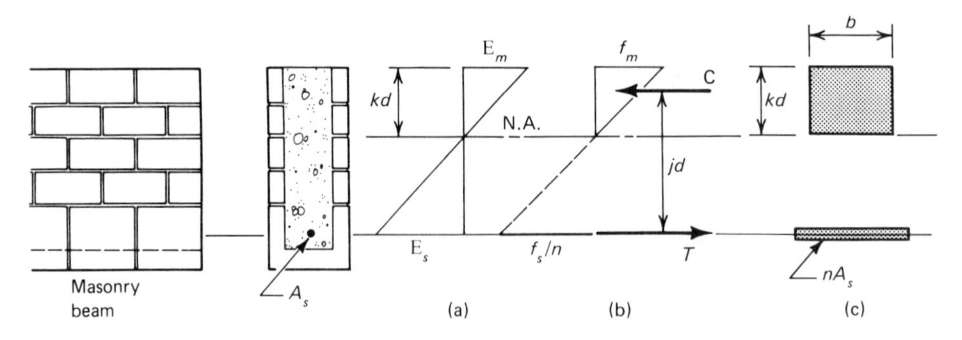

Fig. 9-8. Stress-strain relationships for reinforced flexural masonry section: (a) strain diagram, (b) stress diagram, (c) transformed section.

and E_s = Modulus of elasticity of steel

E_c = Modulus of elasticity of masonry.

As the steel must carry all the tension, the stress in the steel will be (n) times the tensile stress that would have existed in the masonry at that level. Referring to Figure 9-8(c), the neutral axis for the section can be located by taking moments of the areas in the transformed section about the neutral axis.

$$b \times kd \times \frac{kd}{2} = nA_s \times (d - kd)$$

Dividing by bd^2

$$\frac{k^2}{2} = \frac{nA_s}{bd^2}(d - kd)$$

Substituting $\rho = \dfrac{A_s}{bd}$

$$\frac{k^2}{2} = \rho n(1 - k)$$

Solving the quadratic

$$k = [(n\rho)^2 + 2n\rho]^{1/2} - n\rho \qquad \text{Eq. (9-5)}$$

Since d is known, kd can be calculated and the location of the neutral axis can be established. From the stress diagram the moment arm between the two forces T and C is

$$jd = d - \frac{kd}{3} \text{ and } j = 1 - \frac{k}{3} \qquad \text{Eq. (9-6)}$$

The magnitude of the two forces T and C can now be calculated as:

$$C = \tfrac{1}{2} f_m bkd \qquad \text{Eq. (9-7)}$$

and

$$T = A_s f_s = \rho bdf_s \qquad \text{Eq. (9-8)}$$

Using the moment arm, jd, the internal resisting moment of the section becomes

$$M = C \times jd$$
$$= \tfrac{1}{2} f_m kjbd^2 \qquad \text{Eq. (9-9)}$$

or

$$M = T \times jd \qquad \text{Eq. (9-10)}$$
$$= \rho f_s jbd^2$$

The above formulas allow the calculation of the bending moment that can be resisted by the flexural section.

The optimum condition in the design is achieved when balanced conditions are obtained; that is, the maximum allowable stress in the masonry is reached at the same time the steel reaches its maximum allowable stress.

Notation

A_s Cross-sectional area of tensile reinforcement.

b Width of rectangular beam.

d Effective depth of flexural member; distance from compression face to centroid of tensile reinforcement.

f_m Compressive stress in masonry.

f_s Stress in tensile reinforcement.

j Ratio of distance between centroid of tension and centroid of compression to the effective depth, d.

k Ratio of depth to the neutral axis to the effective depth.

M Bending moment at any cross section.

ρ Ratio of steel area to the cross-sectional area, bd.

Glossary of Terms

Seismic Movement of the earth as caused by an earthquake.

Working Load A load due to the actual dead and live loads acting on a building.

Factored Load A working load to which a factor has been applied as required by ultimate strength design methods.

Homogeneous A body that is composed of a material of uniform consistency.

Review Questions

1. Using the empirical design method, calculate the maximum allowable vertical force per meter on a wall having hollow 390 mm block units with 50 percent voids. The concrete block ultimate strength is 10 MPa and type S mortar is to be used.
2. For the wall in Question 1, calculate the maximum height allowed between lateral supports using the empirical design method. Assume the wall is load bearing.
3. Using the empirical design method, calculate the minimum thickness of a non-load-bearing wall having a horizontal distance of 8 m (26 ft) between pilasters and 3 m (9.8 ft) between lateral supports in the vertical direction.
4. Using the rational design method, calculate the maximum compressive load per meter on a 290 mm (12 in) wall built of hollow concrete block units. Assume 50 percent voids, f'_m = 15 MPa (or 2175 psi) and e_{max} = 50 mm (2 in) and wall height 4 m. $e_1/e_2 = 0$.

Selected Sources of Information

Alberta Building Code, Edmonton, Alberta.
Brick Institute of America, McLean, Virginia.
Clay Brick Association of Canada, Willowdale, Ontario.
National Building Code of Canada, Ottawa, Ontario.
National Concrete Masonry Assn., McLean, Virginia.
National Concrete Producers Assn., Downsview, Ontario.

Chapter 10

PLAIN MASONRY DESIGN

In the design of plain masonry, the basic materials brick, stone, tile and concrete block, may be used individually or in conjunction with one another. This usually depends on the loading conditions or the final appearance that is desired. In modern construction, brick and stone in many instances, are used as exterior veneers. Concrete block is normally used for load-bearing walls and may be either hollow or solid. In some cases, brick and concrete block are tied together to form composite walls that have both good load-bearing capacities and an attractive finish. Composite construction is not limited to brick and block. There are various combinations that may be used, depending on the conditions that may arise.

The more common units of clay brick and concrete block will be used in various situations to illustrate the basic concepts of the rational design method as applied to plain masonry.

Various organizations such as the Brick Institute of America, the National Concrete Masonry Association in the United States, the National Concrete Producers Association and the Ontario Masonry Contractors Association in Canada have developed various tables to relieve the designer of many of the more routine calculations.

The following examples deal with three areas of plain masonry design; *columns, load-bearing walls* and *non-load-bearing walls.*

COLUMNS

Example 10–1

Calculate the capacity of a nonreinforced brick masonry column as shown in Figure 10–1. The following conditions are given: t = 190 mm (7.6 in), b = 390 mm (15.6 in), h = 3.5 m (11.5 ft), f'_m = 16 MPa (2300 psi), and e_1/e_2 = 0.

Solution

As there are no accumulative loads above the column, the virtual eccentricity is equal to the actual eccentricity.

$$e_{top} = 38\text{mm}$$

$$e_{bottom} = 0$$

As bending occurs about one principal axis only, 4.4.3.28.(1) of the code must be complied with:

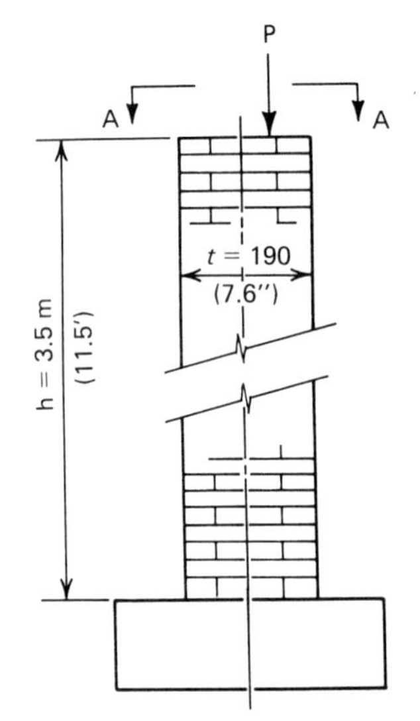

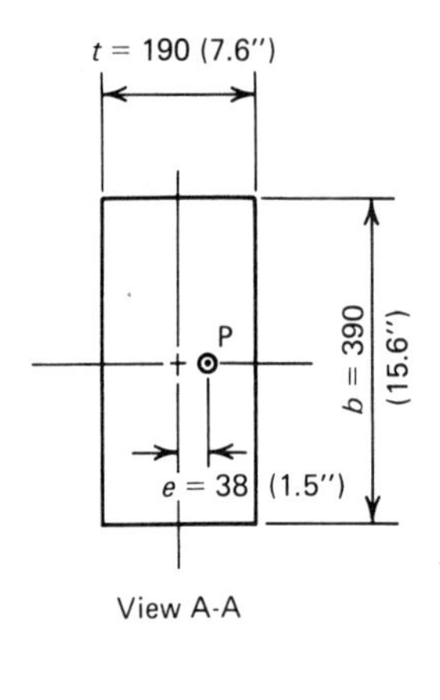

Fig. 10-1. Brick masonry column.

t/20 = 190/20 = 9.5 mm which is less than 38 mm.

$$\frac{t}{3} = \frac{190}{3}$$

= 63.3 mm (or 2.5 in) which is greater than 38 mm (or 1.5 in).

Part (b) of 4.4.3.28 must be considered, which states that the allowable capacity of the column is calculated by;

$$P = C_e \times C_s \times f_m \times A_n$$

Before any additional calculations are made, the slenderness ratio, h/t, of the column should be compared to the maximum allowable as given in 4.4.3.19.(1), which states that for columns, the slenderness ratio (h/t) shall not exceed $5(4 - e_1/e_2)$.

$$\frac{h}{t} = \frac{3500}{190}$$

$$= 18.42$$

$$5(4 - \frac{e_1}{e_2}) = 5(4 - 0)$$

$$= 20$$

As 18.42 is less than 20, the column slenderness is acceptable.

To establish the eccentricity coefficient, C_e, the ratios e/t and e_1/e_2 must be applied to Table 9 of the code. In this case, e_1/e_2 is given as zero and e/t = 38/190 = 0.2, giving C_e = 0.66.

To establish C_s, the quantities h/t and e_1/e_2 are applied to Table 8. For e_1/e_2 = 0 and h/t = 18.42, C_s = 0.71. For plain brick masonry:

$$f_m = 0.20 f_m'$$

$$= 0.20 \times 16$$

$$= 3.2 \text{ MPa (or 460 psi)}$$

The cross-sectional area of the column:

$$A_n = 190 \times 390$$

$$= 74\,100 \text{ mm}^2 \qquad (115 \text{ in}^2)$$

The maximum allowable load on the column is:

$$P = 0.66 \times 0.71 \times 3.2 \times \frac{74\,100}{1000 \times 1000}$$

$$= 0.110 \text{ MN}$$

$$= 110 \text{ kN (24.75 kips)}$$

Example 10-2

Calculate the capacity of a circular nonreinforced brick masonry column given the following conditions: Diameter = 400 mm (16 in), effective height h = 4 m (13 ft), f'_m = 16 MPa (2300 psi), and e_t = 75 mm (3 in). Assume $e_1/e_2 = 0$.

Solution

Since the column is circular, in order to calculate the h/t ratio, an effective thickness must be calculated. The effective thickness, (t), in relation to the principal axis is assumed as 3.5 times the column radius of gyration. (See Appendix, sentence 4.4.3.24.(2). The radius of gyration for circular sections is equal to $D/4$, where D is the diameter of the column. The effective thickness then becomes:

$$t = 3.5 \times \frac{D}{4}$$

$$= 3.5 \times \frac{400}{4}$$

$$= 350 \text{ mm (or 14 in)}$$

For columns, $h/t \leqslant 5(4 - e_1/e_2)$. For $e_1/e_2 = 0$,

$$h/t = 5(4 - 0)$$

$$= 20$$

For column effective height, h = 4 m (13 ft):

$$\frac{h}{t} = \frac{4000}{350}$$

$$= 11.43 \text{ which is less than 20.}$$

Column is within slenderness limits.

Check for e_t = t/3 for given e_t = 75 mm.

$$t/3 = \frac{350}{3} = 117 > 75$$

and $$\frac{t}{20} = \frac{350}{20} = 17.5 < 75$$

Therefore, the allowable load on the column will be:

$$P = C_e \times C_s \times f_m \times A_n$$

as per 4.4.3.28.(1) of the code.

For $e_1/e_2 = 0$ and $e/t = 75/350 = 0.214$, from Table 9, $C_e = 0.64$. For $e_1/e_2 = 0$ and $h/t = 11.43$, from Table 8, $C_s = 0.90$. For plain brick masonry, $f_m = 0.20\ f'_m$ for allowable axial compressive stress.

$$f_m = 0.20 \times 16$$

$$= 3.2 \text{ MPa (460 psi)}$$

The cross-sectional area for the column is:

$$A_n = \frac{\pi D^2}{4}$$

$$= \frac{\pi \times 400^2}{4}$$

$$= 125\ 663 \text{ mm}^2 \text{ (195 in}^2\text{)}$$

The resulting allowable load on the column is:

$$P = 0.64 \times 0.90 \times 3.2 \times \frac{125\ 663}{1000 \times 1000}$$

$$= 0.232 \text{ MN}$$

$$= 232 \text{ kN (52.2 kips)}$$

Example 10-3

A concrete block column of plain masonry is built of 30 MPa (4350 psi) solid concrete block units and type M mortar. The column effective height is 5 m (16.4 ft). The load P is offset 25 mm (1 in) from the X-X axis and 15 mm (0.6 in) from the Y-Y axis as shown in Figure 10-2. Assume $e_1/e_2 = 0$.

Calculate the maximum axial load P that can be applied under the given conditions.

Solution

The first requirement is to determine the value of f'_m of the block units. Referring to Table 3 of the code, for a 30 MPa (4350 psi) concrete unit and Type M mortar, f'_m = 15 MPa (2175 psi).

Next, check the h/t requirements.

For columns, $h/t \leqslant 5(4 - e_1/e_2)$.

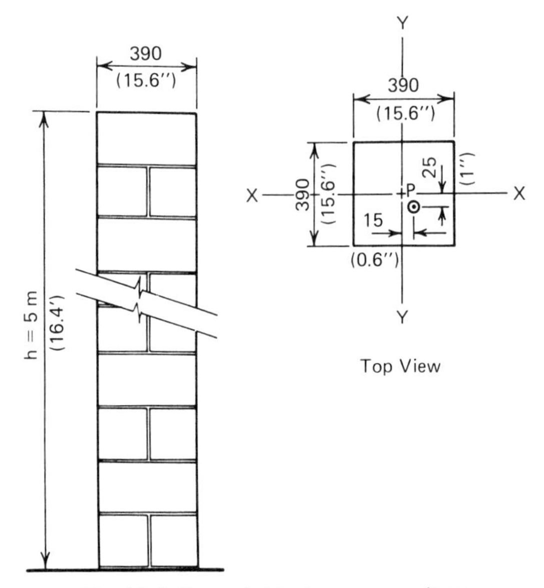

Fig. 10-2. Concrete block masonry column.

$$\text{Maximum allowable } \frac{h}{t} = 5(4 - 0)$$
$$= 20$$

Actual h/t = 5000/390 = 12.82. The actual slenderness ratio does not exceed that allowable; therefore, the column is within limits.

As the load is offset from both major axes, 4.4.3.28(3) of the code must be considered. Part (b) requires that $(e_t \times b) + (e_b \times t) \leqslant bt/3$ where b and t are the cross-sectional dimensions of the column and e_b and e_t are the respective eccentricities as shown in Figure 10-2. Since the column is square, $b = t = 390$ mm.

$$e_t = 25 \text{ mm}$$
$$e_b = 15 \text{ mm}$$
$$(e_t \times b) + (e_b \times t) = (25 \times 390) + (15 \times 390)$$
$$= 15{,}600$$
$$\frac{bt}{3} = \frac{390 \times 390}{3}$$
$$= 50\ 600$$

As part (b) is satisfied, the allowable load that can be carried by the column is:

$$P = C_e \times C_s \times f_m \times A_n$$

where f_m = the allowable compressive stress taken from Table 5 of the code.

$$f_m = 0.20 \times f'_m$$
$$= 0.20 \times 15$$
$$= 3 \text{ MPa}$$

A_n = Cross-sectional area.

$$A_n = 390 \times 390$$
$$= 152\ 100 \text{ mm}^2$$
$$= 0.1521 \text{ m}^2$$

C_s = Eccentricity coefficient, from Table 9 of the code.

$$\text{For } e_1/e_2 = 0 \text{ and } \frac{e_t b + e_b t}{bt} = 0.102$$
$$C_e = 0.84$$

C_s = Slenderness coefficient from Table 8.

$$\text{For } h/t = 12.82 \text{ and } e_1/e_2 = 0$$
$$C_s = 0.875$$

The maximum allowable load is:

$$P = 0.84 \times 0.875 \times 3.0 \times 0.1521$$
$$= 0.335 \text{ MN}$$
$$= 335 \text{ kN (75.4 kips)}$$

LOAD-BEARING WALLS

Load-bearing walls are walls that are subjected to vertical loads due to floors or roofs. In addition to the vertical loads, load-bearing walls may be subjected to horizontal forces such as wind. This produces a combined load situation that could cause overstressing in the wall due to the compounding of the stresses.

Example 10-4

Calculate the capacity of one meter of wall of a nonreinforced brick masonry wall section as shown in Figure 10-3, given the following: t = 190 mm (7.6 in), h = 2.4 m (7.8 ft), f'_m = 19 MPa (2755 psi), M_{top} = 10 kN · m (7.38 ft-k) and M_{bot} = 2 kN · m (1.5 ft-k).

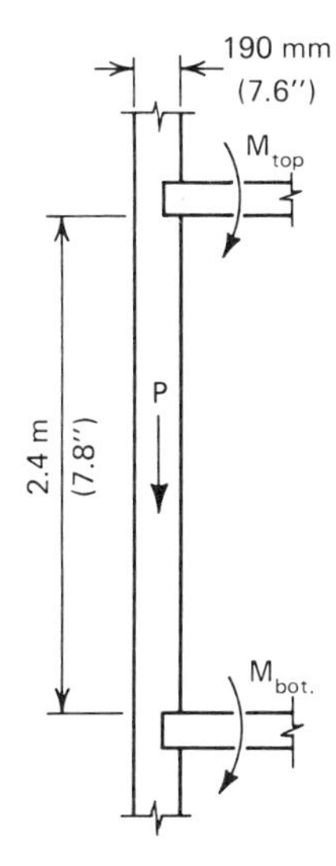

Fig. 10-3. Wall moments.

Solution

Ignoring the mass of the wall, calculate the eccentricities at the top and bottom of the wall:

$$e_{top} = \frac{M_{top}}{P} = \frac{10 \times 1000}{P} \text{ mm}$$

$$e_{bot} = \frac{M_{bot}}{P} = \frac{2 \times 1000}{P} \text{ mm}$$

Next, calculate the ratio e_1/e_2 for the given conditions:

$$\frac{e_1}{e_2} = \frac{2/P}{10/P} = -0.20 \text{ (double curvature)}$$

Calculate the ratio, h/t for the wall:

$$\frac{h}{t} = \frac{2.4 \times 1000}{190} = 12.63$$

The allowable axial load for the wall is given by:

$$P = C_e \times C_s \times f_m \times A_n$$

From Table 8 of the code, C_s = 0.90.
From Table 4, $f_m = 0.25 f'_m$.

$$f_m = 0.25 \times 19 = 4.75 \text{ MPa} = 4750 \text{ kPa}$$

For one meter of wall, $A_n = 1000 \times 190 = 190 \times 10^3 \text{ mm}^2 = 0.19 \text{ m}^2$

In order to calculate the allowable load, the factor C_e must be established. To calculate C_e, the ratio e/t must first be determined. Using the larger virtual eccentricity, calculate the ratio e/t:

$$\frac{e}{t} = \frac{\frac{10 \times 1000}{P}}{190}$$

Try P = 600 kN, $\frac{e}{t} = \frac{\frac{10\,000}{600}}{190} = 0.0877$

From Table 9 of the code, C_e = 0.88
Allowable load, $P = 0.88 \times 0.9 \times 4750 \times 0.19$
= 715 kN as compared to 600 kN.

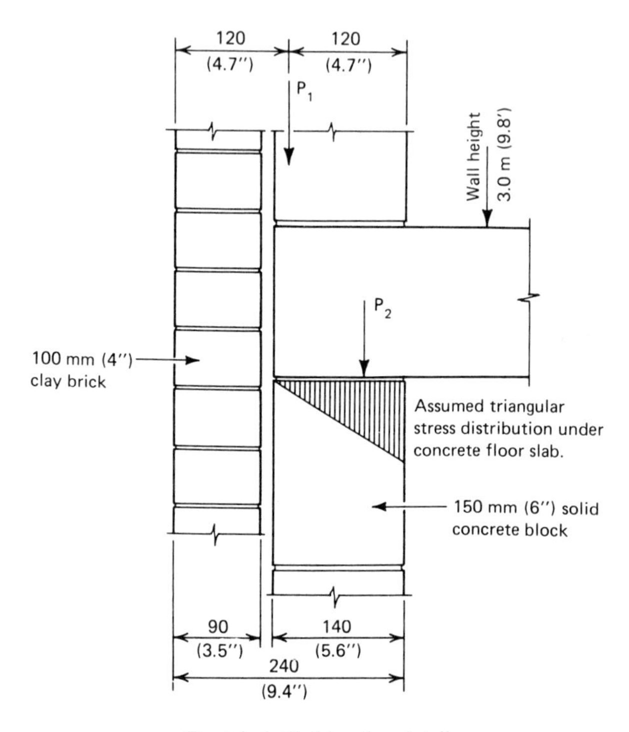

Fig. 10-4. Wall-loading detail.

As the assumed P must equal the allowable P, try $P = 763$ kN:

$$\frac{e}{t} = \frac{\frac{10\,000}{600}}{190}$$

$$= 0.069 \text{ and } C_e = 0.939$$

Allowable load, P

$$= 0.939 \times 0.9 \times 4750 \times 0.19$$

$$= 763 \text{ kN } (171.7 \text{ kips})$$

Example 10-5

A 250 mm (10 in) wall of composite construction is composed of 100 mm (4 in) clay brick and 150 mm (6 in) solid concrete block units. The clay brick has a compressive strength of 40 MPa (5800 psi), and the concrete block compressive strength is 20 MPa (2900 psi). Type M mortar is to be used. The wall effective height is 3 m (9.8 ft).

Neglecting the mass of the wall, compute the wall stresses for the vertical loads shown in Figure 10-4 across the wall section, where P_1 = 100 kN/m (6 k/ft) and P_2 = 30 kN/m (2 k/ft).

Solution

As the wall is of composite construction, the sectional properties of each material in the wall must be established. From Table 2, f'_m (clay brick) = 15 MPa (2175 psi) and from Table 3, f'_m (solid block) = 12 MPa (1740 psi). Calculating the modular ratio of the two materials:

$$n = \frac{E_{\text{brick}}}{E_{\text{block}}}$$

$$= \frac{15\,000}{12\,000}$$

$$= 1.25$$

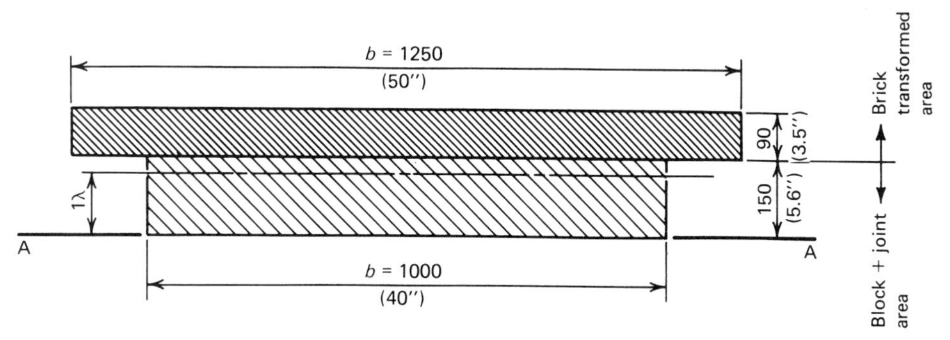

Fig. 10-5. Transformed wall section.

With the modular ratio, a transformed section for one meter of wall can be developed as shown in Figure 10-5.

For the transformed section, calculate the location of the centroidal axis (x-x):

$$\bar{y} = \frac{\Sigma A \times y}{\Sigma A}$$

$$\bar{y} = \frac{(1000 \times 150)75 + (1250 \times 90)195}{(1000 \times 150) + (1250 \times 90)}$$

$$= 126 \text{ mm}$$

Calculate the moment of inertia about the axis (x-x),

$$I_{x-x} = \frac{bd^3}{12} - 2\left[\frac{b_1 d_1{}^3}{12}\right]$$

where $b = 1250$, $d = 240$, $b_1 = (1250 - 1000)/2 = 125$ and $d_1 = 150$:

$$I_{x-x} = \frac{1250 \times 240^3}{12} - 2\left[\frac{125 \times 150^3}{12}\right]$$

$$= 1.44 \times 10^9 - 7.031 \times 10^7$$

$$= 1.37 \times 10^9 \text{ mm}^4$$

Calculate the section modulus of the wall with respect to the block face and the brick face.

$$S_{\text{(block face)}} = \frac{I}{c}$$

$$= \frac{1.37 \times 10^9}{126}$$

$$= 1.087 \times 10^7 \text{ mm}^3$$

$$S_{\text{(brick face)}} = \frac{I}{c}$$

$$= \frac{1.37 \times 10^9}{114}$$

$$= 1.202 \times 10^7 \text{ mm}^3$$

As there are two loads acting on the wall, the resulting virtual eccentricity (e) is calculated as:

$$e = \frac{M}{P} = \frac{P_1 e_1 + P_2 e_2}{P_1 + P_2}$$

P_1 is located at the center of the composite wall or 120 mm from the brick face. The stress distribution on the concrete block portion of the wall due to the floor slab is assumed as triangular as shown in Figure 10-4. The force P_2 occurs 140/3 = 43 mm (1.7 in) from the inner face of the block wall or 240 - 43 = 197 mm (7.8 in) from the exterior brick face. e_1 *and* e_2 are the eccentricities of P_1 and P_2 with respect to the centroidal axis (x-x). The resulting virtual eccentricity is:

$$e = \frac{(100 \times 13) + (30 \times 43)}{(100 + 30)}$$

$$= 20 \text{ mm } (0.8 \text{ in})$$

Due to the eccentricity, compressive stresses are produced by the vertical loads plus the induced moment.

Check $t/6$:

$$\frac{t}{6} = \frac{240}{6}$$

= 40 mm (1.6 in) which is greater than the resulting virtual eccentricity of 20 mm; as a result, the entire wall section will be subjected to compressive stress.

The total load on the composite wall is:

$$100 + 30 = 130 \text{ kN/m } (8 \text{ k/ft}).$$

The stress produced by this load acting on the transformed section is:

$$f_a = \frac{P}{A}$$

$$= \frac{130 \times 1000 \times 1000}{(1250 \times 90) + (1000 \times 150)}$$

$$= 0.495 \text{ MPa } (72 \text{ psi})$$

For (e) less than $t/3$, the maximum allowable compressive load is:

$$P = C_e \times C_s \times f_m \times A_n$$

Divide by the area A_n, and the allowable compressive stress becomes:

$$\frac{P}{A_n} = C_e \times C_s \times f_m$$

C_e = 0.90 for e_1/e_2 = 0 and e/t = 20/240 = 0.83
C_s = 0.87 for e_1/e_2 = 0 and h/t = 3000/240 = 12.5

$$f_m = 0.2 f'_m$$

$$= 0.2 \times 12$$

$$= 2.4 \text{ MPa } (348 \text{ psi})$$

The maximum allowable stress then becomes:

$$F_a = \frac{P}{A_n}$$

$$= 0.90 \times 0.87 \times 2.4$$

$$= 1.88 \text{ MPa } (273 \text{ psi})$$

Calculate the compressive stress due to bending on the block face and the brick face:

for the block face $f_b = \dfrac{P \times e}{S_{(\text{block face})}}$

$$= \frac{130 \times 20 \times 10^6}{1.087 \times 10^7}$$

$$= 239.2 \text{ kPa}$$

$$= 0.239 \text{ MPa } (35 \text{ psi})$$

for the brick face $f_b = \dfrac{P \times e}{S_{(\text{brick face})}}$

$$= \frac{130 \times 20 \times 10^6}{1.202 \times 10^7}$$

$$= 216.31 \text{ kPa}$$

$$= 0.216 \text{ MPa } (31 \text{ psi})$$

The allowable compressive stress due to bending as per the code for the concrete block is:

$$F_b = 0.3 \times f'_m$$

$$= 0.3 \times 12$$

$$= 3.6 \text{ MPa } (522 \text{ psi})$$

For the above results, check the concrete block for the combined stresses using the interaction equation:

$$\frac{f_a}{F_a} + \frac{f_b}{F_b} \leqslant 1.0$$

$$\frac{0.495}{1.88} + \frac{0.239}{3.6} = 0.329 \leqslant 1.0$$

Therefore, the wall is not overstressed.

To determine the stress distribution over the cross section of the wall, the combined stresses must be calculated at each face of the brick and the block. Calculating the stress at the extreme block fiber:

$$f_a + f_b = 0.495 + 0.239$$

$$= 0.734 \text{ MPa } (106 \text{ psi})$$

The combined stress at the extreme brick fiber is:

$$n(f_a - f_b) = 1.25(0.495 - 0.216)$$
$$= 0.349 \text{ MPa (51 psi)}$$

Calculate the combined stresses at the block-brick interface:

for the block,

$$(f_a - f_b) = 0.495 - \frac{\dfrac{130 \times 20 \times 10^6}{1.37 \times 10^9 \times 10^3}}{(150 - 126)}$$
$$= 0.495 - 0.044$$
$$= 0.451 \text{ MPa (65 psi)}$$

Where f_b is the stress due to bending in the block.

For the brick, $n(0.451) = 1.25 \times 0.451$
$= 0.564$ MPa (82 psi)

The resulting stress distribution across the wall is as shown in Figure 10-6.

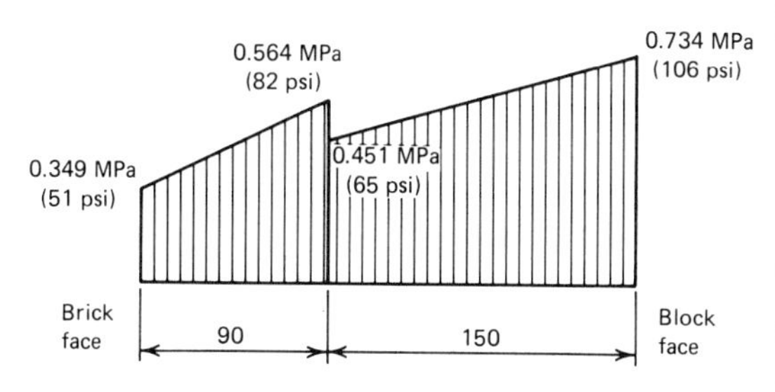

Fig. 10-6. Wall stress diagram.

NON-LOAD-BEARING WALLS

Non-load-bearing walls are walls that carry very little vertical load. They can be either exterior walls, where they are primarily subjected to horizontal wind loads, or interior partition walls that carry only their own dead weight.

As interior partition walls do not support any load as such, the empirical design method is used to establish minimum dimensions. Exterior walls that must carry loads due to wind are designed by the rational design method.

Example 10-6

An interior partition wall composed of hollow concrete block units is to be 3 m (9.8 ft) high. The horizontal distance between pilasters is 6 m (19.7 ft) . Using the empirical design method, calculate the minimum thickness of the wall.

Solution

From clause 4.4.4.17 of the code, the maximum distance between horizontal or vertical supports shall be not more than 36 times the thickness of the wall.

For the vertical conditions:

$$t = \frac{3000}{36}$$
$$= 83.3 \text{ mm (3.28 in)}$$

For the horizontal conditions:

$$t = \frac{6000}{36}$$
$$= 167 \text{ mm (6.57 in)}$$

The larger of the two values governs, and the minimum block size that should be used is 190 mm (8 in).

In the design of non-load-bearing walls that must resist wind loads, it is assumed that most of the lateral load is transmitted in the vertical direction with no end fixity at the lateral supports. This assumption may be conservative where the ratio of horizontal to vertical distances between lateral supports is relatively small and end fixity does exist. The curves in Figure 10-7 can be used to determine the proportion of wind load transmitted in the vertical and horizontal directions. The curves

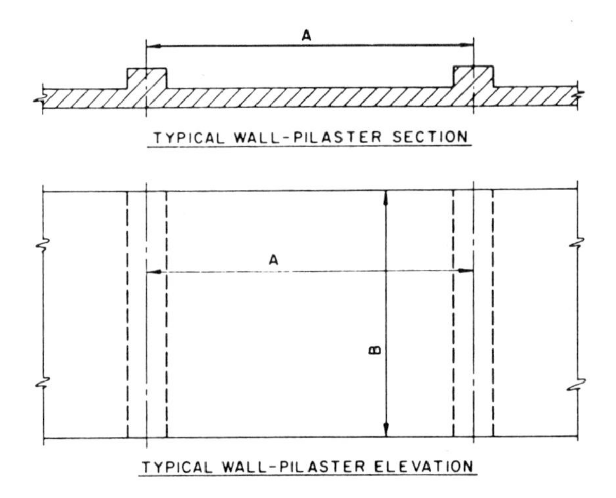

CASE 1: WALLS FIXED AT PILASTERS

A. FIXED AT BOTTOM, FREE AT TOP
B. SUPPORTED TOP AND BOTTOM
C. FIXED AT BOTTOM, SUPPORTED AT TOP

CASE 2: WALLS SUPPORTED AT PILASTERS

A. FIXED AT BOTTOM, FREE AT TOP
B. SUPPORTED TOP AND BOTTOM
C. FIXED AT BOTTOM, SUPPORTED AT TOP

CASE 3: WALLS FIXED AT ONE END, SUPPORTED AT OTHER

A. FIXED AT BOTTOM, FREE AT TOP
B. SUPPORTED TOP AND BOTTOM
C. FIXED AT BOTTOM, SUPPORTED AT TOP

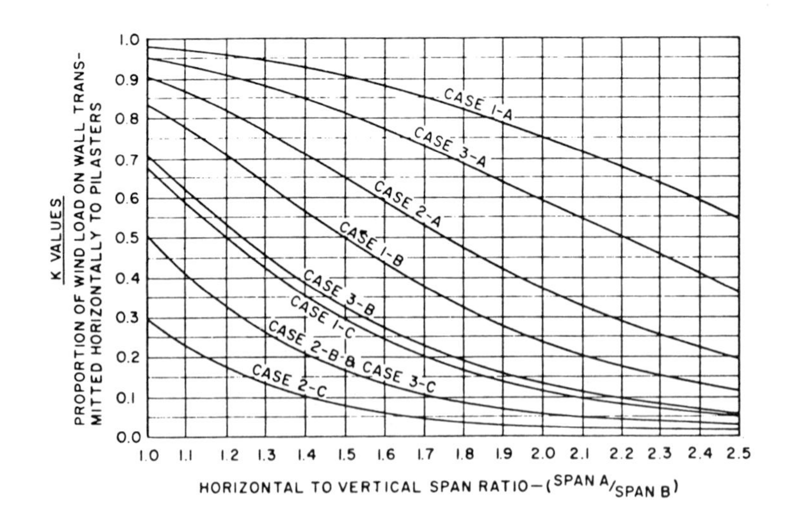

Fig. 10-7. Approximate wind load distribution to horizontal and vertical supports. (*Courtesy of National Concrete Masonry Association*)

are based on the assumption that the moment of inertia and moduli of elasticity of the walls are the same in both directions and that the wall has no openings that may affect the stiffness of the wall.

Example 10–7

A building has exterior walls spanning 4 m (13 ft) vertically between floor and ceiling. The walls are to span 6 m (19.7 ft) horizontally from center-to-center of pilasters which are built integrally with the wall. The roof loads will be carried by trusses supported by the pilasters, so the walls will be considered as free at the top and fixed at the bottom and at the pilasters.

The wall construction consists of 300 mm (12 in) hollow concrete masonry units laid up in running bond with face-shell mortar bedding, using type M mortar. Assume block shell thickness to be 32 mm (1¼ in).

Determine the resulting bending moments and flexural stresses developed if the wall must resist a horizontal pressure due to wind of 0.75 kN/m² (16 psf).

Solution

Refer to Figure 10–7; span A/span B = 6/4 = 1.5. For the conditions described above, the wall is represented in Figure 10–7 by Case 1-A. Using the curve for Case 1-A, the proportion of the wind load on the wall transmitted to the horizontal span, represented by K, is equal to 0.91 or 91 percent. With the wall having fixed ends at the pilasters, the maximum moment per meter of wall in the horizontal span is;

$$M = \frac{wl^2}{12} \text{ (See Fig. 16–10)}$$

$$= \frac{0.75 \times 0.91 \times 6^2}{12}$$

$$= 2.05 \text{ kN} \cdot \text{m (1.5 foot kips} = 1500 \text{ ft-lb)}$$

The flexural tensile stress developed is calculated by:

$$f_t = \frac{M}{S_{y-y}}$$

where S is the section modulus of one meter of wall in the vertical direction as shown in Figure 10–8.

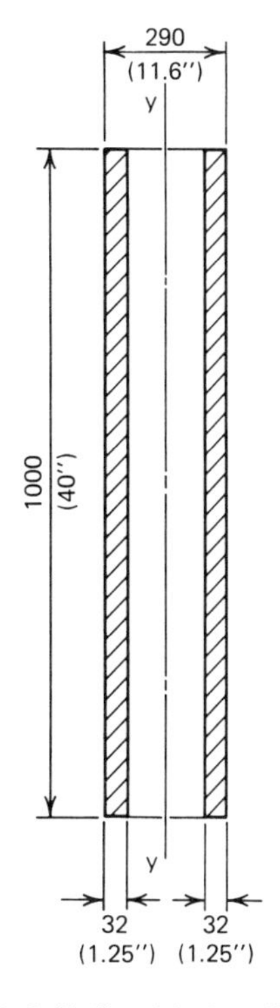

Fig. 10–8. Hollow block wall section.

$$S_{y-y} = \frac{b(h^2 - h_1^2)}{6}$$

$$= \frac{1000(290^2 - 226^2)}{6}$$

$$= 5.504 \times 10^6 \text{ mm}^3$$

$$f_t = \frac{2.05 \times 1000}{5.504 \times 10^6}$$

$= 0.372 \times 10^{-3}$ kN/mm^2

$= 0.372 \times 10^{-3} \times 10^6$ kN/m^2

$= 372$ kPa (54 psi)

The allowable tensile unit stress parallel to the bed joints is 320 kPa (46 psi). For wind this may be increased by one-fourth (as per the National Building Code) giving an allowable stress of 400 kPa (58 psi), which is greater than the actual stress of 372 kPa (54 psi).

In the vertical span, the wall described carries 9 percent of the loading. Since the wall is free at the top and fixed at the base, the maximum bending moment is as shown in Figure 15-10.

$$M = \frac{wh^2}{2}$$

$$= \frac{0.75 \times 0.09 \times 4^2}{2}$$

$= 0.54$ kN · m (0.40 foot kips = 400 ft-lb)

The flexural tensile unit stress developed is:

$$f_t = \frac{M}{S_{y-y}}$$

$$= \frac{0.54 \times 1000}{5.504 \times 10^6}$$

$= 0.098 \times 10^{-3}$ kN/mm^2

$= 98$ kPa (14.2 psi)

This value may be reduced by the dead load stress on the wall at the point of maximum moment, which in this case occurs at the bottom of the wall. Assuming the wall exerts a dead load force of 2.5 kN/m^2 of wall, the compressive unit stress developed is equal to:

$$\frac{\text{weight of the wall}}{\text{wall cross-sectional area}}$$

$$\frac{2.5 \times 4}{0.064 \times 1.0} = 156.3 \text{ kN/m}^2 \text{ (48 psi)}$$

The net result is a compressive stress of:

$156.3 - 98 = 58.3$ kPa (8.5 psi)

Glossary of Terms

Transformed Section An assumed section having the same elastic properties as a section composed of two different materials.

Centroidal Axis That axis which passes through the center of area of a section.

Review Questions

1. Calculate the capacity of a nonreinforced concrete block masonry column, given the following: $t = 190$ mm, $b = 190$ mm, $h = 3$ m, $f'_m = 15$ MPa and zero eccentricity top and bottom. Assume the block units are 40 percent void.
2. A solid column built of clay brick masonry units has dimensions of 190 × 290 mm. The column effective height is 3.25 m. What is the maximum load the column will be able to support if the load is to be offset 50 mm from the weak axis? Assume $f'_m = 19$ MPa and $e_1/e_2 = +0.5$.
3. A circular nonreinforced column of clay brick masonry is to carry a load of 60 MN that acts through the centroid of the column. Assume the column effective height is to be 4 m, calculate the strength of brick masonry and mortar combination required to support the given load.

4. A solid nonreinforced masonry wall has the following specifications: t = 240 mm, h = 4 m, f'_m = 10 MPa and e_1/e_2 = −0.5. Determine the maximum thickness of the wall and calculate the maximum load per meter that the wall can support assuming a maximum eccentricity of 15 mm.
5. Calculate the minimum thickness of a nonreinforced solid wall having a height of 6 m. If no lateral support is provided along the top of the wall, calculate the maximum load per meter of wall if 55 MPa units with type M mortar are to be used. Assume zero eccentricity.
6. A nonreinforced composite wall of 90 mm brick and 290 mm concrete block units has an effective height of 6 m. Calculate the maximum load per meter if the load is to be concentrated 175 mm from the outer face of the block. Assume 20 MPa solid concrete block and 40 MPa brick units, type M mortar and e_1/e_2 = +1. Allow for a 10 mm mortar joint between the brick-block interface.

Selected Sources of Information

Alberta Building Code, Edmonton, Alberta.
Brick Institute of America, McLean, Virginia.
Clay Brick Association of Canada, Willowdale, Ontario.
National Building Code of Canada, Ottawa, Ontario.
National Concrete Masonry Assn., McLean, Virginia.
National Concrete Producers Assn., Downsview, Ontario.

Chapter 11

REINFORCED MASONRY DESIGN

When applying the working stress design method to reinforced masonry, you should remember that the allowable stresses used are set at some predetermined value. Clause 4.4.3.15 of the code gives the maximum allowable stresses in the reinforcing steel when it is used in conjunction with masonry units. From the given values, you can see that the allowable stresses are well below the stress at which yielding occurs.

To introduce the working stress design method as applied to reinforced masonry, we will investigate typical building components such as *beams, columns* and *pilasters* and walls on an individual basis in a series of practical examples.

BEAMS

Beams in reinforced masonry are used primarily as lintels over doors and windows and as bond beams. Walls that must support loads and span relatively large openings may also be designed as deep beams.

The following examples will deal with the basic requirements that must be considered to develop a good design, namely:

1. Tension reinforcing to absorb the tensile stresses due to bending.

2. Shear reinforcing to absorb the shear stresses due to vertical shear.

3. Bond and anchorage requirements for the reinforcing steel.

Example 11-1

A 190 mm (7.6-in) four-course deep masonry beam composed of 15 MPa (2000-psi) concrete block units filled with concrete grout, has an effective depth of 700 mm (27.5 in) as shown in Figure 11-1.

Calculate the required steel area, A_s, to produce balanced conditions. Allowable steel stress 125 MPa (18 ksi) and type M mortar is to be used.

Solution

With the aid of Figure 9-8, a stress diagram can be drawn for the beam as shown in Figure 11-2. To produce balanced conditions, the

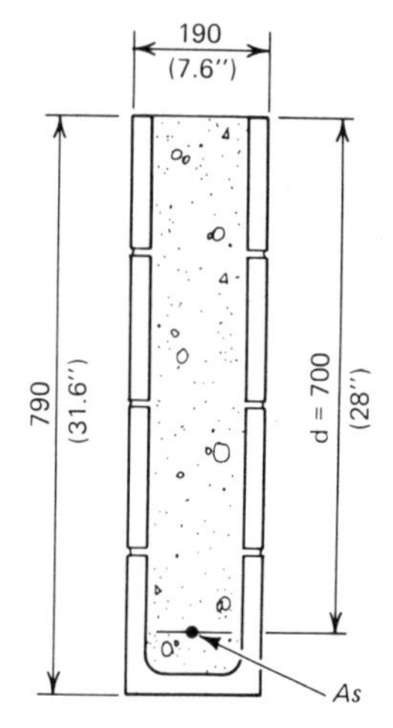

Fig. 11-1. Masonry block beam cross-section.

internal resisting forces, T and C, must be equal. In equation form:

$$T = C$$

To calculate the value of the compression force C, the location of the neutral axis must be determined. Referring to the stress diagram, the values of f_m and f_s/n must first be established. From Table 7 of the code:

$$f_m = 0.33 f'_m$$

From Table 3 of the code, for 15 MPa (2000-psi) block and type M mortar, f'_m = 10 MPa (1450 psi).

$$f_m = 0.33 \times 10$$
$$= 3.3 \text{ MPa (478 psi)}$$

Calculating the modular ratio:

$$n = \frac{E_{\text{steel}}}{E_{\text{masonry}}}$$

$$E_{\text{steel}} = 200 \times 10^3 \text{ MPa}$$

From Table 7 of the code, $E_{\text{masonry}} = 10 \times 10^3$ MPa.
The resulting modular ratio becomes:

$$n = \frac{200 \times 10^3}{10 \times 10^3}$$
$$= 20$$

The value for $f_s/n = \dfrac{125}{20}$
$$= 6.25$$

Using the similar triangles of the stress diagram:

$$\frac{f_m}{kd} = \frac{f_s/n + f_m}{d}$$

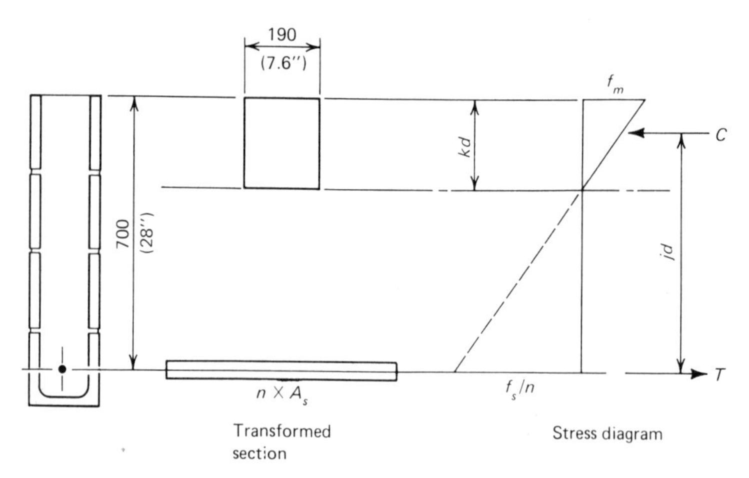

Fig. 11-2. Masonry beam transformed section and stress diagram.

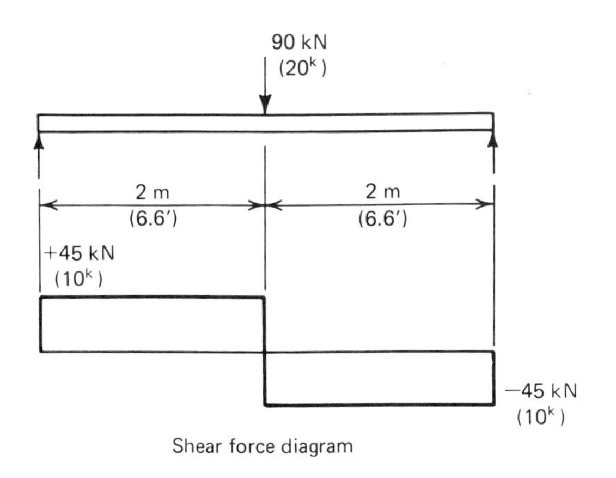

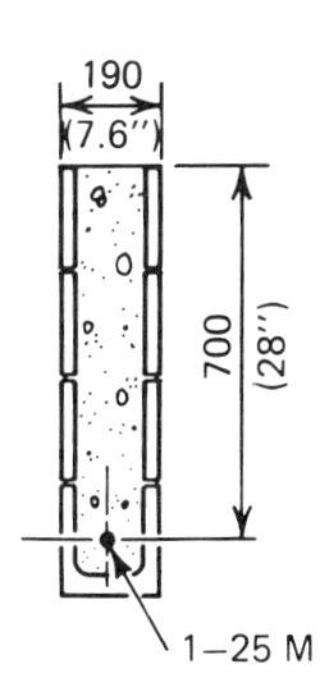

Fig. 11-3. Beam shear force diagram.

$$\frac{3.3}{k} = \frac{6.25 + 3.3}{1}$$

$$k = 0.35$$

$$kd = 0.35 \times 700$$

$$= 245 \text{ mm}$$

Apply Eq. (9-7); $C = \frac{1}{2} \times 3.3 \times \frac{190 \times 245}{10^6}$

$$= 0.077 \text{ MN}$$

$$= 77 \text{ kN (17.3 kips)}$$

Using Eq. (9-8) and knowing that tension must equal compression:

$$T = A_s \times f_s$$

$$A_s \times f_s = 77 \text{ kN}$$

$$A_s = \frac{77 \times 1000 \times 1000}{125 \times 1000}$$

$$= 616 \text{ mm}^2$$

(Equivalent to one 30M bar)

Example 11-2

For the beam in Figure 11-3, neglecting the beam dead weight, calculate the reinforcing steel required for the shear stresses developed. Check the bond requirements for the main tension steel. Assume 15 MPa (2000 psi) concrete block units with type M mortar, and the allowable steel stress as 125 MPa (18 ksi).

Solution

Consider the shear requirements first. In Table 7, the maximum allowable shear stress is given as $v = 0.02f'_m$, but not to exceed 350 kPa (51 psi). For 15 MPa (2000 psi) units with type M mortar, $f'_m = 10$ MPa (1450 psi).

$$v = 0.02 \times 10$$

$$= 0.20 \text{ MPa}$$

$$= 200 \text{ kPa (29 psi)}$$

From clause 4.4.3.39 of the code, the actual shear stress due to the applied load is calculated as:

$$v = \frac{V}{b \times d}$$

where V = shear force at distance (d) from the support (clause 4.8.3.2), b = beam width and d = beam effective depth. From the shear force diagram in Figure 11-3, V = 45 kN at distance (d), and the resulting actual shear stress becomes:

$$v = \frac{45 \times 1000 \times 1000}{190 \times 700}$$

$$= 338 \text{ kPa (49 psi)}$$

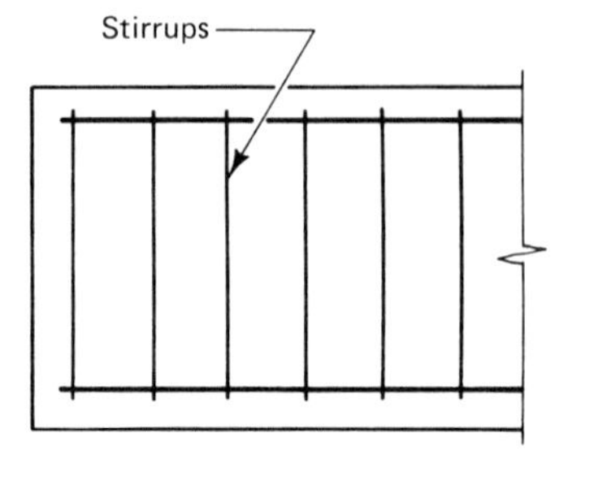

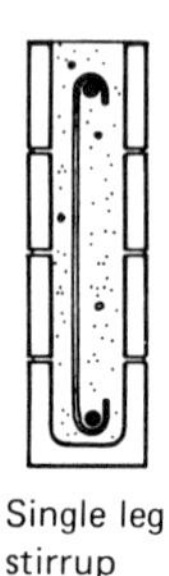

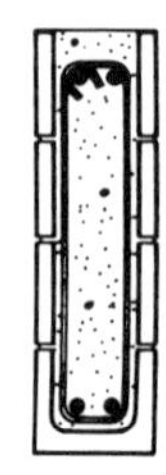

Fig. 11-4. Beam stirrups.

As the actual shear stress is greater than the allowable, the beam requires shear reinforcing steel. Clause 4.4.3.39.(4) of the code requires that if the actual stresses in shear exceed the allowable stresses, the web reinforcing must be designed for the total actual shear stress. In this case, the shear reinforcing steel must be designed for 338 kPa (49 psi).

For shear reinforcing (stirrups perpendicular to the tension steel—see Fig. 11-4) the area of steel required as per 4.8.4.3 of the code is:

$$A_v = \frac{V \times S}{f_v \times d}$$

where A_v = Cross-sectional area of the stirrup.
V = Vertical shear force in beam at particular section.
S = Stirrup spacing.
f_v = Allowable stress in steel.
d = Beam effective depth.

Using 15M bars (see Table 13-2), A_v = 200 mm^2, the required spacing for a single leg stirrup is;

$$S = \frac{A_v \times f_v \times d}{V}$$

$$= \frac{200 \times 125 \times 10^3 \times 700}{45 \times 1000 \times 1000}$$

$$= 388 \text{ mm } (15.25 \text{ in})$$

Use 15M stirrups, single leg, at 300 mm (12 in) o.c. or 15M stirrups, double leg, at 600 mm (24 in) o.c. However, maximum spacing as per 4.4.3.43 is $d/2$ = 700/2 = 350 mm (14 in). Therefore, single leg stirrups at 300 mm (12 in) are required.

Checking the bond requirements for the main steel, 4.4.3.44 of the code gives the following formula for the calculation of bond stresses:

$$u = \frac{V}{\Sigma_o\, jd}$$

where u = The bond stress developed.
V = Vertical shear force due to the applied load.
jd = Distance between internal forces as calculated by Eq. (9-6).
Σ_o = Sum of the perimeters of the bars being considered.

For the given steel:

$$u = \frac{45 \times 10^6}{79.2 \times 625}$$

$$= 909 \text{ kPa } (131 \text{ psi})$$

The maximum allowable bond stress, as per Table 7, is 1100 kPa (160 psi), and so, the actual bond stress is within limits.

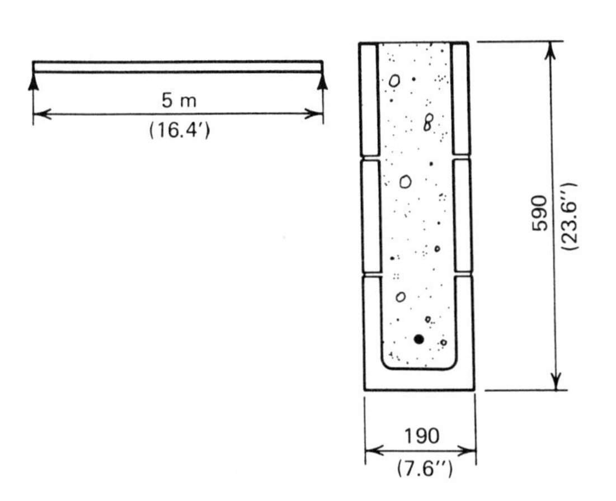

Fig. 11-5. Beam span and cross-section.

Example 11-3

Calculate the required tension steel in a beam constructed of 200 mm (8 in) concrete block masonry units, three courses high, spanning 5 m (16.4 ft) and carrying a total design load of 600 N/m (40 plf) as shown in Figure 11-5. Type M mortar is to be used with units having an f'_m = 7.70 MPa (1100 psi). The reinforcing steel has a yield strength of 400 MPa (58 ksi).

Solution

Calculate the maximum bending moment due to the applied load.

$$M = \frac{w \times l^2}{8}$$

$$= \frac{600 \times 5^2}{8}$$

$$= 1875 \text{ N} \cdot \text{m (1383 ft-lb)}$$

In order to establish an area of steel, assume a minimum steel ratio (ρ_{min}) as given in the code

$$\rho_{min} = \frac{0.55}{f_y}$$

where f_y is expressed in MPa:

$$\rho_{min} = \frac{0.55}{400{,}000}$$

$$= 0.0014$$

Also, knowing that $\rho = A_s/bd$, and selecting an effective depth of 500 mm (20 in), the required steel area becomes:

$$A_s = 0.0014 \times 190 \times 500$$

$$= 133 \text{ mm}^2 \text{ (0.2 in}^2\text{)}$$

Using a 15M bar, A_s = 200 mm² (0.31 in²), and the resulting steel ratio becomes:

$$\rho = \frac{200}{190 \times 500}$$

$$= 0.0021$$

Applying the basic theory discussed in Chapter 9, and using Eq. (9-5):

$$k = [(n\rho)^2 + 2n\rho]^{1/2} - \rho n$$

Solving for the modular ratio (n):

$$n = \frac{E_{steel}}{E_{masonry}}$$

$$= \frac{200 \times 10^3}{7.7 \times 10^3}$$

$$= 26$$

$$k = [(26 \times 0.0021)^2 + 2 \times 26 \times 0.0021]^{1/2} - 26 \times 0.0021$$

$$= (0.0030 + 0.1092)^{1/2} - 0.0546$$

$$= 0.280$$

The distance from the top of the beam to the neutral axis becomes:

$$kd = 0.280 \times 500$$

$$= 140 \text{ mm (5.51 in)}$$

Using Eq. (9-6):

$$jd = d - \frac{kd}{3}$$

$$= 500 - \frac{140}{3}$$

$$= 453.3 \text{ mm } (17.85 \text{ in})$$

The maximum internal moment developed, using the capacity of the masonry in compression as per Eq. (9-9) is:

$$M = \tfrac{1}{2} f_m \times kjbd^2$$

where $f_m = 0.33 f'_m$ as per Table 7 of the code.

$$f_m = 0.33 \times 7.70$$
$$= 2.54 \text{ MPa } (368 \text{ psi})$$

The resulting moment is:

$$M = \frac{\tfrac{1}{2} \times 2.54 \times 140 \times 453.3 \times 190}{1000 \times 1000}$$
$$= 15.31 \text{ MN} \cdot \text{mm}$$
$$= 15.31 \text{ kN} \cdot \text{m } (11.30 \text{ ft-kips})$$

which is greater than the 1875 N · m bending moment due to the applied load. The capacity of the beam in compression due to bending is not exceeded. Checking the internal moment developed by the reinforcing steel in tension as per Eq. (9-10):

$$M = \rho f_s \times jbd^2$$

For steel with a yield strength of 400 MPa (58 ksi), the maximum allowable steel stress (f_s) allowed by the code is 165 MPa (24 ksi). The stress developed in the steel due to the applied load may be calculated by:

$$f_s = \frac{M}{\rho \times jbd^2}$$
$$= \frac{1875 \times 10^9}{0.0021 \times 140 \times 453.3 \times 190}$$
$$= 74.0 \times 10^6 \text{ N/m}^2$$
$$= 74.0 \text{ MN/m}^2$$
$$= 74.0 \text{ MPa } (10.73 \text{ ksi})$$

Which is less than the 165 MPa (24 ksi) allowed, and one 15M bar is sufficient.

Example 11-4

Using the beam in Example 11-3, check the conditions for bond and anchorage.

Solution

Clause 4.4.3.44 of the code gives the formula for bond as:

$$u = \frac{V}{\Sigma_o \, jd}$$

where u = Bond stress developed in the steel.
V = Vertical shear force developed by the applied load.
Σ_o = Sum of the perimeters of the reinforcing bars.
jd = Distance between the internal tension and compression forces developed in the beam due to bending.

From the previous example,

$$V = \frac{w \times l}{2}$$
$$= \frac{600 \times 5}{2}$$
$$= 1500 \text{ N } (340 \text{ lb})$$

Σ_o = 50.1 mm (Table 13-2) and jd = 453.3 mm

$$u = \frac{1500}{50.1 \times 453.3}$$
$$= 0.066 \text{ N/mm}^2$$
$$= 66.0 \times 10^3 \text{ N/m}^2$$
$$= 66.0 \text{ kPa } (9.57 \text{ psi})$$

The allowable bond stress as per Table 7 for deformed bars is 1100 kPa (160 psi).

The anchorage length of a bar may be checked by determining the length of a bar required to resist the maximum allowable tensile stress of the bar by the maximum allowable bond stress. Consider the bond force developed between the bar and the masonry in equation form:

$$F_u = u \times \pi d_b \times L_d$$

where u = Maximum allowable unit bond stress.

d_b = Reinforcing bar diameter.

L_d = Bar length available beyond the point of maximum stress.

This force must be equal to the maximum tensile force developed in the steel at maximum allowable stress. In equation form:

$$f_t = \frac{\pi d_b^2}{4} \times f_s \qquad \text{Eq. (11-2)}$$

Equating the two equations:

$$\frac{\pi d_b^2}{4} \times f_s = u \times \pi d_b \times L_d$$

Solving for the length (L_d):

$$L_d = \frac{f_s \times d_b}{4 \times u} \qquad \text{Eq. (11-3)}$$

Applying this to the preceding example where f_s = 165 MPa, d_b = 16 mm and u = 1100 kPa,

$$L_d = \frac{165 \times 1000 \times 16}{4 \times 1100}$$

$$= 600 \text{ mm (24 in)}$$

The actual length of bar available from point of maximum stress is one half the beam span or 2500 mm (8.2 ft).

COLUMNS AND PILASTERS

In many masonry buildings, pilasters or columns are used to carry concentrated roof or floor loads. These loads rarely pass through the

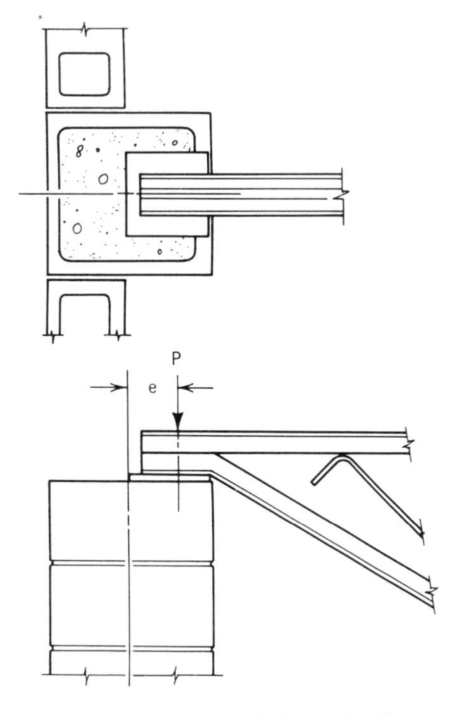

Fig. 11-6. Joist and pilaster detail.

centroid of the column, thereby producing an eccentric loading condition. This condition can occur in buildings that use steel joists for the roof-framing system. A typical detail is shown in Figure 11-6. Where the eccentricity is greater than one-third the lateral dimension of the supporting pilaster or column, tensile stresses develop and reinforcing steel must be added.

Example 11-5

The pilaster shown in Figure 11-7 is to be 4 m (13 ft) high and support a truss load of 75 kN (17 kips) with an eccentricity of 100 mm. The pilaster consists of concrete block units filled with concrete grout. Type M mortar is to be used, and f'_m = 15 MPa (2.18 ksi), f_s = 140 MPa (20 ksi), and e_1/e_2 = +1. The vertical reinforcing is four 20M bars placed as

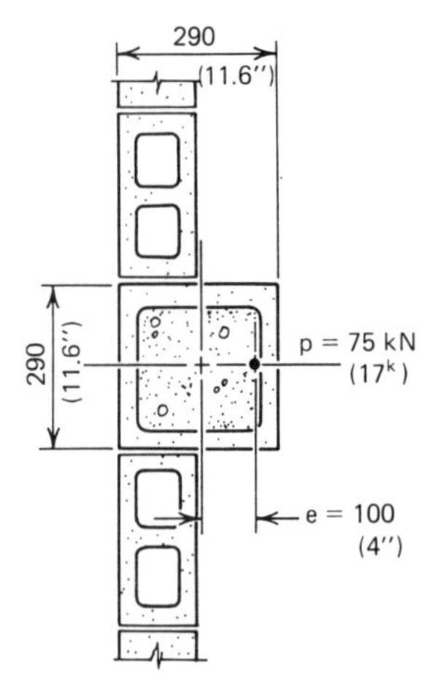

Fig. 11-7. Grouted masonry block pilaster.

shown, with no horizontal ties. Will the pilaster have sufficient capacity to support the load due to the roof truss? Assume $n = 13$.

Solution

To determine which clause in the code will govern the design of this particular pilaster, the eccentricity (e) must be checked to find out whether or not it exceeds the value of $t/3$. In this case, $t/3 = 290/3 = 96.7$ mm, which is smaller than the eccentricity of 100 mm. In this case, 4.4.3.31.(2) is the governing clause, and the capacity of the pilaster must be determined from a transformed section. Since lateral ties have not been provided, the steel in compression is ignored when calculating the transformed section properties.

The cross-sectional area of the two 20M bars shown in Figure 11-8 is 600 mm² (0.93 in²). To establish the section properties of the transformed section, assume a value for the distance kd and calculate the location of the centroidal axis.

By trial and error, $kd = 154$ mm. Let $\bar{y}$ be the distance from the compression side of the pilaster to the centroidal axis. In equation form:

$$\bar{y} = \frac{\Sigma A \times y}{\Sigma A}$$

Calculate the areas:

$$b(kd) = 290 \times 154 = 44\,660 \text{ mm}^2$$

$$n(A_s) = 13 \times 600 = 7800 \text{ mm}^2$$

Total area is 52 460 mm²

Next, calculate $A \times y$;

$$b(kd)\,(kd/2) = 44\,660 \times \frac{154}{2} = 3.439 \times 10^6 \text{ mm}^3$$

$$n(A_s)\,(d) = 7800 \times 230 = 1.794 \times 10^6 \text{ mm}^3$$

$$\Sigma A \times y = 5.233 \times 10^6 \text{ mm}^3$$

$$\bar{y} = \frac{5.288 \times 10^6}{52\,460}$$

$$= 99.8 \text{ mm (4 in)}$$

The moment of inertia of the transformed section about the centroidal axis (CA) is:

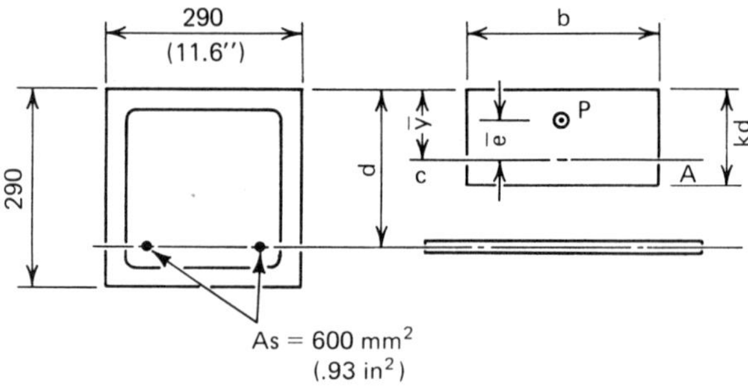

Fig. 11-8. Pilaster transformed section.

$$I_{CA} = \frac{b(kd)^2}{12} + A(y_1)^2 + (nA_s)(y_2)^2$$

(comp. area) (ten. area)

$$= \frac{290 \times (154)^3}{12} + 44\,660 \times \left(99.8 - \frac{154}{2}\right)^2$$

$$+ 7800(230 - 99.8)^2$$

$$= 8.826 \times 10^7 + 2.322 \times 10^7 + 13.223 \times 10^7$$

$$= 24.371 \times 10^7 \text{ mm}^4$$

The distance between the centroidal axis and the point of load application is:

$$\bar{e} = \bar{y} - (t/2 - e)$$

$$= 99.8 - (290/2 - 100)$$

$$= 54.80 \text{ mm (2.15 in)}$$

$$\text{and } d - \bar{y} = 230 - 99.80$$

$$= 130.2 \text{ mm (5.12 in)}$$

The combined stress due to the axial compressive stress plus the compressive stress due to the load eccentricity developed in the masonry can be written as:

$$f_m = \frac{P}{A} + \frac{P\bar{e}\bar{y}}{I_{CA}}$$

$$= \frac{75 \times 10^6}{(290)^2} + \frac{75 \times 10^6 \times 54.8 \times 99.8}{24.371 \times 10^7}$$

$$= 891.8 + 1683.1$$

$$= 2574.9 \text{ kPa}$$

$$= 2.575 \text{ MPa (373 psi) compression}$$

The combined stress in the reinforcing steel can be written as:

$$\frac{f_s}{n} = \frac{P}{A} - \frac{P\bar{e}(d - \bar{y})}{I_{CA}}$$

$$= 891.8 - \frac{75 \times 10^6 \times 54.8 \times 130.2}{24.371 \times 10^7}$$

$$= 891.8 - 2180$$

$$= -1288.2 \text{ kPa}$$

$$= 1.288 \text{ MPa (187 psi) tension}$$

Now, the assumed (kd) may be checked.

$$kd = \frac{f_m}{f_m + \frac{f_s}{n}} \times (d)$$

$$= \frac{2.575}{2.575 + 1.288} \times (230)$$

$$= 153 \text{ mm (6 in) which checks with the assumed.}$$

The modified stress allowed by the code for the masonry in compression can be stated as:

$$f_m = C_s\,(0.28 f'_m)$$

$$\text{and } f'_m = \frac{f_m}{0.28 \times C_s}$$

For $h/t = 4000/290 = 13.8$ and $e_1/e_2 = +1$, the value for $C_s = 0.648$.

$$f'_m = \frac{2.575}{0.28 \times 0.648}$$

$$= 14.19 \text{ MPa (2.06 ksi).}$$

The maximum available was given as 15 MPa (2.18 ksi). The tension stress in the steel can be calculated as:

$$f_s = \frac{f_s}{n} \times n$$

$$= 1.288 \times 13$$

$$= 16.744 \text{ MPa (2.42 ksi)}$$

The maximum allowable tensile stress available is:

$$C_s \times f_s = 0.648 \times 140$$

$$= 90.72 \text{ MPa (13.15 ksi)}$$

From the foregoing figures it becomes apparent that the stress in the steel is quite low, but the masonry units have been stressed in compression to very near the allowable maximum.

Example 11-6

A nominal 250 by 250 mm or 10 by 10 in (240 by 240 mm or 9.6 by 9.6 inches actual size) reinforced brick masonry column is supported in both directions, top and bottom, at 4m (13 ft) intervals. The column is to be constructed of 30 MPa (4000 psi) units, type M mortar and have minimum reinforcing. Determine the allowable concentric load.

Solution

The minimum area of reinforcement (see clause 4.4.3.32) for reinforced masonry columns is 0.5 percent of the gross cross-sectional area of the column. The area of steel required is:

$$A_s = (0.005)(240 \times 240)$$
$$= 288\ \text{mm}^2\ (0.4\ \text{in}^2)$$

Four 10M bars provide 400 mm^2 of steel area. Use four 10M bars with a steel f_y = 400 MPa (58 ksi).

For 30 MPa units with type M mortar, f'_m = 15.0 MPa. As the column is to be loaded concentrically top and bottom, the end eccentricities e_1 and e_2 will be zero. Thus the virtual eccentricity (e) will also be zero.

$$e_1/e_2 = 0$$
$$t = 240\ \text{mm}$$
$$e/t = 0$$
$$h = 4000\ \text{mm}$$
$$h/t = 4000/240$$
$$= 16.67$$

From Table 8, C_e = 1.00
From Table 9, C_s = 0.75
From Table 7, $f_m = 0.20 f'_m$

$$= 0.20 \times 15$$
$$= 3.00\ \text{MPa}$$

For steel with f_y = 400 MPa, f_s = 165 MPa.

The actual steel ratio (ρ) can now be calculated.

$$\rho = \frac{\text{Area of steel}}{\text{Gross area of masonry}}$$
$$= \frac{400}{(240 \times 240)}$$
$$= 0.007$$

If we assume that the vertical reinforcing is tied as per clauses 4.4.32.(1) and (2), the allowable axial load can be calculated by:

$$P = C_e \times C_s (f_m + 0.80\, \rho n f_s) A_n \quad \text{(clause 4.6.7.5)}$$

$$n = \frac{E_{steel}}{E_{masonry}}$$
$$= \frac{200\,000}{15\,000}$$
$$= 13.3$$

The resulting allowable compressive load is:

$$P = 1.0 \times 0.75(3.0 + 0.8 \times 0.007 \times 13.3 \times 165)57\,600/10^6$$
$$= 0.66\ \text{MN}$$
$$= 660\ \text{kN (48.5 kips)}$$

WALLS

Walls that are to be reinforced must have minimum reinforcing in the vertical and horizontal directions. Clause 4.4.3.30.(1) gives the requirements for minimum amounts of steel for walls reinforced in the CSA code. Non-load-bearing walls have two different requirements depending on the seismic zone in which they are to be located.

Example 11-7

Calculate the minimum allowable steel requirements in the vertical and horizontal directions for a 250 mm (10 in) non-load-bearing concrete block wall.

Solution

Where reinforcing is to be placed in both directions, clause 4.4.3.30 of the code specifies that maximum spacing shall be not more than 1.2 m o.c. for the principal reinforcing. The amount of steel to be used for vertical and horizontal reinforcement can be specified by the following two formulas.

$$\text{Vertical reinforcement } A_v = 0.002\, A_g \alpha$$

$$\text{Horizontal reinforcement } A_h = 0.002\, A_g(1 - \alpha)$$

where $\alpha = 0.33$ to 0.67, as decided on by the designer.

A_g = Gross cross-sectional area of the wall.

A_v = Vertical reinforcing steel area.

A_h = Horizontal reinforcing steel area.

The required steel area for one meter of wall becomes:

$$A_g = 240 \times 1000 = 240 \times 10^3 \text{ mm}^2$$

For $\alpha = 0.4$, the vertical reinforcing area becomes:

$$A_v = 0.002 \times 240 \times 10^3 \times 0.4$$
$$= 192 \text{ mm}^2 \text{ per meter of wall.}$$

For bars spaced at 1.2 m o.c. maximum, the required bar area becomes:

$$A_v = \frac{1200}{1000} \times 192$$
$$= 230.4 \text{ mm}^2 \ (0.36 \text{ in}^2)$$

Use 20M bars at 1200 mm o.c., $A_s = 300 \text{ mm}^2$. For the horizontal reinforcing steel area:

$$A_h = 0.002 \times 240 \times 10^3 \ (1 - 0.4)$$
$$= 288 \text{ mm}^2 \text{ per meter of wall}$$

At maximum spacing, the required area is:

$$A_h = \frac{1200}{1000} \times 288$$
$$= 345.6 \text{ mm}^2 \ (0.54 \text{ in}^2)$$

Use 15M bar at 600 mm o.c., $A_s = 200 \text{ mm}^2$.

Example 11-8

Design a reinforced concrete masonry cantilever retaining wall that is to be 3 m in height above the top of the footing, given the following:

Reinforced concrete masonry: $f'_m = 15$ MPa
Flexural compressive stress: $f_m = 4.95$ MPa
Shearing stress: $V_m = 0.3$ MPa
Bond stress: $u = 1.10$ MPa
Reinforcement: $f_s = 140$ MPa
Earth: Equivalent fluid density, 480 kg/m^3
Allowable soil pressure, 120 kPa.
Friction coefficient, 0.55.
Densities: Earth backfill, 1000 kg/m^3
Reinforced masonry, 1000 kg/m^3
Reinforced concrete, 2400 kg/m^3
Safety factors: Overturning, 2.0
Sliding, 1.5

Solution

The stability of a cantilever retaining wall depends on the wall's ability to resist the horizontal forces produced by the backfill. To determine the wall stability, trial dimensions must be selected for the wall and the resulting moments calculated.

Try: $T = 290 \text{ mm} = 0.29 \text{ m}$

$t = 300 \text{ mm} = 0.3 \text{ m}$

$d = 2000 \text{ mm} = 2.0 \text{ m}$

$a = 500 \text{ mm} = 0.5 \text{ m}$

First, the soil pressure at the underside of the footing must be calculated.

$$p = 480 \times 9.81 \times (3 + 0.3)$$
$$= 15\ 539 \text{ N/m}^2$$
$$= 15.54 \text{ kN/m}^2$$

The horizontal soil force (H) becomes:

$$H = \tfrac{1}{2} \times 15.54 \times 3.3$$

$$= 25.64 \text{ kN per meter of wall.}$$

The resulting moment produced by the force (H) calculated with respect to the toe of the footing is:

$$M = H \times \frac{3.3}{3}$$

$$= 25.64 \times \frac{3.3}{3}$$

$$= 28.20 \text{ kN} \cdot \text{m}$$

The resisting moment with respect to the toe of the footing is:

	Force (kN)	Arm (m)	Moment (kN ∘ m)
Stem: (3 × 0.29) (1600 × 9.81)	13.66	0.645	8.74
Earth: (1.21 × 3) (1000 × 9.81)	56.98	1.40	79.77
Footing: (0.3 × 2) (2400 × 9.81)	14.13	1.00	14.13
	84.77		102.64
	Overturning moment		-28.20
			74.44

Check safety factors ($S.F$):

$$S.F._{(\text{overturning})} = \frac{102.64}{28.20} = 3.64 > 2.0$$

$$S.F._{(\text{sliding})} = \frac{84.77 \times 0.55}{25.64} = 1.82 > 1.5$$

Had the safety factor for sliding been less than 1.5 a key at the bottom of the footing would be required. Had the safety factor for overturning been less than two, a larger footing would be required.

The location of the resultant of the footing pressure with respect to the toe of the footing may now be calculated (Fig. 11-9.)

$$x = \frac{74.44}{84.77}$$

$$= 0.878 \text{ m}$$

(The resultant falls within the middle third of the footing.)

The resulting pressures at the toe and the heel of the footing can now be calculated using the formula for combined stresses:

$$p = \frac{P}{A} \pm \frac{Mc}{I}$$

$$= \frac{P}{b \times d} \pm \frac{6 \times Pe}{bd^2}$$

where P = Total load of wall plus backfill.

e = Distance from the footing center line to the resultant of the footing pressure.

b = 1 m length of wall.

d = Footing width.

Calculate: (e):

$$e = \frac{2000}{2} - 878$$

$$= 122 \text{ mm}$$

For the pressure at the toe:

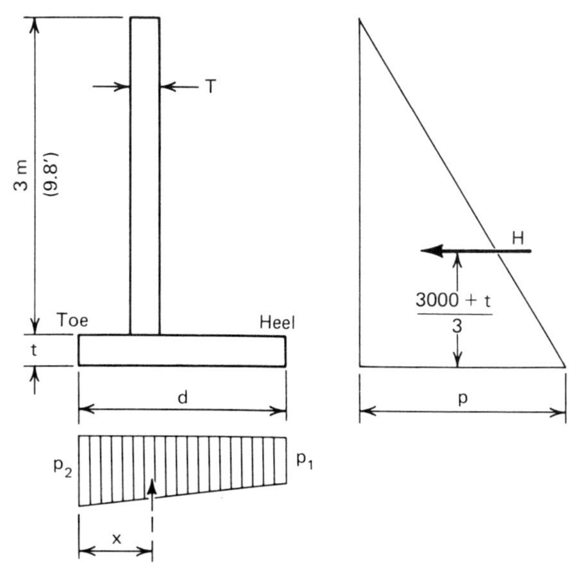

Fig. 11-9. Cantilever retaining wall.

$$p_1 = \frac{84.77}{1 \times 2} + \frac{6 \times 84.77 \times 0.122}{1 \times 2^2}$$

$$= 42.39 + 15.51$$

$$= 57.9 \text{ kPa} < 120 \text{ kPa}$$

For the pressure at the heel:

$$p_2 = 42.39 - 15.51$$

$$= 26.88 \text{ kPa}$$

The design of the stem can now be approached. Calculate the forces on the stem due to the backfill. (Fig. 11-10.) The shear due to the backfill pressure is:

$$V = (480 \times 9.81) \times \frac{3^2}{2}$$

$$= 21.2 \text{ kN}$$

The maximum bending moment produced by the backfill is:

$$M = 21.2 \times \frac{3}{3}$$

$$= 21.2 \text{ kN} \cdot \text{m}$$

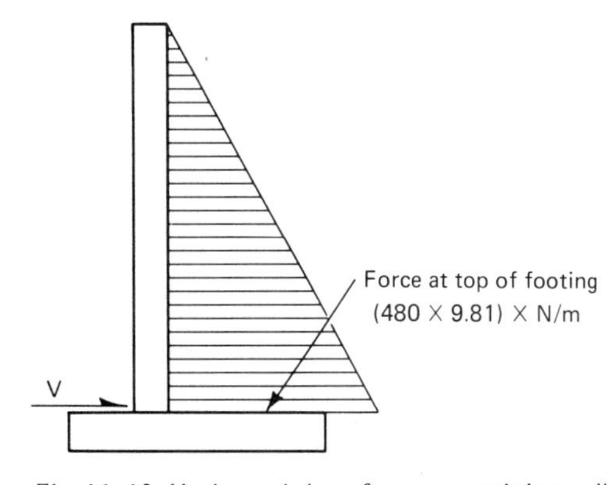

Fig. 11-10. Horizontal shear force on retaining wall.

Assume an effective depth of d = 290 - 100 = 190 mm. A transformed section as shown in Figure 11-11 can be drawn. The value for jd is calculated using an assumed value of j = 0.875:

$$jd = 0.875 \times 190$$

$$= 166 \text{ mm}$$

The internal resisting moment based on the steel area can now be calculated.

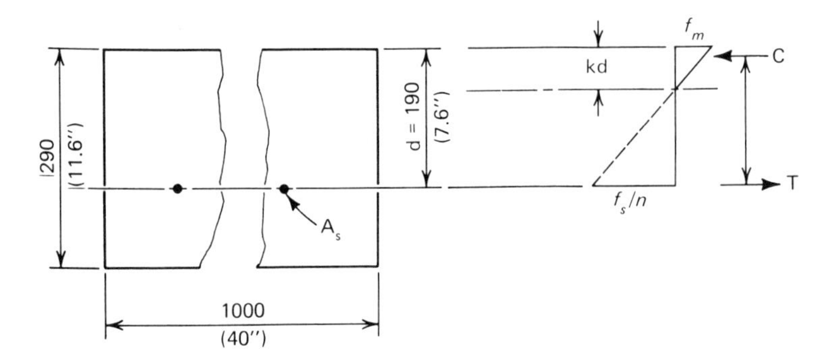

Fig. 11-11. Internal forces on wall section.

$$M = A_s \times f_s \times jd$$

In terms of A_s, the equation becomes:

$$A_s = \frac{M}{f_s \times jd}$$

$$= \frac{21.2 \times 10^6}{140 \times 166}$$

$$= 912 \text{ mm}^2 \text{ per meter of wall.}$$

Use 20M bars at 300 mm o.c. Actual area of steel is:

$$A_s = 300 \times \frac{1000}{300}$$

$$= 1000 \text{ mm}^2$$

From Example 11-7 the vertical and horizontal steel areas can be calculated. Use $\alpha = 0.67$:

$$A_v = 0.002\, A_g\, \alpha$$

$$= 0.002 \times 190 \times 1000 \times 0.67$$

$$= 254.6 \text{ mm}^2 < 1000 \text{ mm}^2$$

For the horizontal steel:

$$A_h = 0.002\, A_g(1 - \alpha)$$

$$= 0.002 \times 190 \times 1000 \times 0.33$$

$$= 125.4 \text{ mm}^2$$

Use 10M bars at 600 mm o.c.; actual steel area becomes:

$$A_h = 100 \times \frac{1000}{600}$$

$$= 167 \text{ mm}^2$$

The stresses for the selected reinforcing can now be checked.

$$\rho = \frac{A_s}{b \times d}$$

$$= \frac{1000}{290 \times 1000}$$

$$= 0.0034$$

$$n = \frac{200\,000}{10\,340}$$

$$= 19.34$$

$$k = [(\rho n)^2 + 2\rho n]^{1/2} - \rho n$$

$$k = [(0.0034 \times 19.34)^2 + 2 \times 0.0034 \times 19.34]^{1/2} - 0.0034 \times 19.34 = 0.303$$

$$j = 1 - \frac{k}{3}$$

$$= 1 - \frac{0.303}{3}$$

$$= 0.899$$

The stress in the steel is:

$$f_s = \frac{M}{A_s \times jd}$$

$$= \frac{21.2 \times 1000 \times 10^6}{1000 \times 0.899 \times 190}$$
$$= 1.24 \times 10^5 \text{ kN/m}^2$$
$$= 124 \text{ MPa},$$

which is less than the maximum allowable stress of 140 MPa.

The stress in the masonry is:

$$f_m = \frac{2M}{kbjd^2}$$
$$= \frac{2 \times 21.2 \times 1000 \times 10^6}{0.303 \times 0.899 \times 1000 \times (190)^2}$$
$$= 4.31 \times 10^3 \text{ kN/m}^2$$
$$= 4.31 \text{ MPa},$$

which is less than the allowable of 4.95 MPa.

Check the shear stress:

$$v = \frac{V}{b \times d}$$
$$= \frac{21.2}{1 \times 0.19}$$
$$= 111.6 \text{ kPa}$$

which is less than the allowable.

Check bond stress:

$$u = \frac{V}{\Sigma_o jd}$$

For 20 M bars at 300 mm o.c.

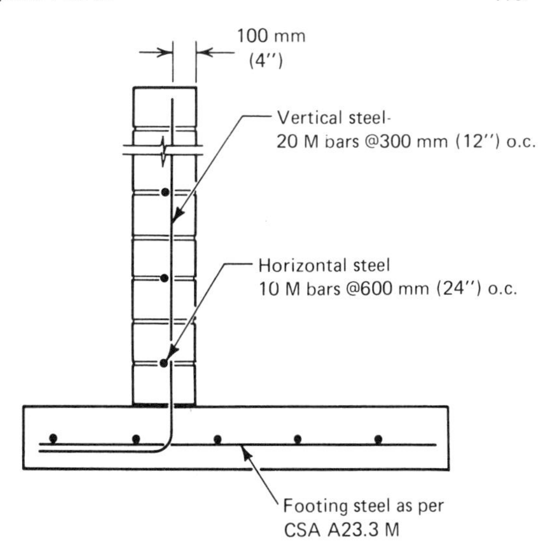

Fig. 11-12. Wall reinforcing.

$$\Sigma_o = 61.3 \times \frac{1000}{300}$$
$$= 204.3 \text{ mm}$$
$$u = \frac{21.2 \times 10^6}{204.3 \times 0.891 \times 190}$$
$$= 613 \text{ kN/m}^2$$
$$= 613 \text{ kPa}.$$

which is less than the allowable.

All the actual stresses are below the allowable limits and the resulting wall detail is as shown in Figure 11-12. The footing must be designed as per CSA A23.3M for reinforced concrete.

Glossary of Terms

Deep Beams Walls that must span large openings and support relatively large loads requiring more than the normal wall steel to prevent wall cracking.

Review Questions

1. Calculate the required steel area for a grouted beam of 300 mm, 15 MPa hollow concrete block units to produce balanced conditions. The effective depth is 1200 mm, f_y = 165 MPa and type M mortar is to be used.

2. A 400 mm clay brick beam is to span 4 m and carry a total design load of 30 kN/m. If the available effective depth is 800 mm, calculate the amount of tension steel required to carry the given load. Use $f_m' = 24$ MPa, $f_s = 140$ MPa and $n = 8$.
3. For the beam in Problem 2, calculate the shear force and check for stirrup requirements. If stirrups are required, assume 10M bars and calculate the appropriate spacing.
4. A steel frame building is to have 4.5 m between floors. Masonry concrete blocks filled with concrete grout are to be used as infill panels between the floors. Calculate the amount of reinforcing steel required to resist a wind load of 1 kN/m^2 if the walls are to be anchored at the top and bottom only. Use 250 mm block units, type M mortar, $f_m = 15$ MPa and $f_s = 125$ MPa.
5. A wall 4 m in height must support a load of 60 N/m. The wall is in single curvature with $e_{top} = 75$ mm and $e_{bot} = 37.5$ mm. Using 20 MPa, 200 mm concrete block units and type M mortar, calculate the amount of steel required in the wall. Ignore the mass of the wall and use $f_s = 125$ MPa.

Selected Sources of Information

Alberta Building Code, Edmonton, Alberta.
Brick Institute of America, McLean, Virginia.
Clay Brick Association of Canada, Willowdale, Ontario.
National Building Code of Canada, Ottawa, Ontario.
National Concrete Masonry Assn., McLean, Virginia.
National Concrete Producres Assn., Downsview, Ontario.

Chapter 12

MASONRY SYSTEMS

WALL TYPES

A number of methods or *systems* are used in the construction of masonry walls, the choice depending on the *use* for which the wall is designed, its *geographic location,* the desired *appearance,* and so forth. Masonry walls generally are classified as *solid, hollow, cavity, composite, veneered* or *reinforced.*

Solid Masonry Walls

Solid masonry walls are built up of *solid* units laid up in mortar, and since solid units are those whose net area is equal to or greater than 75 percent of the gross area (see Fig. 2-3), most solid walls will be built of brick or, in some cases, of stone.

Such walls may be either *single* or *multiple* wythe. Single-wythe, solid masonry walls are generally used as partitions in a non-load-bearing capacity (see Fig. 12-1), although 150 mm (6 in) single-wythe walls are also used in a load-bearing capacity in residential construction.

Multiple-wythe, solid walls are used extensively in load-bearing construction for fire walls and in reinforced brick masonry. (See Fig. 12-2.) The *facing* wythe and backup wythe or wythes are bonded together by *masonry headers* or by *metal ties* (see Chapter 14), and in many cases, the facing wythe may be decorative in nature.

Hollow Masonry Walls

Hollow masonry walls are built with hollow masonry units, which may be brick (see Fig. 1-8), concrete masonry or structural tile. Hollow units are those whose net area is less than 75 percent of the gross area. (See Fig. 2-6.) As in the case of solid units, all horizontal and vertical edges are embedded in mortar.

Hollow walls may be used in either load-bearing or non-load-bearing capacities, depending on the type of unit and the particular circumstances. (See Fig. 12-3.)

Cavity Walls

Cavity walls consist of two wythes of masonry, separated by an air space of 50 to 75 mm (2 to 3 in) and joined together by horizontal metal ties for structural strength. The outside or *face* wythe and the backup wythe may be of similar materials; or, as in many cases, the face wythe will be of brick or

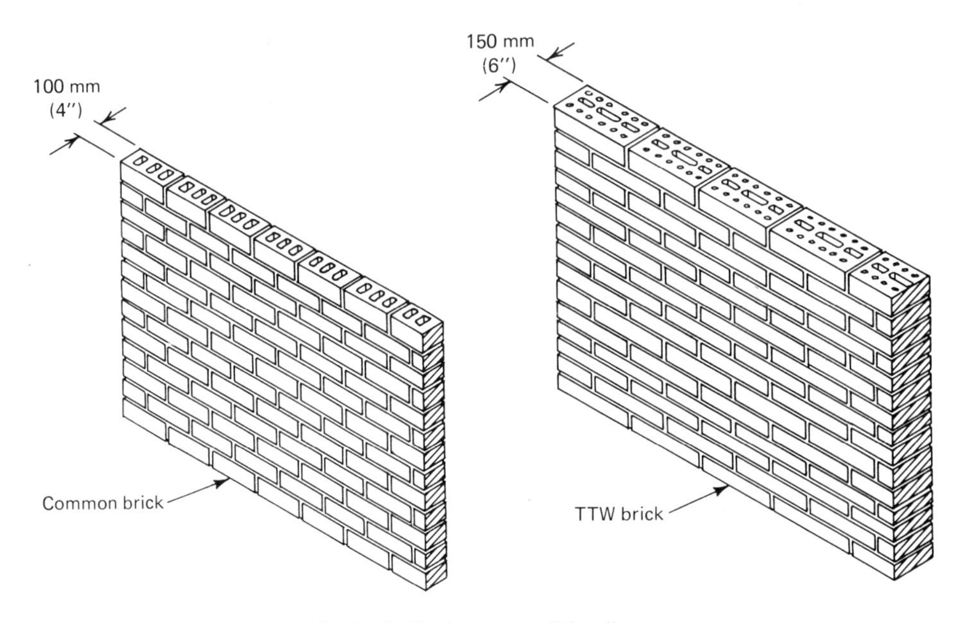

Fig. 12-1. Single-wythe solid walls.

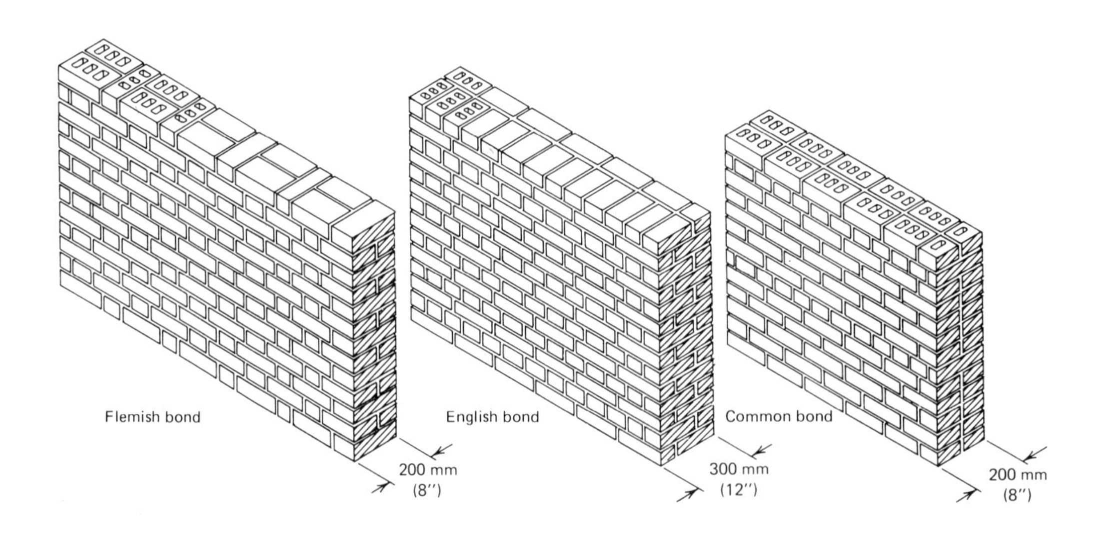

Fig. 12-2. Multiwythe solid walls.

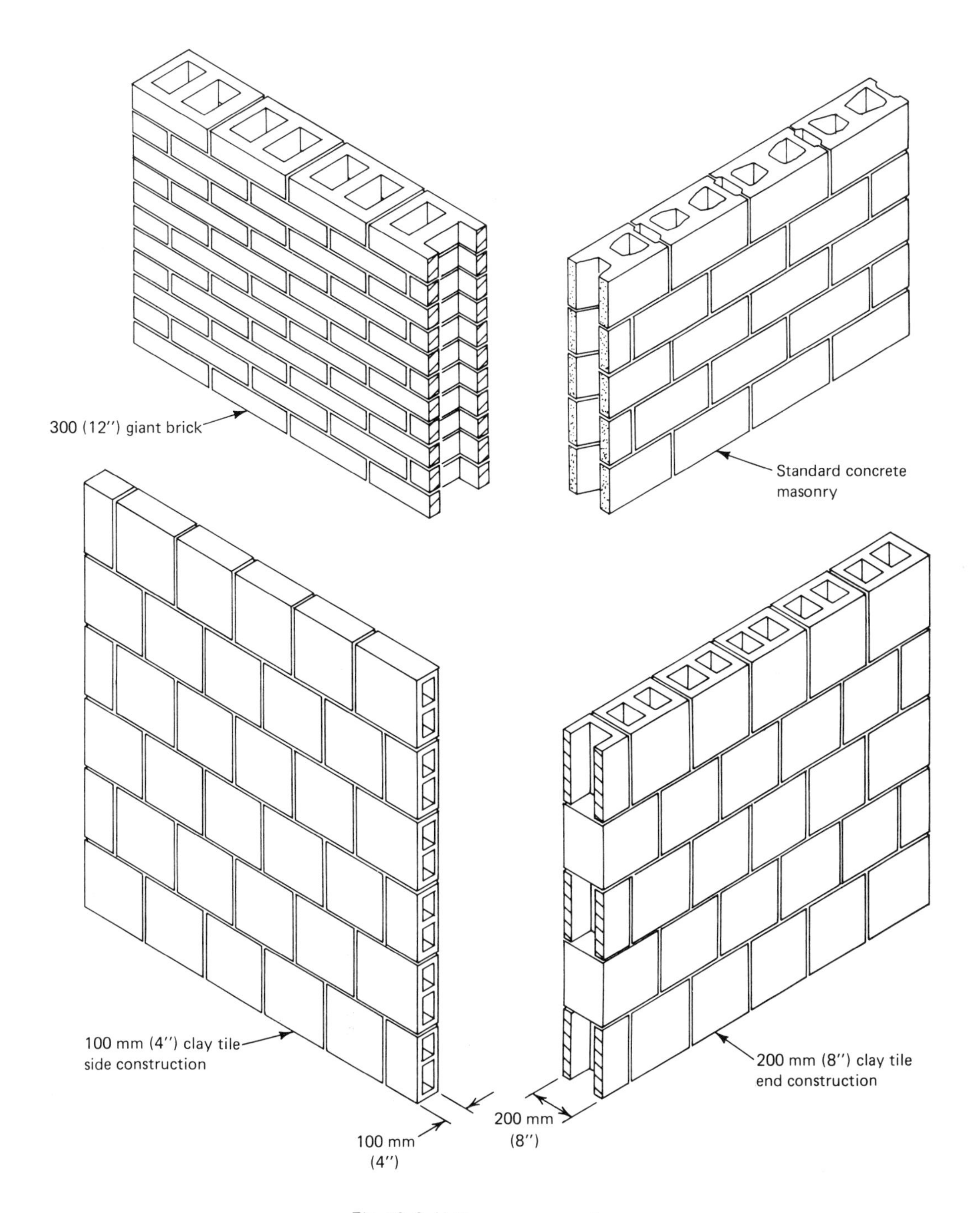

Fig. 12-3. Hollow masonry walls.

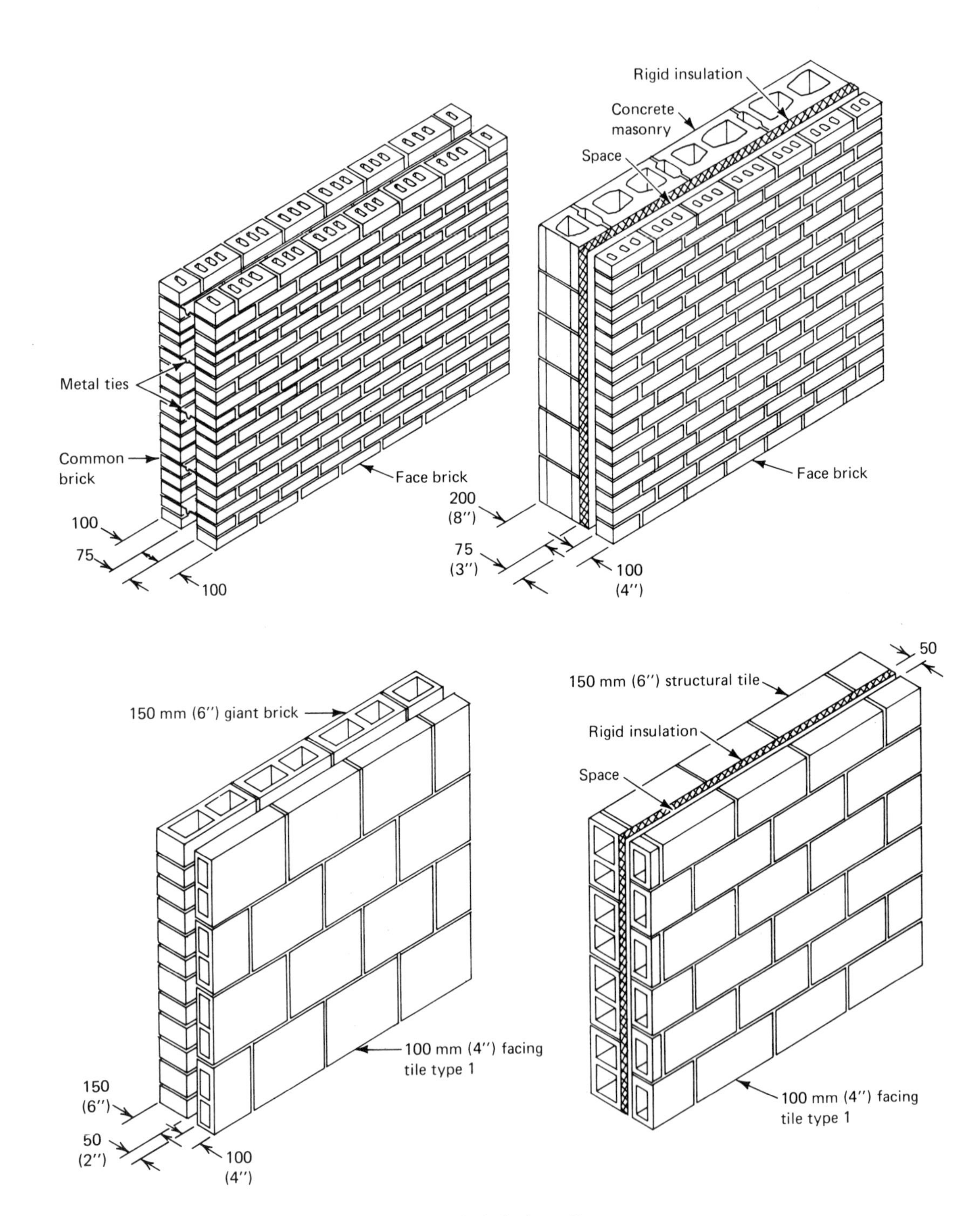

Fig. 12-4. Cavity walls.

facing tile and the backup wythe will be concrete masonry or structural tile and will serve to support the vertical loads of the building. (See Fig. 12-4.)

The important advantages of cavity walls over those of solid masonry are the positive protection against rain penetration that cavity walls can provide and the fact that insulation can be introduced into the cavity to provide thermal protection.

Rain may penetrate the outer wythe of a cavity wall, but the water trickles down the inner face of the outer wythe and cannot traverse the cavity. *Metal flashing* and *weep holes* (see Chapter 14) are provided at the base of the wall and above openings to direct the water to the exterior of the building again.

It is important that the cavity be of sufficient width to facilitate the introduction of insulation–usually of the rigid type (see Fig. 12-4)–and to allow for construction inaccuracies without the two wythes touching. (See Fig. 7-37.)

Composite Walls

A composite wall is a multiple-wythe wall made of solid or hollow units, in which the material in one wythe is dissimilar to that in the other. Composite walls are normally used as exterior walls and will usually consist of a brick, stone or tile face bonded to a concrete masonry or structural tile backup. The tie is made between the two by *masonry headers* or noncorrosive *metal ties.* (See Fig. 12-5.)

Three factors must be taken into consideration when choosing between masonry headers and metal ties: *possible moisture penetration, differential movement* and *economy.*

Any movement which creates a crack through the wall will provide a ready path for moisture to follow. Differential movement is due to the fact that concrete masonry tends to shrink, while brick has a tendency to expand; the *rate of movement* due to moisture and temperature changes is different in concrete masonry and brick, which results in uneven pressure on masonry ties. (See Fig. 12-6.) From the point of view of economy, it requires less time to place metal ties than to lay courses of masonry headers. As a result, metal ties have become more popular, due to their economy and flexibility.

Veneered Walls

A veneered wall consists of a brick, stone, tile, or concrete masonry facing over a backup wall of wood, concrete, concrete masonry, common brick or structural tile. The outside wythe or *veneer* is merely a facing, a weather protective coating, non-load-bearing and, in some cases, quite thin. This veneer is *tied* but not *bonded* to the backup wall. (See Fig. 12-7.) The metal ties used are flexible to allow for differential movement.

Reinforced Masonry Walls

When buildings are more than three stories in height and are located in seismic zones or in any situation where greater resistance is required to applied loads and stresses, added strength should be provided for the walls by introducing reinforcement into them.

Reinforced masonry walls are walls in which the interior space or spaces are filled or partially filled with grout and reinforcing steel. This may involve a single-wythe concrete masonry wall with some or all of the cells filled or a cavity wall with the space between the two wythes filled with a reinforced grout. Reinforced masonry acts very similarly to reinforced concrete, due to the flexibility added when reinforcing is introduced.

Reinforced masonry walls are generally of three types: a single-wythe wall with only *some of the cells filled,* a single-wythe wall *fully*

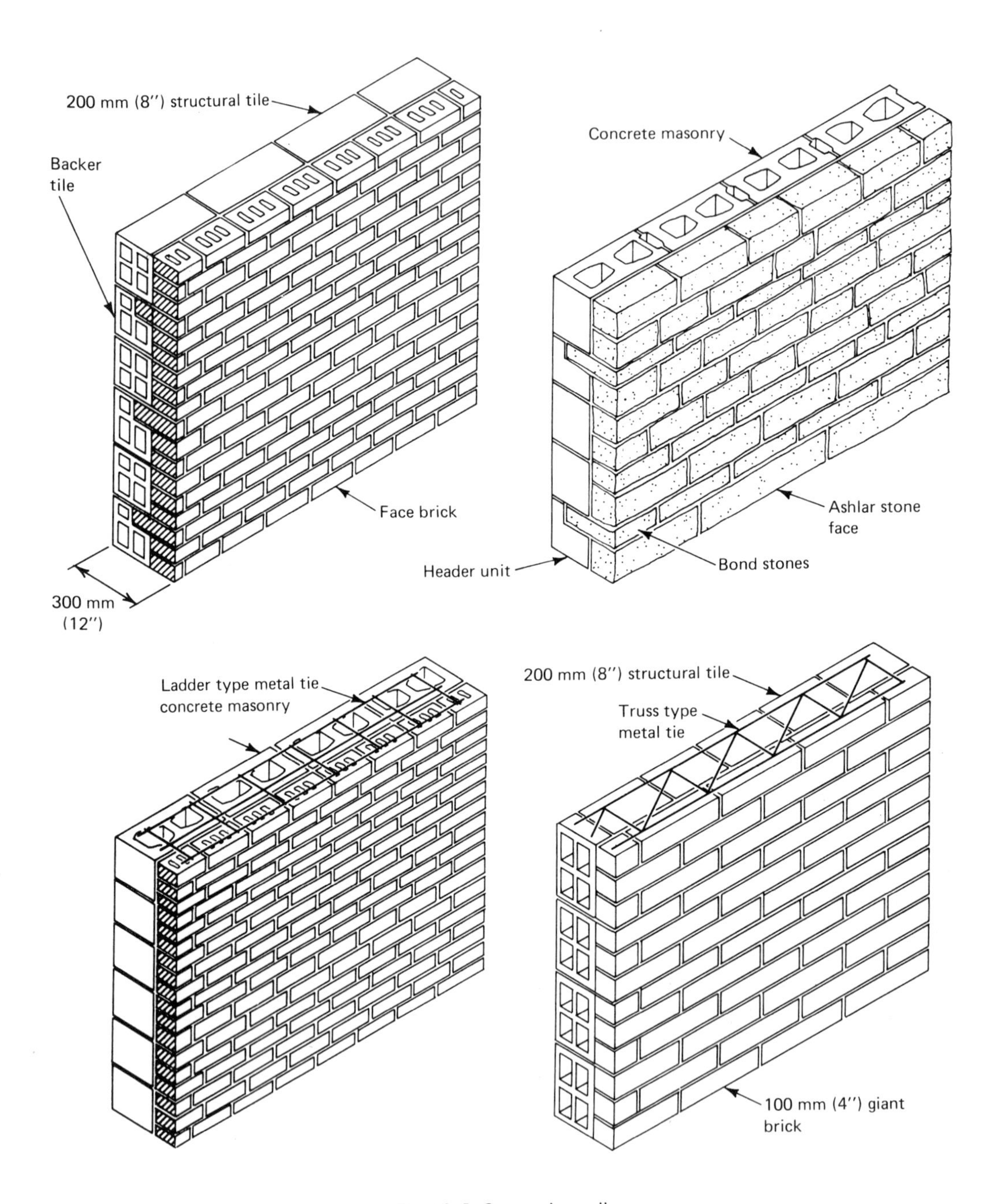

Fig. 12-5. Composite walls.

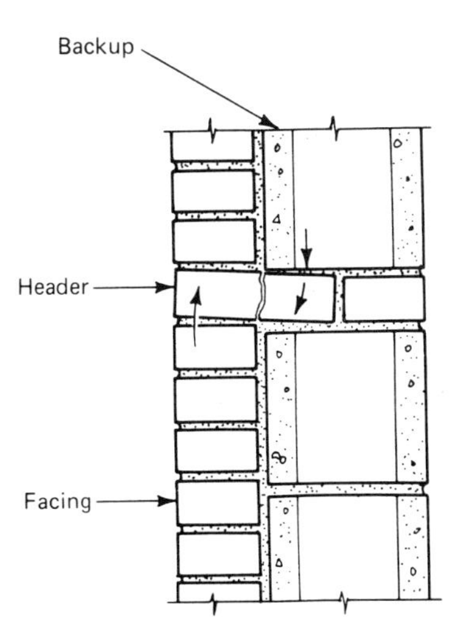

Fig. 12-6. Differential movement.

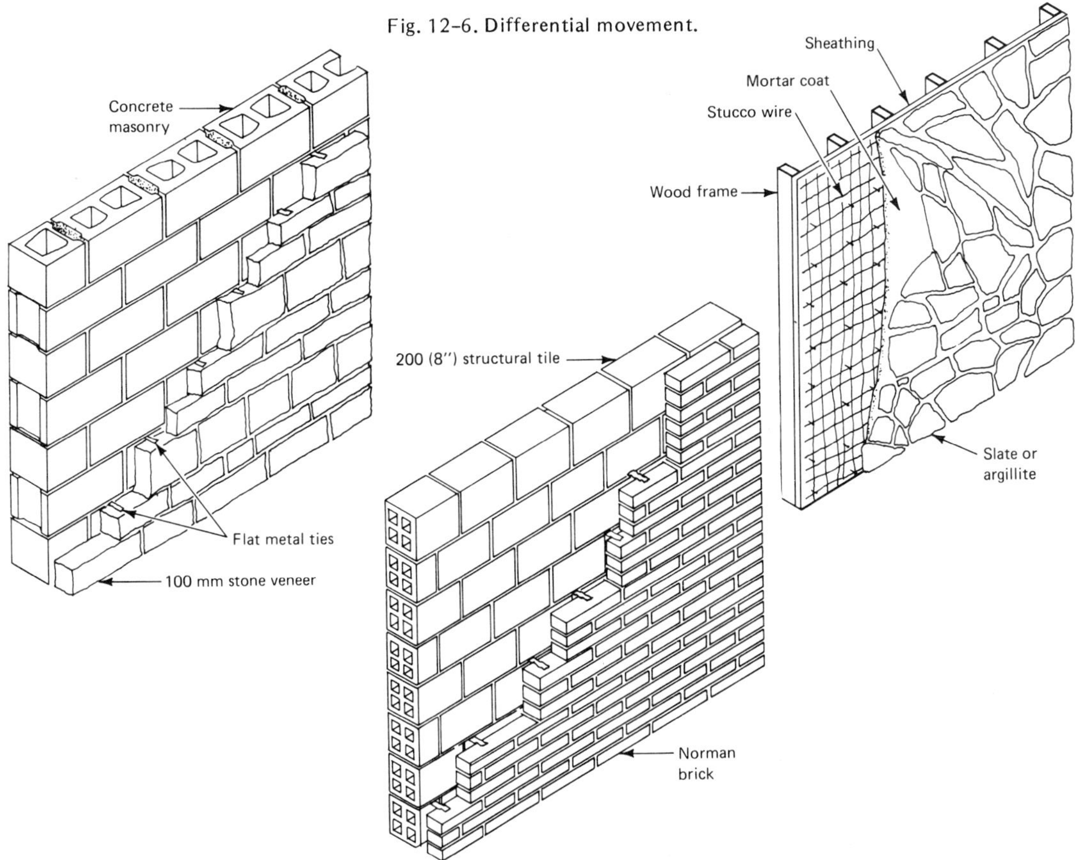

Fig. 12-7. Veneered walls.

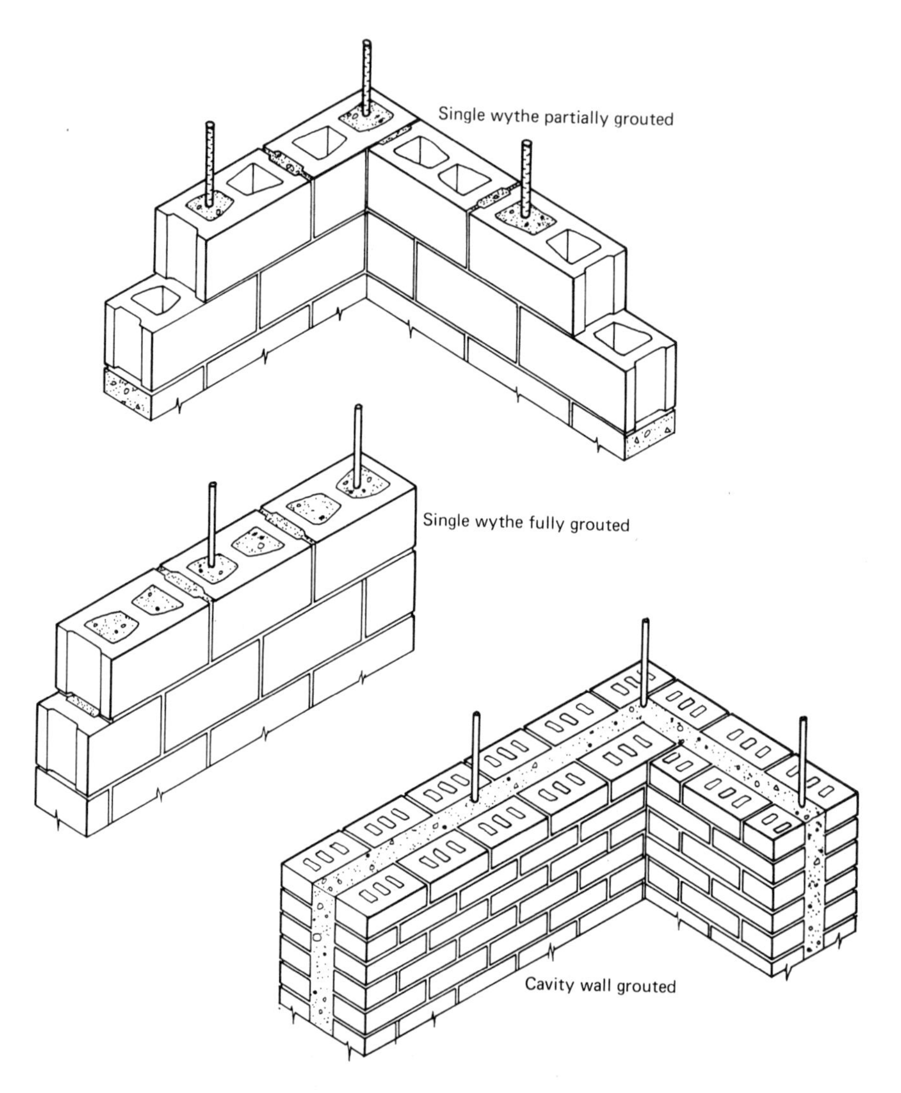

Fig. 12-8. Reinforced masonry walls.

grouted and a cavity wall with the *space between the wythes grouted.* (See Fig. 12-8.)

For reinforced, single-wythe walls, the vertical reinforcement is usually set before the units are placed, since otherwise it is difficult to tie to the foundation steel. In that case, when working with concrete masonry, H-blocks are used at the bar locations since they do not have to be let down over the ends of the steel, as would be the case with standard units.

Horizontal reinforcement is placed in a single-wythe wall when it reaches the bond beam level. (See Fig. 12-9.) Joint reinforcement should not be considered as part of the horizontal reinforcement.

Care must be taken with the partially grouted wall to make sure that the cross-joints on either side of the cavity to be grouted are well bedded

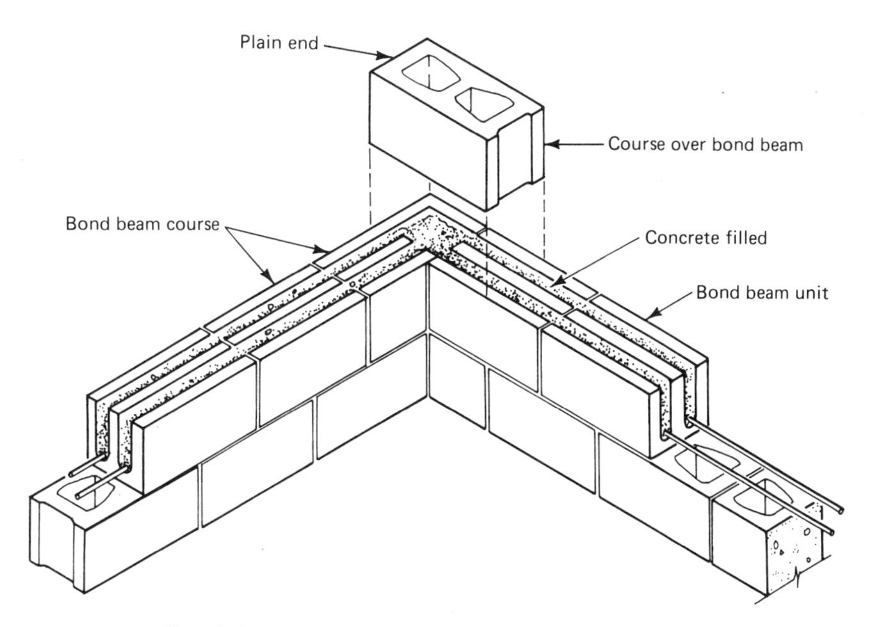

Fig. 12-9. Bond beam course in concrete masonry wall.

with mortar. The cross-sectional area of the space into which the reinforcement is placed should not be less than 50 × 75 mm (2 × 3 in) as adequate bonding will not occur in a smaller space.

Grouting

There are generally two methods used to grout a masonry wall. One is known as *low lift* grouting and the other as *high lift* grouting.

Low Lift Grouting. The low lift grouting method is generally used for walls up to 1200 mm (48 in) in height. After the wall has cured sufficiently (twenty-four hours for single-wythe walls and three days for cavity walls), the grout is placed in the wall, usually by pump. (See Fig. 12-10.) Grouting should stop 40 mm (1½ in) short of the top of the last course in order to create a key with the next lift. (See Fig. 12-11.)

Reinforcing can be placed before or after the masonry is laid. If it is placed prior to laying, in the case of concrete masonry, special

Fig. 12-10. Pumping grout into masonry wall. (*Courtesy Portland Cement Association*)

Fig. 12-11. Grout stopped below top of wall. (*Courtesy Portland Cement Association*)

open-ended blocks (see Fig. 2-18) are used, whereas if reinforcing is placed after the masonry is laid, regular units can be used. The reinforcing bars are generally quite short, since they do not need to extend more than 30 bar diameters past the top course of the wall being grouted.

When one section of wall is completed, the whole operation is then repeated with the next section above. One disadvantage of this method is that it creates a *cold joint* between the two wall sections.

High Lift Grouting. High lift grouting involves grouting a story height wall continuously in 1200 mm (48 in) layers, which are consolidated and reconsolidated through the layer above.

The first 1200 mm (48 in) layer is grouted, consolidated and then allowed to stand for from 20 to 30 minutes. The second 1200 mm layer is then grouted and consolidated; then the first layer is reconsolidated through the second, blending the two and providing a denser grout through the entire wall. Care must be taken not to wait too long before the second operation is begun, as problems may occur due to the fact that the wall can fail with vibration after the grout has set up.

Preparation for Grouting

Before starting to grout, the space at the bottom of the wall must be cleaned out, through *cleanouts* that have been left in the beginning course of masonry. These can be provided by cutting a piece out of the face shell of a concrete masonry unit, by using an inverted bond beam block and omitting a stretcher unit at intervals along the wall. (See Fig. 12-12.) The purpose of a cleanout is to make it possible to clear the debris and mortar droppings from the bottom of the cavity, as this is the area that is subjected to the greatest pressure.

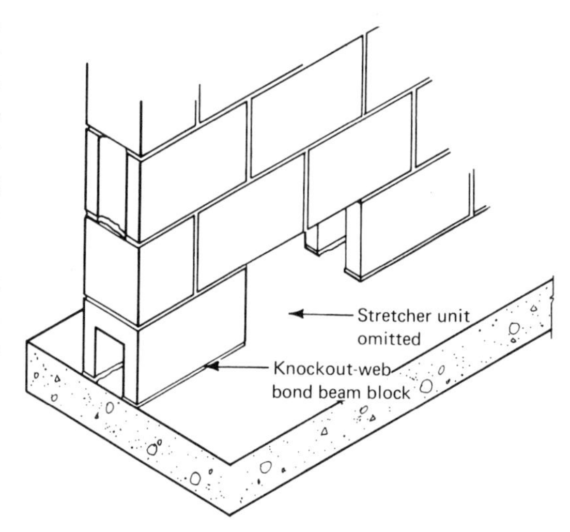

Fig. 12-12. Cleanout detail.

Placing sand in the bottom of the cavity can act as a bond breaker. When placing grout into a cavity wall, the outer wythe or facing should be left 300 to 400 mm (12 to 16 in) higher than the backup to stop spilling on the face of the units. In grout spaces the mortar bed should be angled back so that the mortar does not stick into the cavity and impede the grouting of the space.

Infill Panels

Infill panels are masonry walls that are built into a structural steel or reinforced concrete structural frame. There are two methods that can be used: first, the panel is built in contact with the frame, thus contributing to the strength and stiffness of the frame; second, the panel is isolated from the frame by up to 25 mm (1 in). (See Fig. 12-13 and 12-15.)

Using the first method, it has been estimated that the frame may be strengthened up to $4\frac{1}{2}$ times by the panel and the stiffness may be increased by 30 times. This is not necessarily a good procedure for earthquake zones, where stiffness is not the main prerequisite.

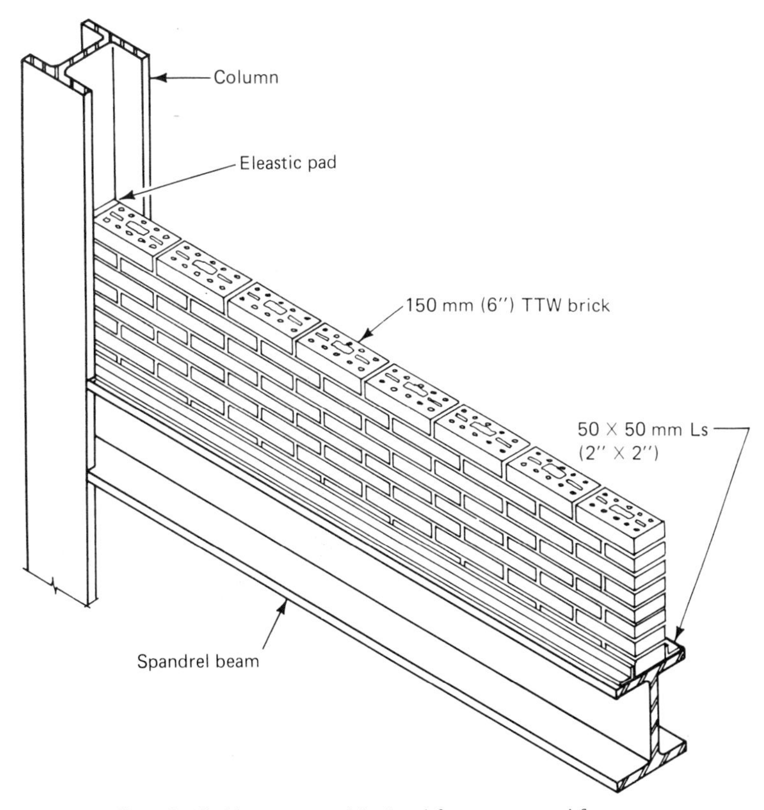

Fig. 12-13. Masonry panel isolated from structural frame.

With the second method, the panel does not contribute to the strength and stiffness of the frame. This may be a wasteful method of construction, since the panel is acting only as a curtain, rather than contributing to the frame in the form of a strut—a contribution that could be even greater if the panel was reinforced and grouted.

Prefabricated Masonry Panels

The construction industry is continually seeking ways to reduce *costs* and *on-site construction time,* and one way of doing this has been the introduction of *prefabricated masonry panels.* The success of precast concrete panels, the development of new and improved masonry units and mortar and the introduction of the rational design method in reinforced masonry have all had an effect on the development of large, prefabricated masonry panels.

Prefabrication began in Europe in the early 1950s and was introduced into North America in the mid-1960s. Most of the early methods were aimed at reducing labor costs, but now prefabricated products are often considered as good or superior to those built in place.

Panel Fabrication

Wall panels are fabricated by two different methods. In one—a manual system—masons position the units in the panels, while in the other a *block-laying machine* is used. The manual system can be employed either in a plant or at the jobsite in all-weather enclosures, on platforms called *launch pads,* but the machine operates in a prefabricating plant or factory.

Manually Built Panels. Under the manual system, units may be laid vertically or horizontally, with *conventional mortar,* with *organic mortar* or with *no mortar.* In the latter case, *surface bonding* will be employed to tie the units into a single panel.

In one surface-bonding method, the units are set face down in the bottom of a form and a backing of concrete or grout is placed over the top—a procedure known as *casting.* Another surface-bonding method involves stacking the units dry in a vertical position and coating both sides with a plaster coat made up of a mixture of portland cement and glass fibers. This coating, from 2 (3/32 in) to 3 mm (1/8 in) thick, will increase the tensile strength of the panel and provide a protective waterproof shield. When color is added, it also serves as a decorative coating.

Regardless of the mortar system used, panels will be reinforced with steel and grouted. Normally, panels up to 2.5 by 9 m (8 by 29.5 ft) will be laid up by this method. Such panels can show improved quality if the work is done where temperature and humidity can be controlled.

Machine-Built Panels. A masonry-laying machine will lay units in either running or stack bond, make the head and bed joints, insert the joint reinforcement and tool the joints. It thus assembles panels about 16 times as fast as they could be built in place, using the same manpower in the machine operation. Such a machine will handle units from 150 to 300 mm (6 to 12 in) thick and is capable of producing panels up to 3.65 by 6 m (12 by 19.7 ft) in face dimensions. (See Fig. 12-14.)

Blocks are arranged in the proper sequence on a feed belt according to the selected design plan and are thus fed onto the machine's layout table. Mortar is fed to the head joint of each individual unit by vibrating fingers. At the same time the mortar is vibrated onto the horizontal surface to form a complete and uniform bed joint, including cross-webs as well as face shells.

Fig. 12-14. Masonry panel-making machine.

The block is then advanced and another takes its place, the complete cycle taking about five seconds per unit. The procedure is repeated until the desired length of panel is on the layout table. A carriage clamp then raises the entire course and places it on the *panel pallet.*

Succeeding courses are placed on top of the first one, with vibration to control the height, and the process continues until the desired panel height is reached.

The panel pallet, a wheeled metal beam mounted on rails, is then moved down the track to make room for another pallet and the fabrication of another panel. Vertical reinforcement then has to be inserted in the panel cores and the cores grouted. After 24 hours, the panels are removed from the pallet by forklift or crane and taken to a storage area.

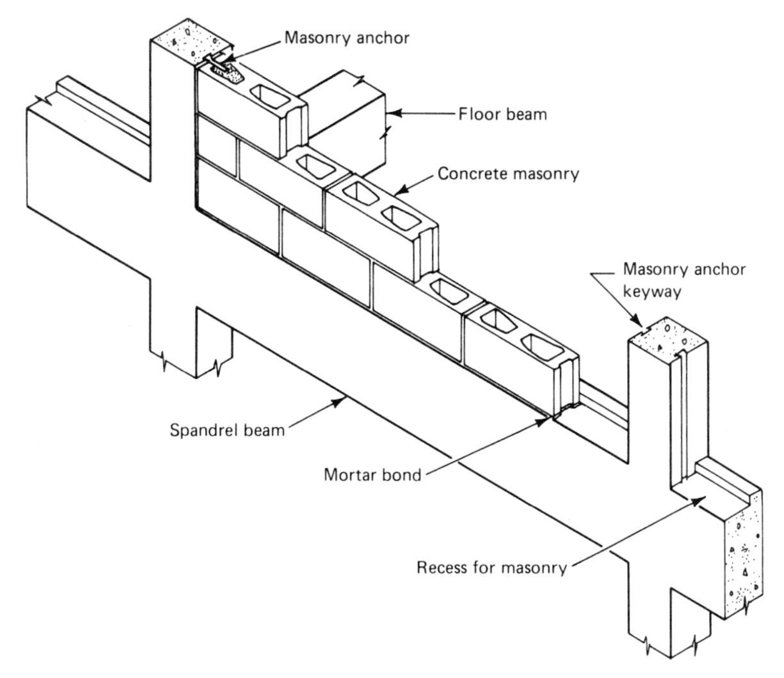

Fig. 12-15. Simple masonry curtain wall connections.

Mortars

In some cases, fabricated joints are used, eliminating the necessity of applying mortar of any kind. (See Chapter 5.)

Panel Anchors

With conventional masonry panels, the tying systems between panels and floors or panels and columns have been relatively simple. (See Fig. 12-15.) The development of prefabricated panel construction has necessitated the development of strong connections between adjoining panels and between panels and floors or columns. Horizontal edges (panel-to-floor edges) may be connected by *spliced reinforcement,* by *tension connections,* by *bolted tension connections,* by *bolting* or by *welding.* Vertical edges (panel-to-panel or panel-to-column edges) may be connected by *steel plates,* by *lapped reinforcement* or by *formed lapped reinforcement,* among others.

Many of these panel-to-panel and panel-to-floor connections are made through *access pockets* left in the panel and in the floor. After the connections are complete, the pockets are filled with units or grout.

See Chapter 14 for details on these connection systems.

CONTROL OF WALL MOVEMENT

All walls and all materials that go to make up a building are in a constant state of movement. The movement of masonry walls is brought about by *changes in temperature, fluctuation in moisture content, carbonation of concrete and mortars* and *movement due to loads* supported by the building.

Table 12-1. Thermal Expansion

Materials	Coefficient of Linear Expansion per °C
Clay or shale brick masonry	6.5×10^{-6}*
Heavyweight concrete masonry	9.4×10^{-6}
Lightweight concrete masonry	7.7×10^{-6}
Granite	8.5×10^{-6}
Limestone	5.4×10^{-6}
Marble	13.1×10^{-6}
Gravel aggregate concrete	10.8×10^{-6}
Lightweight aggregate concrete	8.1×10^{-6}
Structural steel	11.0×10^{-6}

Since this movement cannot be stopped, consideration must be given to allowing for freedom of movement without a failure in the structure. When masonry units of different coefficients of thermal expansion are used together, the design should allow for differential movement. Ordinarily thermal movements can be estimated from the given coefficients of expansion. Table 12-1 indicates the average values for various building materials.

It seems apparent that temperature variations will cause greater stresses in an exterior wall than changes in moisture content. A temperature change of 30°C (54°F) will cause as much change in volume as the maximum change expected due to a moisture variation from wet to dry. Masonry units tend to expand when there is an increase in moisture content and some of them will shrink when there is a decrease. Clay masonry will expand with an increase in water content, but the process is not reversible under normal atmospheric conditions. Concrete masonry units, on the other hand, will shrink with a loss of moisture.

It is recommended that provisions be made in the design of clay masonry for some moisture expansion (coefficient of moisture expansion of 0.0002). With concrete masonry, the initial drying or shrinkage is of greatest concern; a shrinkage coefficient of 0.005 is recommended for concrete masonry units exposed to average weather conditions.

If freezing should occur, the units used for exterior walls will usually possess sufficient strength to resist the expansive forces, due to the fact that they do not usually reach a critical degree of saturation.

Although it is known that the surface concrete and mortar will shrink due to carbonation, little is known about the amount of shrinkage.

There are two techniques used to accommodate movement that may cause cracking in wall systems—*horizontal reinforcing* and *control joints.*

Horizontal Reinforcing

Horizontal reinforcement of masonry walls is usually accomplished by the introduction of *horizontal reinforcing steel into the collar joints* of multiwythed walls, by the use of *bond beams* and by *joint reinforcement.*

The steel used for horizontal reinforcing is the deformed variety, similar to that used in reinforced concrete walls. When used in collar joints (see Fig. 12-16), it may be necessary to widen the joint in order to properly accommodate the diameter of steel required.

Bond beams are horizontal members in a wall that integrate the components into a structural unit. Such a beam may be incorporated at the base of a wall, at floor and roof levels and over large openings. In concrete masonry walls, they may be constructed of special units (see Fig.

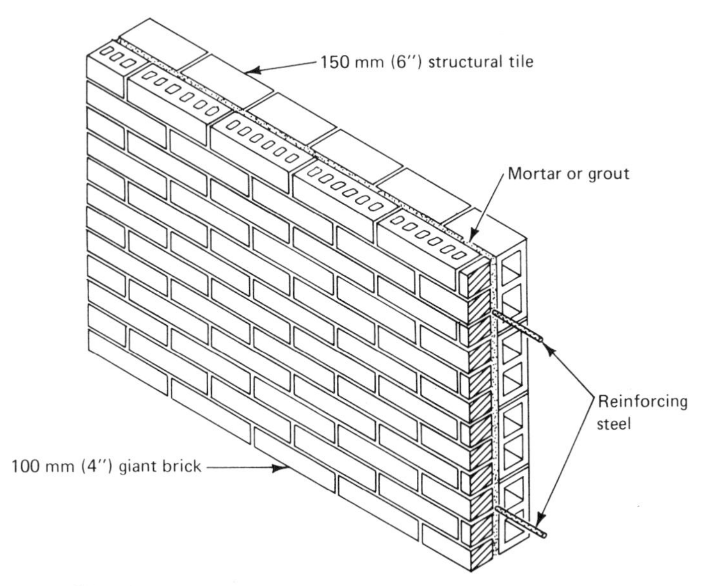

Fig. 12-16. Horizontal reinforcing steel in collar joint.

2-11) capable of receiving horizontal steel and concrete. In clay masonry walls, bond beams may be constructed in several ways, some of which are illustrated in Fig. 12-17.

Bond beams will also tend to distribute the vertical concentrated loads that might be placed on a wall. As a means of crack control, the bond beam will resist cracking due to horizontal movement over an area 600 mm (24 in) above and below the beam's location in the wall. The rebar used may include both smooth and deformed bars.

The size of these bars is generally specified by a number corresponding to the diameter in mm. Table 12-2 shows some of the typical bar sizes, diameters and weights.

Horizontal joint reinforcing was developed to help control cracking in walls. It does not stop the cracking due to tensile stress, but it does distribute the stress so that there will be a number of fine cracks, as opposed to one or two large ones. It consists of prefabricated reinforcing (See Fig. 14-19) embedded in the horizontal joints, which adds reinforcing to a wall but is not considered in the steel area

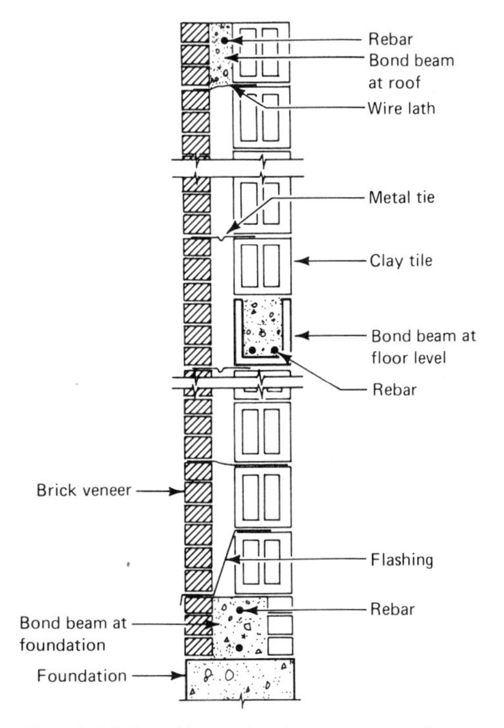

Fig. 12-17. Bond beams in clay masonry wall.

Table 12-2. Metric Reinforcing Bars

Bar Designation	Mass kg/m	Nominal Dimensions		
		Diameter mm (in)	*Cross-Sectional Area mm^2 (in^2)*	*Perimeter mm (in)*
10M	0.785	11.3 (.45)	100 (.16)	35.5 (1.42)
15M	1.570	16.0 (.64)	200 (.31)	50.1 (2.0)
20M	2.355	19.5 (.78)	300 (.47)	61.3 (2.45)
25M	3.925	25.2 (1.00)	500 (.78)	79.2 (3.17)
30M	5.495	29.9 (1.20)	700 (1.09)	93.9 (3.76)
35M	7.850	35.7 (1.43)	1000 (1.55)	112.2 (4.49)
45M	11.775	43.7 (1.75)	1500 (2.33)	137.3 (5.49)
55M	19.625	56.4 (2.26)	2500 (3.88)	177.2 (7.09)

(Courtesy of the Ontario Reinforcing Manufacturers Association)

The nominal dimensions of a deformed bar are equivalent to those of a plain round bar having the same mass per meter as the deformed bar.

calculations for the wall. The joint reinforcing is placed at vertical distances of from 200 to 600 mm (8 to 24 in), depending on wall height and the spacing of control joints. It should have a minimum mortar cover of 10 mm (3/8 in), and the pieces should overlap at the ends by 150 mm (6 in).

Control Joints

Control joints are vertical joints built into a wall to relieve stresses that may be either tensile or compressive in nature, by reducing restraint and permitting movement to take place. They are incorporated into a wall in locations at which stress concentration might occur; for example, at locations where *abrupt changes occur in wall height* or *in wall thickness.* Control joints may be located at pilasters and at pipe or duct chases, *above joints in foundation or floor, below joints in roofs or floors bearing on the wall, at a distance not over one-half the allowable joint spacing from bonded intersections or corners* and at *one or both sides of all door and window openings,* depending on the size.

A control joint is needed only at one side of an opening if it is less than 1700 mm (68 in) wide; otherwise there should be one on both sides. Above the opening, the joint should coincide with the end of the lintel. (See Fig. 12-18.) Care should be taken to avoid joints between openings that are close together. (See Fig. 12-19.)

The distance between control joints will

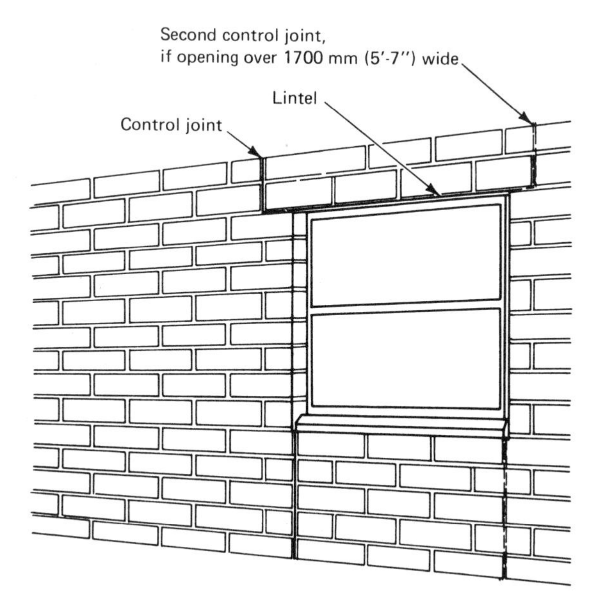

Fig. 12-18. Control joint coincides with end of lintel.

depend on the wall height-to-length relationship and the amount of horizontal reinforcing in the wall. For the spacing in nonreinforced walls, in nonearthquake zones, the recommendations in Table 12-3 may be followed.

The spacing in reinforced walls should conform to local building code restrictions, where they apply. Otherwise, a spacing of about 15 m (49 ft) between joints is reasonable.

If masonry is carried on a structural frame,

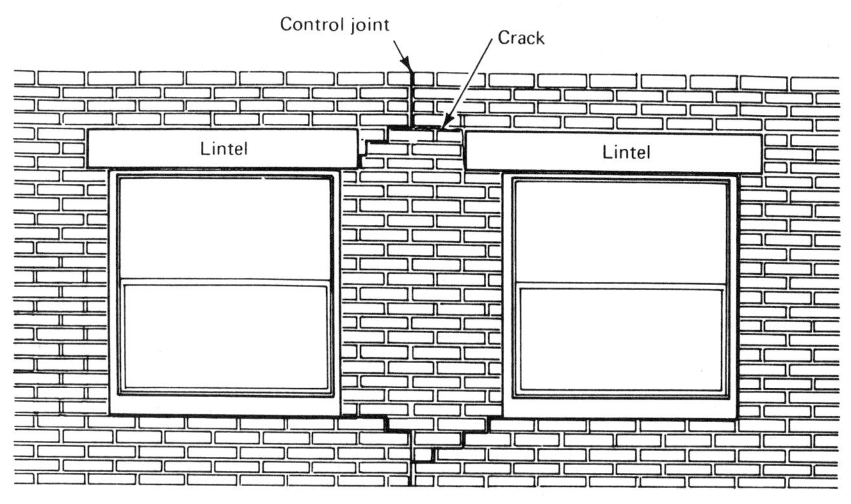

Fig. 12-19. Incorrectly located control joint.

consideration should be given to horizontal control joints, as there is differential movement in such a situation, particularly with a reinforced concrete frame. With a steel frame, this joint will occur at the bearing angles.

A control joint must not only allow for movement but must also tie the two sections together and provide a watertight seal. For details on control joint-forming methods, see Chapter 14.

Table 12-3. Control Joint Spacing

Joint Reinforcing Spacing		Maximum Spacing of Control Joints		
		Panel Length/ Height	*Panel Length*	
mm	*(in)*		*m*	*(ft)*
None		2	12	(39)
600	(24)	2.5	13.5	(44)
400	(16)	3	15	(49)
200	(8)	4	18	(59)

(Courtesy Portland Cement Association)

Glossary of Terms

Carbonation The process of converting into a carbonate.

Chase A regularly shaped groove.

Coefficient A number expressing the amount of some change or effect under certain specified conditions.

Deformed Steel Steel bars with ridged surfaces for better bond in concrete.

Firewall A wall built of noncombustible materials, to stop the spread of fire from one part of a building to another.

Pallet A portable platform, used for transporting materials.

Seismic Pertaining to an earthquake.

Tensile Stress The stress caused by a force tending to cause extension.

Review Questions

1. When can a solid unit be used in a single-wythe reinforced wall?
2. What are two advantages of masonry cavity wall construction?
3. Discuss the effect that inaccuracies in cavity wall construction will have on the size of cavity.
4. What effect will a single-wythe block wall, that has every other vertical void grouted, have on the method of placing mortar?
5. How can the amount of separation of the grout fill from the inside of the unit, due to shrinkage, be minimized?
6. Why is it so critical that the grout space be cleaned at the bottom before the grouting procedure starts?

7. What are the advantages and disadvantages of using prefabricated masonry panels as a facing for a high-rise building?
8. When organic mortar is used to build a prefabricated panel, why is the size of the unit so critical?
9. How can allowances be made for the differences in movement of a brick facing and a block backup wall and where should they occur?
10. What affects the distance between control joints?

Selected Sources of Information

Alberta Masonry Institute, Calgary, Alberta.

Brick Institute of America, McLean, Virginia.

IXL Industries Ltd., Medicine Hat, Alberta.

Leuchars, J. & Schrivener, J. "Infill Panels," *Reader in Civil Engineering,* Christchurch, New Zealand: University of Canterbury, 1975.

National Concrete Masonry Association, McLean, Virginia.

National Research Council, Ottawa, Ontario.

Ontario Reinforcing Manufacturer's Assn., Toronto, Ontario.

Portland Cement Association, Skokie, Illinois.

Tomax Corporation, Phoenix, Arizona.

Chapter 13

PLANNING AND COORDINATING CONSTRUCTION

WALL PLANNING AND LAYOUT

Economic wall planning and layout usually involve the use of *modular coordination* in the size of units and in their placement in a wall. This is extremely significant when placing and choosing *openings* for doors and windows. The advantages of this system will become evident as one becomes familiar with it.

Modular Coordination

At one time most masonry walls were designed or planned on a grid of 4 inches or multiples of 4 inches. With the shift to the metric system, the grid or module that appears to be used is the 100 mm (or multiples of 100) module. (See Fig. 13-1.) If the building is planned to that module, there is a saving in labor because unit cutting or fitting is minimized. (See Fig. 13-3.) All horizontal, vertical, and thickness dimensions should be in multiples of 100 millimeters as this is the internationally accepted grid system. Figures 13-1 and 13-2 illustrate how standard units fit into this grid.

The typical coordinating metric panel is designated at 600 mm (24 in) square. This will require 9 courses of standard brick or 27 bricks for the panel. Three courses of metric block (a total of $4\frac{1}{2}$ blocks) or 2 courses of tile or 4 units would be used to fill the panel. Some stone is split to a specific size that would require 6 courses to reach the panel height. As the lengths are often random, the number of units needed to fill a panel varies. All units are used with a 10 mm ($\frac{3}{8}$ in) mortar joint to fit the panel shown in Figure 13-2.

Openings

If modular planning is to be followed, then doors and windows that fit the same grid must be used. Figure 13-3 illustrates what will happen if modular planning is followed as well as the difficulties that will arise if it is not followed. As window openings will be specified by the outside dimension, it will be easier to choose the units that fall into the metric grid without worrying about the actual glass measurement.

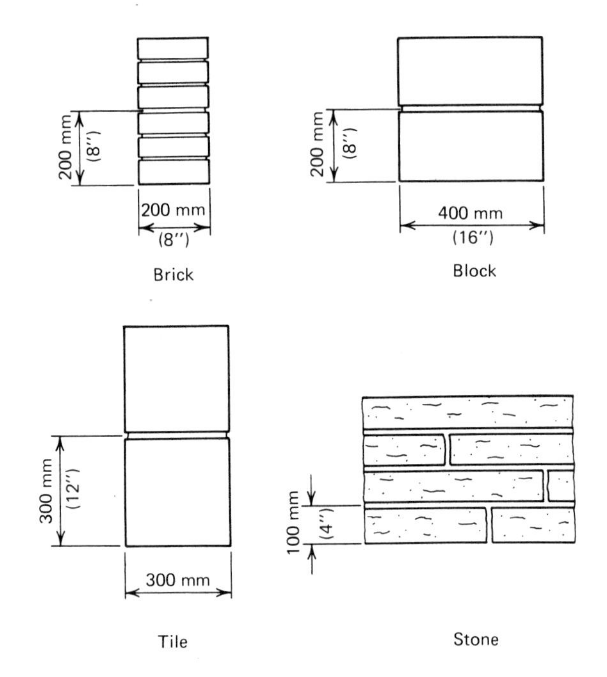

Fig. 13-1. 100 mm grid.

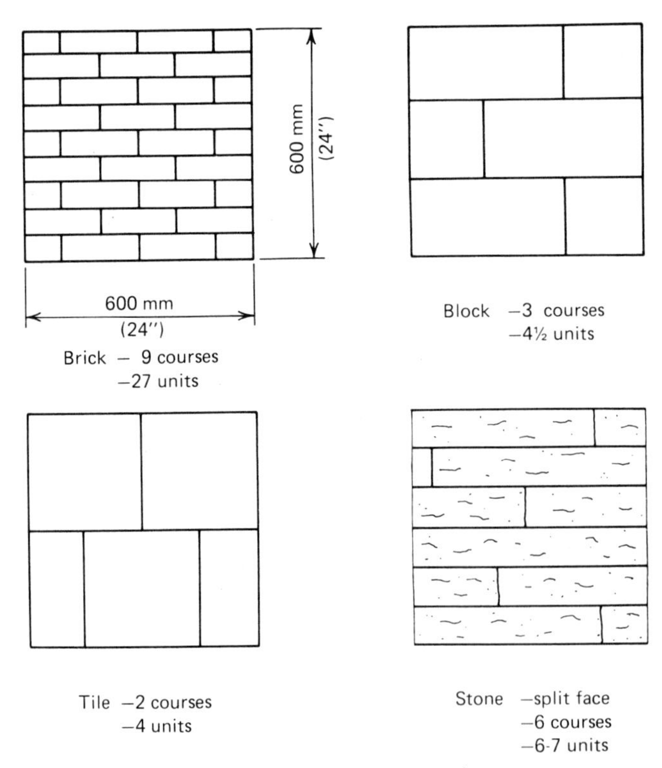

Fig. 13-2. 600 by 600 mm panel.

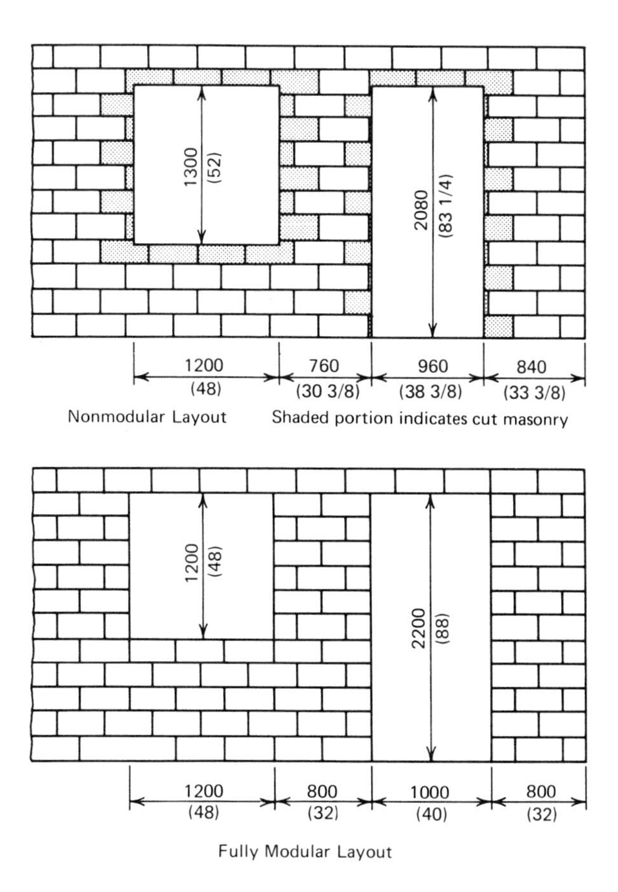

Fig. 13-3. Modular and nonmodular coordination. (*Courtesy National Concrete Producers Assn.*)

COORDINATING CONSTRUCTION

Consideration must be given to many items during the construction of a building. Even before the actual building starts, a detailed *estimation* of the costs involved is required. When the contract has been awarded, a careful *planning and scheduling* of the construction procedure is necessary. During the actual construction procedure, careful *inspection* will result in a building that meets all expectations.

Estimation

Estimation of a building involves two steps: *quantity survey* and *labor costs.* Quantity survey determines the number of units, the quantity of mortar and grout, the amount of reinforcing and also other specialty items where required. Labor costs are costs incurred while installing and completing the masonry elements. They constitute a vital part of the estimate and should be given serious consideration. These costs will include not only the mason's time but also that of all helpers and supervisory staff. The combination of materials, labor and overhead will form the elements of a *typical estimate.*

Quantity Survey. Calculations to find the quantity of *units* in a wall are usually done by finding the gross area of each individual wall and then deducting the area of the openings to find the net area of the wall. This net area is then multiplied by the number of units that are required to complete a square meter of wall area as shown in Tables 13-1, 13-2, and 13-3. Openings with areas of less than 0.5 m^2

Table 13-1. Mortar Quantities For Brick

Width X Height X Length (Nominal) mm (in)	Joint Thickness mm (in)	m^3/1000 Units (yd^3/1000)	Units/m^2 (100 ft^2) Surface Area	m^3/m^2 (yd^3/100 ft^2) Surface Area
Standard Brick				
100 X 67 X 200 (4 X 2-2/3 X 8)	10 (3/8)	0.27 (0.35)	75 (700)	0.020 (0.243)
Norman				
100 X 67 X 300 (4 X 2-2/3 X 12)	10	0.37 (0.48)	50 (465)	0.018 (0.219)
Roman				
100 X 50 X 300 (4 X 2 X 12)	10	0.35 (0.46)	67 (620)	0.023 (0.279)
100 X 67 X 400 (4 X 2-2/3 X 16)	10	0.47 (0.61)	37.5 (350)	0.018 (0.219)
TTW				
150 X 67 X 300 (6 X 2-2/3 X 12)	10	0.55 (0.72)	50 (465)	0.027 (0.328)
200 X 67 X 300 (8 X 2-2/3 X 12)	10	0.73 (0.95)	50	0.036 (0.437)
75 mm Brick				
75 X 67 X 200 (3 X 2-2/3 X 16)	10	0.23 (0.30)	75 (700)	0.017 (0.207)
Monarch				
75 X 100 X 400 (3 X 2-2/3 X 16)	10	0.43 (0.56)	25 (230)	0.011 (0.134)
Saxon				
100 X 100 X 300 (4 X 4 X 12)	10	0.40 (0.52)	33.5 (310)	0.013 (0.158)
Giant (Face-Shell Bedding)				
All Sizes	10	0.46 (0.60)	25 (230)	0.012 (0.146)
Giant (Full Bedding)				
100 X 100 X 400 (4 X 4 X 16)	10	0.49 (0.64)	25	0.013 (0.158)
150 X 100 X 400 (6 X 4 X 16)	10	0.51 (0.67)	25	0.013 (0.158)
200 X 100 X 400 (8 X 4 X 16)	10	0.52 (0.68)	25	0.014 (0.170)
300 X 100 X 400 (12 X 4 X 16)	10	0.59 (0.77)	25	0.015 (0.182)

Note: The void configuration in brick can affect the amount of mortar.

Table 13-2. Mortar Quantities For Block (Face-Shell Bedding)

Height X Length mm (in)	Joint Thickness mm (in)	m^3/1000 Units (yd^3/1000)	Units/m^2 (100 ft^2) Surface Area	m^3/m^2 (yd^3/100 ft^2) Surface Area
200 X 400 (8 X 16)	10 (3/8 in)	0.54 (0.71)	12.5 (116)	0.007 (0.085)
200 X 200 (8 X 8)	10	0.36 (0.47)	25 (232)	0.009 (0.109)
200 X 600 (8 X 24)	10	0.73 (0.95)	8.5 (79)	0.006 (0.073)
100 X 200 (4 X 8)	10	0.27 (0.35)	50 (464)	0.014 (0.170)
100 X 400 (4 X 16)	10	0.46 (0.60)	25 (232)	0.012 (0.146)
300 X 600 (12 X 24)	10	0.82 (1.07)	5.6 (52)	0.005 (0.061)

Note: *Full Bedding*
100 mm thick wall add 7% to face-shell amount;
200 mm thick wall add 14% to face-shell amount;
300 mm thick wall add 28% to face-shell amount.

(5 ft^2) can usually be ignored in the calculation as the waste involved will result in no reduction in the number of units required.

Calculation usually starts with the largest units and then progresses to the smaller ones. After the amount of standard-type units has been calculated, the specialty units will be considered. The walls are generally considered in the order of construction; that is, the outside walls are considered before the interior walls.

Mortar requirements for a masonry job will vary depending on the type of unit used, the thickness of the mortar joint, and the type of wall. The type of mortar will be determined by the load on the wall and its location. (See Table 5-3.) The quantity of mortar can easily be de-

Table 13-3. Mortar Quantities For Structural Tile (Side Construction)

Face-Shell Bedding

Height X Length mm (in)	*Joint Thickness mm (in)*	*m^3/1000 Units (yd^3/1000)*	*Units/m^2 (100 ft^2) Surface Area*	*m^3/m^2 (yd^3/100 ft^2) Surface Area*
300 X 300 (12 X 12)	10 (3/8 in)	0.54 (0.71)	11.5 (107)	0.006 (0.073)
200 X 300 (8 X 12)	10	0.46 (0.60)	17 (158)	0.008 (0.097)
250 X 300 (10 X 12)	10	0.50 (0.65)	13.5 (125)	0.007 (0.085)
135 X 300 (5-1/3 X 12)	10	0.40 (0.52)	25 (232)	0.010 (0.122)

Full Bedding

Thickness of Wall mm	*(in)*	*Percent Increase Over Face Shell*
50	(2)	0%
75	(3)	0%
100	(4)	5%
150	(6)	30%
200	(8)	60%
250	(10)	90%
300	(12)	115%

Table 13-4. Mortar Quantities For Stone

100 mm (4-in) Facing	m^3/m^2 Surface Area ($yd^3/100\ ft^2$)
Coursed ashlar	0.013 (0.158)
Single-course split face	0.012 (0.146)
Rubble	0.015 (0.182)

termined by considering the net area of the wall involved and multiplying it by the factor given in Tables 13-1, 13-2, 13-3, and 13-4.

The amount of each of the ingredients in the mortar should also be determined. Sand should be ordered by mass rather than by volume to minimize the loss due to bulking. (See Fig. 5-12.) The quantity of sand will be equal to the total amount of mortar required. The amount of cement (normal portland or masonry), lime and any admixtures will be based on the ratio that is required for each type of mortar. (See Table 5-1.)

The quantities for individual crews can vary as the waste factor for each crew is not always the same. Mixing and placing of mortars are included in the cost of unit placement.

The amount of *grout* needed is calculated in cubic meters and is usually delivered to the jobsite by a ready-mix supplier, so a breakdown of ingredients is not crucial. The amount of grout will vary with the type of unit being filled and with the size of the cavity. Tables 13-5 and 13-6 indicate quantities of grout that

Table 13-5. Grout Quantities For Hollow Brick and Block

Size of Units mm (in)	Fully Grouted/Unit m^3/unit (yd^3/unit)	Fully Grouted/ m^3/m^2 ($yd^3/100\ ft^2$) Surface Area
Giant Brick		
150 X 100 X 400 (6 X 4 X 16)	0.0026 (0.0036)	0.066 (0.802)
200 X 100 X 400 (8 X 4 X 16)	0.0041 (0.0057)	0.102 (1.239)
300 X 100 X 400 (12 X 4 X 16)	0.0070 (0.0097)	0.174 (2.114)
Block		
150 X 200 X 400 (6 X 8 X 16)	0.0048 (0.0066)	0.060 (0.729)
200 X 200 X 400 (8 X 8 X 16)	0.0074 (0.0103)	0.092 (1.118)
250 X 200 X 400 (10 X 8 X 16)	0.0088 (0.0122)	0.109 (1.324)
300 X 200 X 400 (12 X 8 X 16)	0.0118 (0.0164)	0.148 (1.798)
200 X 200 X 200 (8 X 8 X 8)	0.0017 (0.0024)	0.042 (0.510)
200 X 200 X 600 (8 X 8 X 24)	0.0111 (0.0154)	0.094 (1.142)
200 X 300 X 600 (8 X 12 X 24)	0.0166 (0.0230)	0.093 (1.130)
O-Block		
200 X 200 X 400 (8 X 8 X 16)	0.0088 (0.0122)	0.110 (1.337)
H-Block		
200 X 200 X 400 (8 X 8 X 16)	0.0074 (0.0103)	0.092 (1.118)

Table 13-6. Grout Quantities For Cavity Walls

Cavity Width	m^3/m^2 Surface Area ($yd^3/100\ ft^2$)
25 mm (1 in)	0.025 (0.304)
50 mm (2 in)	0.050 (0.608)
75 mm (3 in)	0.075 (0.911)
100 mm (4 in)	0.100 (1.215)

Table 13-7. Joint Reinforcing and Masonry Tie Quantities

Vertical Spacing of Reinforcing		Lineal Meters of Reinforcing Per m^2 of Wall Area (ft/100 ft^2)
200 mm (8 in)		2.50 (76)
400 mm (16 in)		1.25 (38)
600 mm (24 in)		0.83 (25)
Vertical Spacing of Ties	**Horizontal Spacing of Ties**	**Number of Ties Per m^2 of Wall Area (100 ft^2)**
200 mm (8 in)	300 mm (12 in)	16.80 (156)
200 mm	600 mm (24 in)	8.40 (78)
400 mm (16 in)	300 mm	8.40
400 mm	600 mm	4.20 (39)

will be required to fill the spaces in a masonry wall based on each square meter of surface area. The width of the cavity as shown in Table 13-6 will have considerable effect on quantity. Accuracy is important during construction as it can result in considerable saving.

The cost of grout placement will vary with each situation as some places are inaccessible and therefore more difficult to grout.

The quantity of *joint reinforcing* can be determined by the number of units, but is quite often calculated per square meter of wall. This area has already been calculated when determining the quantity of masonry units. The area is then multiplied by the factor given in Table 13-7 to obtain the lineal meters of reinforcing. The vertical spacing of the reinforcing will dramatically increase the amount of reinforcing. The vertical spacing is usually in multiples of 200 mm (8 in) as this is the module bricks and blocks fit, though some spacings will occur in 100 mm (4 in) increments to facilitate other sizes. The labor required to install reinforcing is often included in the placing cost of the units but in some placements, additional costs may be incurred. (See Table 13-8.)

The number of *masonry ties* is dependent on the horizontal and vertical spacing and is specified as a number of ties per square meter as shown in Table 13-7. This can easily be determined by multiplying the number of ties per square meter by the area of the wall to be tied.

Specialty items include items like cleaning surfaces, placement of expansion and control

Table 13-8. Quantity Requirements For Speciality Items

Rigid insulation adhesive	
Dab	4 m^2/L(200 ft^2/gal)
Trowell and butter	2 m^2/L(100 ft^2/gal)
5% silicone dampproofing	
Brick, tile and stone	2.4 m^2/L(120 ft^2/gal)
Block	1.6 m^2/L(80 ft^2/gal)
Insulation–Verrniculite	
One m^3 will fill	175 (150 mm blocks) (20)
[one bag (4 ft^3) will fill]	140 (200 mm blocks) (16)
	100 (250 mm blocks) (11.5)
	80 (300 mm blocks) (9)
Colored mortar	1 kg/26.7 kg masonry (3 lbs/bag) cement
Caulking	3.125 m/312.5 cc tube (10 ft/11 oz) (10 mm X 10 mm bead) (3/8 in X 3/8 in)

Table 13-9. Labor Required For Speciality Items

Item	Labor
Cleaning	
Face brick, block or tile	0.09 hr/m^2 (0.84 hr/100 ft^2)
Rough textured brick or block	0.11 hr/m^2 (1.02 hr/100 ft^2)
Expansion joints	0.05 hr/m (0.015 hr/ft)
Control joints	0.10 hr/m (0.030 hr/ft)
Back parging	0.08 hr/m^2 (0.74 hr/100 ft^2)
Flashing	0.05 hr/m (0.015 hr/ft)
Insulation	
Pour type (200 mm block in cell)	0.05 hr/m^2 (0.46 hr/100 ft^2)
Foam (200 mm block in cell)	0.19 hr/m^2 (1.77 hr/100 ft^2)
Rigid (50 mm thickness)	0.17 hr/m^2 (1.58 hr/100 ft^2)
Scaffolding	0.15 hr/m^2 (1.39 hr/100 ft^2)
Vapor barrier	0.04 hr/m^2 (0.37 hr/100 ft^2)

joints, flashing, insulation installation, scaffolding and vapor barriers. Tables 13-8 and 13-9 indicate some values that can be used for these costs where they occur.

Labor Costs. Labor costs are generally developed and tabulated with regard to each individual operation. These values will vary with individual companies. Most costs are estimated on the basis of the mason's productivity. This productivity will vary depending on the units being laid (see Table 13-10), the pattern used, whether the units are being laid from a scaffold and weather conditions. A

Table 13-10. Mason's Unit Placement Productivity

Size of Units Nominal	Number of Units/ Eight-Hour Day	Square Meters of Wall/Day (ft^2/day)
Bricks		
Common		
100 X 67 X 200 (4 X 2-2/3 X 8)	550–650	7.3–8.6 (79–93)
Norman		
100 X 67 X 300 (4 X 2-2/3 X 12)	360–440	7.2–8.8 (78–95)
Roman		
100 X 50 X 300 (4 X 2 X 12)	450–550	6.7–8.2 (72–88)
100 X 67 X 400 (4 X 2-2/3 X 16)	320–380	8.5–10.1 (92–109)
Giant		
100 X 100 X 400 (4 X 4 X 16)	220–280	8.8–11.2 (95–120)
150 X 100 X 400 (6 X 4 X 16)	220–280	8.8–11.2 (95–120)
200 X 100 X 400 (8 X 4 X 16)	210–270	8.4–10.8 (90–116)
300 X 100 X 400 (12 X 4 X 16)	200–250	8.0–10.0 (86–108)
TTW		
150 X 67 X 300 (6 X 2-2/3 X 12)	270–330	55.4–6.6 (58–71)
200 X 67 X 300 (8 X 2-2/3 X 12)	240–300	4.8–6.0 (52–65)
75 mm		
75 X 67 X 200 (3 X 2-2/3 X 8)	600–700	8.0–9.3 (86–100)
Monarch		
75 X 100 X 400 (3 X 4 X 16)	230–290	9.2–11.6 (99–125)
Saxon		
100 X 100 X 300 (4 X 4 X 12)	270–330	8.1–9.9 (87–107)

Table 13-10. Cont.

Size of Units Nominal	Number of Units/ Eight-Hour Day	Square Meters of Wall/Day (ft^2/day)
Block (Lightweight)		
100 X 200 X 400 (4 X 8 X 16)	190–230	15.2–18.4 (164–198)
150 X 200 X 400 (6 X 8 X 16)	200–240	16.0–19.2 (172–207)
200 X 200 X 400 (8 X 8 X 16)	180–220	14.4–17.6 (155–189)
250 X 200 X 400 (10 X 8 X 16)	175–205	14.0–16.4 (151–177)
300 X 200 X 400 (12 X 8 X 16)	165–195	13.2–15.6 (142–168)
100 X 100 X 200 (4 X 4 X 8)	360–440	7.2–8.8 (78–95)
100 X 100 X 400 (4 X 4 X 16)	210–270	8.4–10.8 (90–116)
200 X 200 X 200 (8 X 8 X 8)	320–380	12.8–15.2 (138–164)
200 X 200 X 600 (8 X 8 X 24)	130–150	15.3–17.6 (165–189)
200 X 300 X 600 (8 X 12 X 24)	90–110	16.1–19.6 (173–211)
Tile (Unglazed)		
75 X 300 X 300 (3 X 12 X 12)	300–360	26.1–31.3 (281–337)
100 X 300 X 300 (4 X 12 X 12)	270–330	23.5–28.7 (253–309)
150 X 300 X 300 (6 X 12 X 12)	220–280	19.1–24.3 (206–262)
200 X 300 X 300 (8 X 12 X 12)	180–220	15.7–19.1 (169–206)
100 X 200 X 300 (4 X 8 X 12)	330–390	19.4–22.9 (209–246)
150 X 200 X 300 (6 X 8 X 12)	300–360	17.6–21.2 (189–228)
200 X 200 X 300 (8 X 8 X 12)	240–300	14.1–17.6 (152–189)
100 X 135 X 300 (4 X 5-1/2 X 12)	360–440	14.4–17.6 (155–189)
150 X 135 X 300 (6 X 5-1/2 X 12)	330–390	13.2–15.6 (142–168)
200 X 135 X 300 (8 X 5-1/2 X 12)	290–350	11.6–14.0 (125–151)
Stone (100 mm Facing)		
Single-course split		10.8–13.2 (116–142)
Coursed ashlar		7.2–8.8 (78–95)
Rubble		1.8–2.2 (19–24)

common mason to laborer ratio is from two-to-one to three-to-one, so consideration must be made in the cost to allow for this as well as any supervisory help that is necessary, as these items will affect the productivity.

Typical Estimate. An estimator will do his quantity takeoff in a specific way. The method that is used will vary with different estimators and different companies. It should be set out in such a way, however, that the work can easily be checked for accuracy by another individual. This takeoff will show all calculations and will usually be listed in the same order as the construction procedure. The completed takeoff will be included with labor costs, general expenses and the markup on a recap sheet to make a typical estimate. The general expenses will include the items in Table 13-12. The markup will vary depending on the company's need for the job.

Table 13-11. Labor Costs Relative To Masonry Patterns

Type of Pattern	Cost Factor
Running	1.00
Common	1.05
Stack	1.11
Dutch or English	1.25
Flemish	1.18
Basket weave	
Vertical	1.43
Diagonal	2.00
Herringbone	2.86

Table 13-12. General Expense Sheet

Building survey														
Temporary buildings														
Office														
Storage														
Sanitary														
Hoists and forklifts														
Enclosures														
Temporary														
Power														
Heat														
Water														
Office supplies														
Telephone, telex and computer														
Snow removal														
Tool rentals														
Scaffold rentals														
Signs														
Material handling and trucking														
Cleanup														
Pumping														
Salaries														
Superintendent														
Accountant														
Engineer														
Watchman														
First aid														
Layout														
Permits														
Bonds														
Financing														
Insurance														
Room and board														
Travelling expenses														
Wage increase														
Penalty														
Bonus														
TOTAL														

The following will illustrate part of the method used to estimate the cost of a small building. The first step is to become familiar with the drawings and specifications. The first part of takeoff would normally be the excavation. If the job is a subcontract, the excavation and foundation are likely provided by the general contractor. The foundation calculation is found by taking the center line perimeter and multiplying it by the width and thickness for the footing. This same figure is then used for the walls if they are of uniform thickness, and is multiplied by the thickness and height of the wall as shown in Example 13-1(a). The concrete slab on grade is calculated by taking the size and multiplying it by the thickness. All sizes are in meters so the answer will be in cubic meters, the measurement in which concrete is supplied. The mason's work begins when the foundation is completed. The walls are considered individually, calculating all the walls with the same size of block at the same time. The exterior walls are generally taken first. The gross area is calculated, then the size of openings are deducted. The net area is in square meters and values from Tables 13-1, 13-2, and 13-3 can be used to determine the number of units, which is easily converted to a cost. Values from Table 13-8 can be related to mason's rate giving labor costs.

The quantity of mortar, grout and reinforcing is calculated from the net area of the walls as shown in Tables 13-1 through 13-7. Other specialty items are also based on the net area in square meters as shown in Tables 13-8 and 13-9. When all materials are calculated they are transferred to a recap sheet as shown in Example 13-1(b). The quantities are then priced and the total is summed. The labor costs are determined by setting the value for units per man-hour. The mason's hours are calculated by taking the total number of units and dividing it by his productivity. The laborer-to-mason ratio will give the laborer's hours. Multiplying the current rate of pay by these hours will give labor cost.

Once the general expenses, the foreman's rate, and the profit expected on the job have been added on, the final value for the estimate can be determined. The masonry subcontract usually becomes part of the general contractor's summary when the entire job is priced.

Planning and Scheduling

Immediately after the awarding of a contract a master plan is ordered by the contractor for the planning and scheduling of the job. The best person to plan and schedule any project is the superintendent as he has a detailed knowledge of the construction procedure. The prime purpose is to provide a logical system for determining the duration of the project and the cost and manpower requirement factors involved in the project.

The *critial path method* is a common method used for *planning, scheduling*, and *controlling* the work. The critical path network attempts to build the entire project on paper prior to the actual construction. This network can be used as a tool to guide and control the tempo and direction of the job. To be useful, the logic, network and duration must reflect reality as closely as possible.

Planning. In the planning stage, the first thing that must be done is to list the specific activities that must be accomplished in their proper order. (See Fig. 13-4.) Once this has been completed, then the *arrow diagram* is drawn. A carefully prepared network will show the accurate flow of activities and their relationships. The diagram is constructed by taking each activity into account. When considering any specific activity one must take into account the previous activities, subsequent ones and those occurring concurrently.

GENERAL ESTIMATE

BUILDING .... ESTIMATE No. 314
LOCATION .... SHEET No. 1
ARCHITECTS .... ESTIMATOR ....
SUBJECT .... CHECKER ....
DATE ....

DESCRIPTION OF WORK	No. Pieces	DIMENSIONS L	DIMENSIONS W	DIMENSIONS H	Extensions	Extensions	Total Estimated Quantity	Unit Price Mat'l	Total Estimated Material Cost	Unit Price Labor	Total Estimated Labor Cost
Strip Topsoil		30.48	15.24	0.15	69.68		70 m³				
Bldg. Size											
12.80 x 8.50											
Ftg.Proj.0.50 x 0.50											
w/s 0.60 x 0.60											
Cutback 2.00 x 2.00											
15.90 x 11.60											
Excavation											
Bsmt.		15.90	11.60	2.00	368.88		369 m³				
Backfill	As Excavated				368.88						
Deduct		12.80	8.50	2.00		217.60					
					368.88 −	217.60	151 m³				
Centreline Perimeter											
2/12.80 = 25.60											
2/8.50 = 17.00											
42.60											
less 4/0.20 .80											
41.80											
Concrete Ftgs.		41.80	0.70	0.20	5.85						
		7.90	0.70	0.20	1.11						
					6.96		7 m³				
Form Footings	2	41.80		0.20	16.72						
	2	8.10		0.20	3.24						
					19.96		20 m²				

GENERAL ESTIMATE

BUILDING ....

LOCATION ....

ARCHITECTS ....

SUBJECT ....

ESTIMATE No. ....

SHEET No. 2

ESTIMATOR ....

CHECKER ....

DATE ....

DESCRIPTION OF WORK	No. Pieces	DIMENSIONS L	W	H	Extensions	Extensions	Total Estimated Quantity	Unit Price Mat'l	Total Estimated Material Cost	Unit Price Labor	Total Estimated Labor Cost
Concrete Walls		41.80	0.25	2.35	24.56						
		8.10	0.25	2.35	4.76						
					29.32		29.5 m³				
Form Walls	2	49.90	–	2.35	234.53		234.5 m²				
Concrete S.O.G.		12.40	8.10	0.10	10.04		10 m³				
Gravel Under S.O.G.		12.40	8.10	0.13	13.06		13 m³				
Cure & Finish S.O.G.		12.40	8.10		100.44						
S.O.T.		8.10	3.05		24.71						
					125.15		125 m²				
Parge Tank Walls		22.3		2.35	52.41		52.5 m²				
Slab Over Tank		8.10	3.05	0.15	3.71		4 m³				
Form Slab		8.10	3.05		24.71		25 m²				
Blocks - 200mm Standard											
West Elevation		12.80	2.40		30.72		31 m²				
East Elevation		12.80	2.40		30.72						
Deduct - Door		1.00	2.20			2.20					
- Window		1.20	1.20			1.44					
					30.72 –	3.64	27 m²				

GENERAL ESTIMATE

BUILDING .... ESTIMATE No. ....

LOCATION .... SHEET No. 3

ARCHITECTS .... ESTIMATOR ....

SUBJECT .... CHECKER ....

DATE ....

DESCRIPTION OF WORK	No. Pieces	DIMENSIONS L	W	H	Extensions	Extensions	Total Estimated Quantity	Unit Price Mat'l	Total Estimated Material Cost	Unit Price Labor	Total Estimated Labor Cost
South Elevation		8.50	2.40		20.40						
Deduct - Window		1.80	1.20			2.16					
					20.40 −	2.16	18.5 m²				
North Elevation		8.50	2.40		20.40		20.5 m²				
Cross Wall		8.10	2.40		19.44		19.5 m²				
Bricks - Common											
West Elevation		12.80	2.40		30.72		31 m²				
East Elevation		12.80	2.40		30.72						
DEDUCT - DOOR		1.00	2.20			2.20					
- WINDOW		1.20	1.20			1.44					
					30.72 −	3.64	27 m²				
South Elevation		8.50	2.40		20.40						
DEDUCT - WINDOW		1.80	1.20			2.16					
					20.40 −	2.16	18.5 m²				
North Elevation		8.50	2.40		20.40		20.5 m²				
Joint Reinforcing											
300mm Double		97	2.50		242.50		242.5 m				
Flashing	2	12.80			25.60						
	2	8.50			17.00		43 m				
LINTELS											
	1	1.40	100/75	10	1.40						
	1	1.60	100/75	10	1.60						
					3.00		3 m				

RECAPULATION

PROJECT	
LOCATION	
ARCHITECT ENGINEER	
GENERAL CONTRACTOR	
BRANCH	
ESTIMATE NO.	
SHEET NO.	
DATE	
CHECKED BY	

CODE	DESCRIPTION	QUANTITY	UNIT	UNIT PRICE	MATERIAL COST	UNITS PER MH	MASON HR.	LABOUR HR.
	Strip Topsoil	70	c.m.		- -			
	Excavation	369	c.m.		- -			
	Concrete - Footings	7	c.m.		- -			
	- Walls	29.5	c.m.		- -			
	- S.O.G.	13	c.m.		- -			
	- Slab	4	c.m.		- -			
	Forming - Footings	20	s.m.		- -			
	- Walls	234.5	s.m.		- -			
	- Slab	25	s.m.		- -			
	Gravel	13	c.m.		- -	2		6.5
	Cure & Finish Slab	125.5	s.m.		- -			
	Parging	52.5	s.m.		- -	12.5	2.5	1.7
	200x200x400 Block 116.5x12.5	1457	blks.		- -	25	59	24
	100x67x200 Brick 116.5x75	8738	brks.		- -	75	116.5	47
	Joint Reinforcing	242.5	m		- -	---	---	----
	Flashing	43	m		- -	20	2	1
	Lintels	3	m		- -	6	0.5	----
				Total	- - -	Total	- -	- -
	Add Bricklayers hours x rate				- - -			
	Labourers hours x rate				- - -			
	Foremans hours x rate				- - -			
	General Expenses				- - -			
				Total	- - -			
	Add Profit Margin - % of Total				- -			
			Final	Total	- - -			

Activity	Description	Duration	Precedes	Follows
A	Layout	1	–	B
B	Excavation	2	A	C
C	Foundation	6	B	D,R
D	Block walls	8	C	E,F
E	Insulation	1	D	K
F	Roof framing	2	D	G,J,H,I
G	Roofing	2	F	L
H	Rough elect.	1	F	O
I	Rough plumb.	2	F	N
J	Ext. trim	1	F	K
K	Brick facing	15	J	–
L	Ceiling fin.	4	G	M
M	Painting	3	L	O
N	Fin. plumbing	6	I	O
O	Fin. elct.	2	H,M,N	P
P	Fin. floor	1	O	Q
Q	Int. trim	3	P	–
R	Rough grade	1	C	–

Fig. 13-4. Activity list.

The arrow diagram shows at a glance all phases of the job in proper sequence. It is easy to see how much work has to be completed before any specific activity begins.

Scheduling. Once the arrow diagram has been drawn as shown in Figure 13-5, the next operation is to determine the actual job duration. The duration of each activity is found by estimating the time required to finish it. These estimates should include labor (usually in man days), time needed for operations that are not included in the general estimate (e.g., curing or setting time of concrete), and also work done by other firms (subcontractors, suppliers or architects).

Each activity and its duration is indicated on the arrow diagram. The shortest possible completion time of the job is established along the critical or longest sequence of events (critical path), as shown in Figure 13-5. The contractor can then see which events are most critical and which ones have slack or *float* time. He can then distribute his manpower, assemble his materials and dispatch the equipment to get the most efficient use of each. The activities along the critical path will not have any float time. By indicating the earliest start and the latest finish time of an activity, the total float time can easily be calculated and be put to efficient use.

Controlling. The actual progress of the job is easily monitored by hand or by computer, depending on the size of the job. A bar chart is often used to show the actual progress of the job so that adjustments can be made when necessary. (See Fig. 13-6,) These adjustments in the schedule may occur due to changes in actual duration, additions or deletions, and the actual start and finish times of an activity. These changes will result in a revised schedule and will have an effect on the completion date.

If specific emphasis is placed on the critical activities, it is possible to finish the entire job sooner. This *crashing* of the job should be given careful consideration as the extra cost involved may exceed any savings resulting from an early

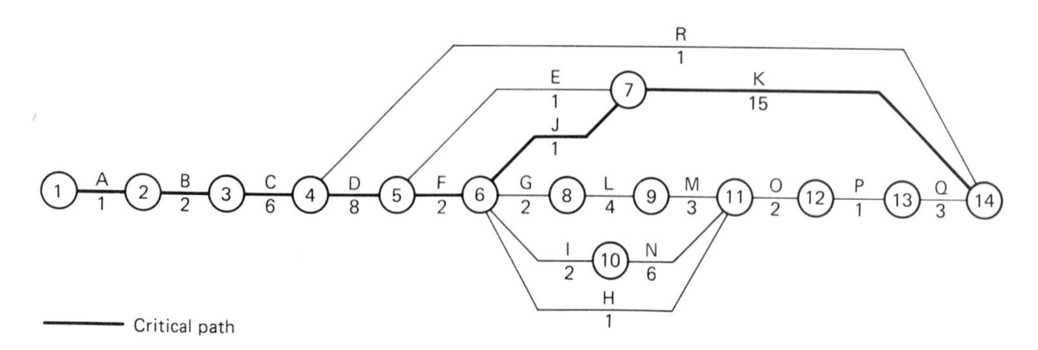

Fig. 13-5. Arrow diagram.

Activity	2	5	8	11	14	17	20	23	26	29	32	35
Layout	x											
Excavation	xx											
Foundation		xxxxxx										
Block walls				xxxxxxxx								
Insulation						x						
Roof framing						xx						
Roofing							xx					
Rough electrical							x					
Rough plumbing							xx					
Exterior trim							x					
Brick facing							xxxxxxxxxxxxxxx					
Ceiling finish								xxxx				
Painting									xxx			
Finish plumbing								xxxxxx				
Finish electrical										xx		
Finish floor											xx	
Interior trim											xxx	
Rough grade				x								

x = One day

Fig. 13-6. Bar chart.

finish. This method can also be used to finish a job on schedule if it has fallen behind due to unforeseen circumstances. This is especially an advantage on jobs that have a penalty clause for a late finish date.

A subcontractor will need to fit into the general contractor's schedule that has been worked out, so he should arrange his activities to meet this schedule. On larger jobs the subcontractor may find it beneficial to take his portion of the arrow diagram and enlarge it, putting greater detail into his activities. He can then set up his own schedule around this diagram, making it easier to finish his portion of the job, so that there is no delay in the finishing date.

This critical path method has a great deal of versatility as it easily allows for changes that result in a continual updating of the schedule, meeting completion deadlines as they occur.

INSPECTION

A project, to be successful, must be handled properly from the beginning. Planning, design, coordination and construction must be done efficiently and effectively to achieve economic, durable and aesthetic buildings. The designer takes the responsibility for the structure, so he must allow for a safety factor. This safety factor allows for discrepancies in materials and inadequacy in construction. It will only come down when the designer is confident that the regulations and the quality expected will be rigidly followed. *Inspection* requires a knowledgeable *inspector* who is called on to check *materials* and *construction methods* on a jobsite when the most critical operations are occurring.

Inspection involves the observation and checking of masonry materials and methods of construction to ensure the presence of desired qualities and adherence to a prepared set of plans and specifications. It includes:

1. Checking, visual inspection and testing of materials.
2. Observing placement of units, mixing and placing of mortar and grout to ensure good workmanship.
3. Examination of work to see that it conforms to plans and specifications.

4. The overseeing of test sample selection and preparation when testing is required.

5. Reporting of results of examinations, observations and test results to appropriate individuals.

Inspector

An inspector must be a diplomat in order to deal efficiently and still command the respect and trust of the people with whom he works. He must have a reasonable understanding of what the designer wants to accomplish, an appreciation for what is possible to accomplish, plus the capability to persuade people to achieve their utmost.

An inspector must deal with people who may not have a clear understanding of design requirements, so he must be an educator and an arbitrator when problems arise in specification requirements. The low bidder may leave out things that are not adequately specified, later trying for extras, complicating the inspector's problems.

Before an inspector leaves for a site inspection, he should have a thorough knowledge of the plans and specifications, relevant CSA and ASTM regulations, and recommended testing procedures. At the outset the inspector should be informed by the engineer and the architect of the parts of the construction that are most important for the creation of the building envisioned. The type of building, its use and wall construction usually will determine inspection priorities. These priorities are usually broken into three areas: materials, construction methods and aesthetic considerations. In order to make an accurate report, the inspector should take along a record book, a check list, building code, specifications and measuring and testing equipment for site tests.

Materials

Masonry Units. All units that are used on a jobsite need to be protected from the elements, clear of contact with mud and water. Units are normally placed without wetting, but clay units are wetted if they have an I.R.A. of over 35 gms (1.23 oz) per minute. The wetting of brick is an expensive item and may not be economically feasible, so consideration should be given to using mortar with high water retentivity. Concrete units are not wetted prior to placing as they will shrink when they dry out.

Masonry units are not so closely controlled that tests on site should be eliminated. If the quality of the units is important enough to specify, it is important enough to ensure compliance. Besides the quality of the units, check the dimensions and squareness; a high degree of perfection cannot be obtained with poor quality units. Clay units should be checked for fire cracking, efflorescence and color blending, as manufacturers have better facilities for blending than does the mason.

Mortar. Engineered masonry requires that the mortar materials and proportions be prequalified before the final specifications are written. The inspector then can ensure that the materials (particularly the sand) meet the specifications at all times. He can make sure that the materials are stored properly to avoid contamination or variation in water content which requires too frequent adjustments for bulking.

Mortar testing and inspection of mixing procedures depend on whether the mortar is specified by proportion or by property. Proportion mixes cannot guarantee compressive strength requirements; they can only guarantee the mix itself. The meeting of these requirements is difficult to prove. The wet sieve analysis could be used, but the test is unable to differentiate between lime and cement. A gauge box (0.03 m^3 or 1 ft^3) can be used as an occasional check against accuracy of shovelfuls. A hopper above the mixer gives a more accurate measurement of volume. With property specifications, the compression test can be used to determine results and then proportions are left to the contractor as the mason is the best judge of mortar. An inspector should not abandon good batch control as he will be depending solely on periodic sampling and a twenty-eight-day waiting period for results.

Compressive strength is not the only quality desired; mortar should resist water penetration, create good bond, and be self-healing. The minimum strength allowed should be used as an increase in strength will tend to diminish other qualities. Mixing should be done with care, about half the water and sand is added first, then all lime and cement, and then the remaining sand and water. Otherwise incomplete mixing can result. The mixing sequence need not be longer than ten minutes as high air content may result.

Grout. Grout usually comes to the job in a ready-mix truck and is pumped into a wall. The grout should be as fluid as possible without causing segregation of materials. Slumps in the range of 200 mm to 250 mm (8 in to 10 in) are preferred. Cleanouts should be provided at the base of all cores being reinforced, and care should be taken that they are of sufficient size (75 cm^2 to 80 cm^2) (11.6 in^2 to 12.4 in^2) to ensure adequate cleaning by pressurized water or air. When spaces are narrow (less than 75 mm or 3 in) only fine aggregate is used in the mixture. Strength requirements can be checked by a compression test, though they are not always specified.

Construction Methods

Foundation Preparation. Before placement begins, check to see that the foundation is level, straight and free of latence, ice and dirt. Foundations that require bedding of more than 20 mm (¾ in) should not be tolerated. A bed joint of less than 6 mm (¼ in) can cause point loading unless it is consistent. Foundations that are out of line can result in unacceptable unit overhang. Corrective action may be indicated before placement proceeds. Miscellaneous materials such as flashing, reinforcing, anchors and insulation should be checked for compliance.

First Masonry Course. All dimensions, location of door frames, position of vertical reinforcing, mortar mix and pattern of wall will be decided with placement of the first course. In some cases a small adjustment of opening position results in a lot fewer units being cut. Dowels that are to be tied to vertical reinforcing need to be accurately located so that the mason does not have a tendency to bend dowels excessively in order to center in the cores. (See Fig. 13-7.) The mortar joint in a cavity wall to be grouted is better beveled away from the cavity to minimize droppings and extrusions into the grout space. (See Fig. 13-8.) If solid units are being placed in a load-bearing wall, see that furrowing is not too deep as illustrated in Figure 13-9. Check this by lifting the unit to see if a furrow remains. All mortar joints should be fully filled, and joint size

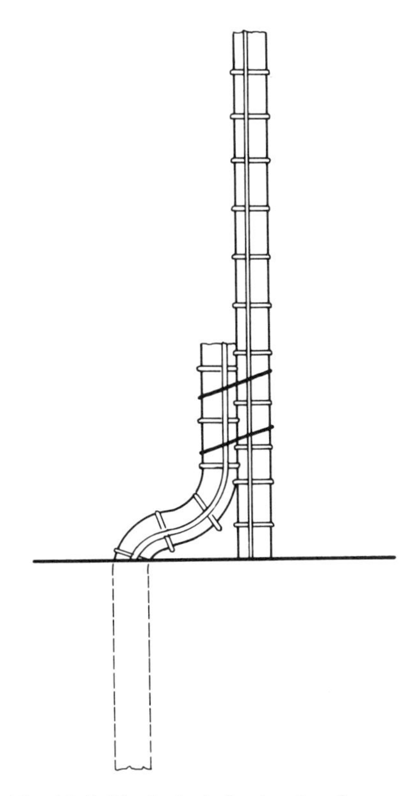

Fig. 13-7. Vertical reinforcing-dowel connection.

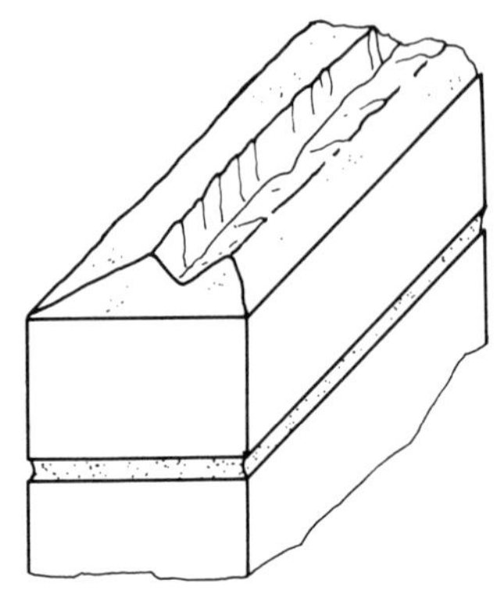

Fig. 13-8. Mortar beveled away from cavity.

should not exceed 13 mm ($^1\!/_2$ in) in thickness. Take care not to allow stringing to the extent that mortar will stiffen before the unit is placed.

Reinforcing. Reinforcing steel needs to be held in place as specified in the plans or, if not specified, in the centerline of the wall. All steel laps should be of sufficient size to ensure that the structural strength is not reduced.

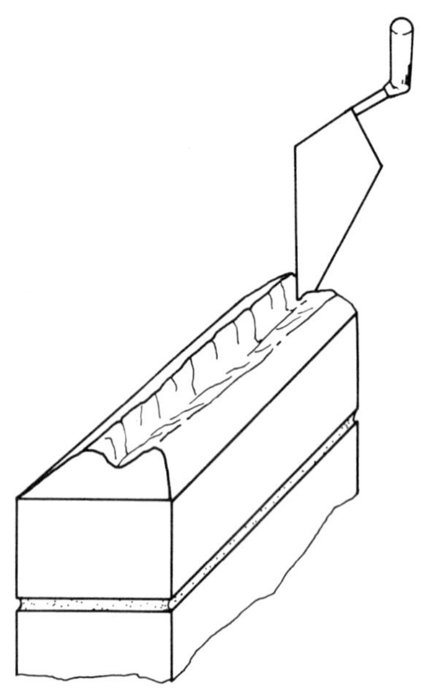

Fig. 13-9. Furrowing.

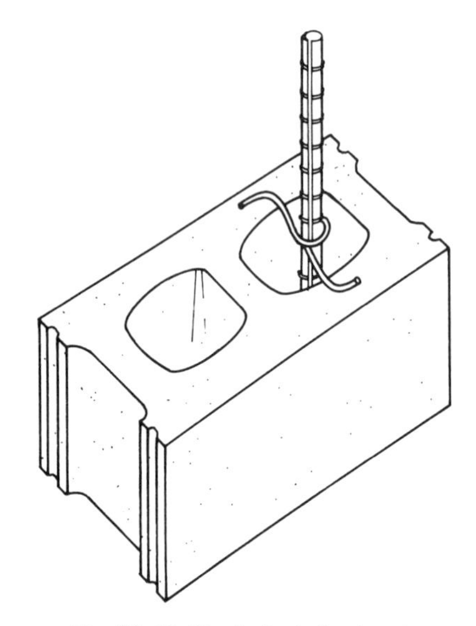

Fig. 13-10. Vertical reinforcing tie.

Vertical reinforcing is held in place by ties spaced not over 192 bar diameters. (See Fig. 13-10.) The placement of joint reinforcing is easily forgotten. Joint reinforcing should be placed in joints at least twice the diameter of the reinforcing wire. This reinforcing should be fully embedded throughout and lapped 150 mm (6 in) where joints occur. All lintel bars across openings should be continuous and extend sufficiently past the edge to provide anchorage, as shown on the plans.

Grouting. Before the grouting of the wall begins, reinforcing is placed in accordance with the plans. Horizontal bars are often placed as work progresses. Where cleanouts are provided, all debris is removed and the hole is filled prior to grouting. The grout is better placed by pumping as crane buckets cause excessive segregation of very fluid mixes. (See Fig. 12-10.) The grout is properly puddled and reconsolidated to avoid voids. Vibration can cause segregation and is difficult to do in narrow spaces. Care should also be taken to see that anchors and reinforcing do not impede grout flow. In low-lift grouting the maximum lift is usually not over 1200 mm (4 ft). The grout should be kept down from the top of the lift to provide a tie to the next lift. High-lift grouting should not be poured in lifts of over 1200 mm (4 ft) as the wall can fail due to the excessive load that might occur. The next lift can proceed as soon as the previous lift has started to stiffen allowing for self-support. Care should be taken that reconsolidation does not occur after initial set has occurred.

Glossary of Terms

ASTM American Society for Testing and Materials.

Bed Joint The first horizontal mortar joint between the masonry and foundation.

CSA Canadian Standards Association.

Float The breathing space in days, for each activity that is not on the critical path.

Furrowing Making a depression in the center of a horizontal brick mortar joint.

Grout A concrete mixture that does not have large aggregate particles in it; used to fill cavities in reinforced masonry walls.

Masonry Tie A metal strip used to tie masonry units together or to a backup wall.

Penalty Clause A clause written into the contract that specifies a penalty for each day the project finishes late.

Recap Sheet A sheet used to price material and labor costs in an estimate.

Water Retentivity The property of mortar that prevents the rapid loss of water by absorption of masonry units.

Review Questions

1. What is the advantage of the 100 mm module, and how is it facilitated in masonry units?
2. Why is it better to use door frames that have been built to multiples of 100 mm in a masonry wall?
3. Why is it critical that an estimate be set down in a neat logical order?
4. Differentiate between quantity surveying and estimating.
5. Discuss the advantages of relating quantities of materials to the square meter of wall area.
6. List four items that affect the productivity of a mason.
7. Explain why the critical path method is often used for planning, scheduling and controlling a job.
8. What is meant by float time?
9. Why does an inspector have to be a special type of individual?
10. Why are masonry units sometimes wetted before placement?

Selected Sources of Information

Alberta Masonry Institute, Calgary, Alberta.

American Society for Testing and Materials, Washington, D.C.

Brick Institute of America, McLean , Virginia.

Canadian Standards Association, Rexdale, Ontario.

Clay Brick Association of Canada, Willowdale, Ontario.

Critical Path Scheduling, New York: Horowitz-Ronald Press, 1967.

CPM in Construction, Washington, D.C.: 1967. Associated General Contractors of America.

National Building Code, Ottawa, Ontario.

National Concrete Masonry Assn., McLean, Virginia.

National Concrete Producers Assn., Downsview, Ontario.

Chapter 14

MASONRY CONSTRUCTION PRACTICES

Men have been building masonry structures for many thousands of years and the methods developed in the past have been carried on from one generation of builders to another. At the same time, new practitioners of the masonry trades have found ways to improve upon their predecessor's techniques. For example, they have learned to lighten the loads imposed by buildings, increase the strength of structural components in relation to their weight, enhance the appearance of masonry structures, increase the speed of erection and integrate masonry with other building materials to provide greater versatility in construction.

Building with masonry involves three types of construction. In one type, the walls transmit the building loads to the foundation and are known as *bearing walls.* In the second, a *structural frame* of some sort carries the loads, and masonry is utilized simply to close in the building around that framework, a system commonly known as *curtain wall* construction. In the third, masonry is used as a protective or decorative skin over a *load-bearing backup wall* of some description, and the result is a *veneered wall.* (See Fig. 14–1.)

The basic conventional technique in masonry construction has been to take individual pieces of some masonry material—tile, brick, stone or concrete masonry—and, using mortar of one kind or another, bond them together into some sort of structure. Masonry construction practices involve the methods and techniques used in achieving that goal.

BRICK CONSTRUCTION

Brick Terms

The first requirement in any discussion of brick construction is to understand the terms used in connection with it.

A single thickness of brick, laid up to form a wall, is called a *wythe* and, if the wall is composed of two or more wythes, it is said to be a *multiwythe* wall. Bricks are laid in rows or *courses* in a wythe, with one or more surfaces exposed.

If the brick is laid flat, with its longest dimension parallel to the face of the wall, it is called a *stretcher.* (See Fig. 14–2, bottom.) If it is laid flat with its longest dimension perpendicular to the wall face, it is a *header.* (Fig.

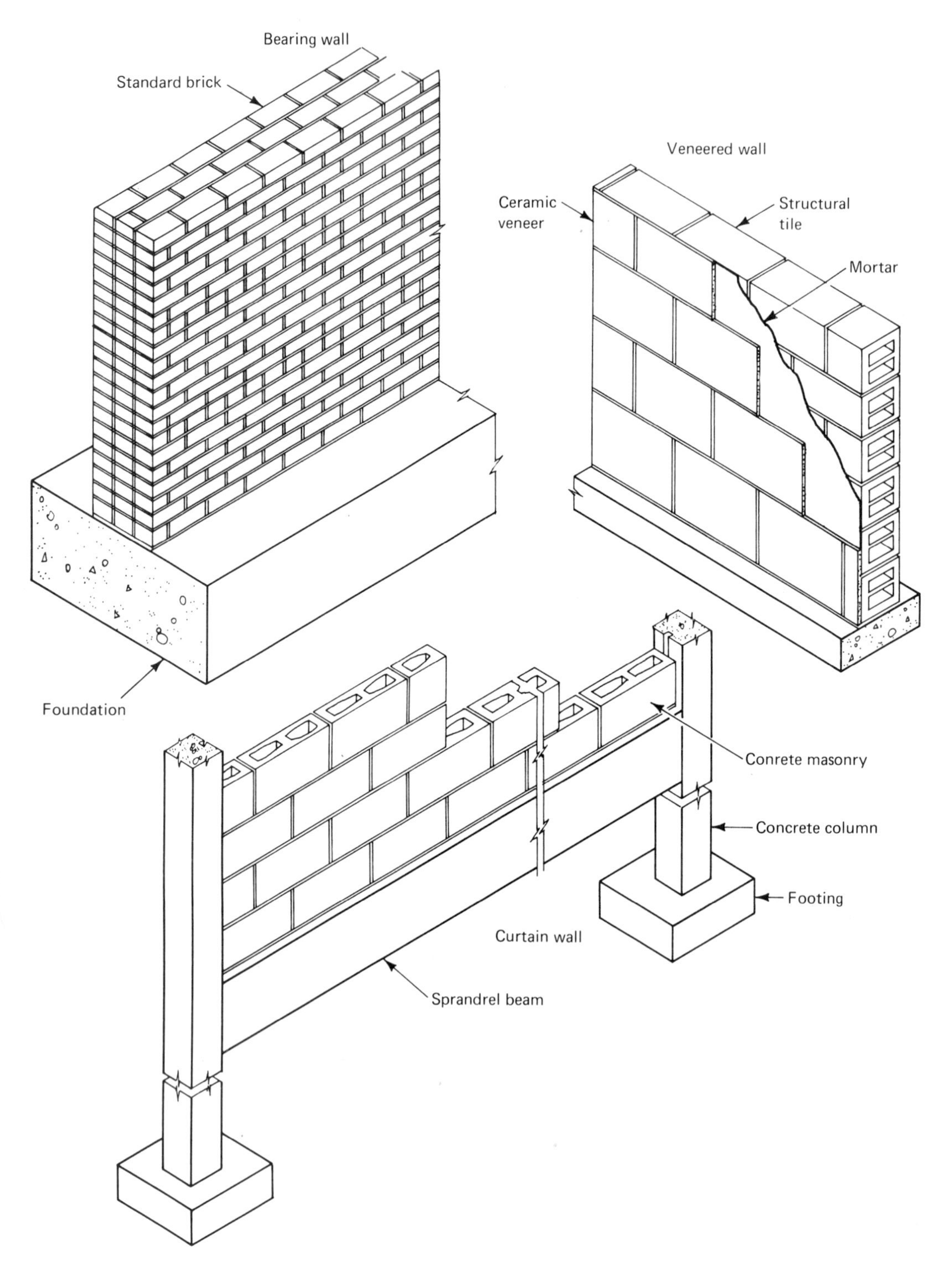

Fig. 14-1. Wall types.

Fig. 14-2. Basic brick courses.

14-2, center.) When a brick is laid on its edge, it is called a *rolok,* and a rolok with its longest dimension perpendicular to a stretcher course is a *soldier.* (Fig. 14-2, top.) Brick are normally laid from both ends of a course towards the center, and the brick that is used to finish the course is called a *closure* brick. Normally, bricks are laid by following a level *chalk line* strung tightly from one end of the wall to the other; thus, brick are said to be laid *to the line.*

Wall Patterns

Bricks are arranged in their courses to form some sort of *pattern* or recurring design on the completed wall face. This may be done by altering the positions of bricks in a course or in succeeding courses or by a change in the color or texture of the units being used. For example, the use of two or more colors of brick may produce an interesting pattern. Regularly recessing or protruding a unit and the use of a number of different joint treatments also produce interesting effects.

The arrangement of bricks in a course and in succeeding courses is called the *pattern,* a number of the conventional ones being illustrated in Figure 14-3. Recently many modifications of those traditional patterns have been developed, and these modifications are known as *contemporary patterns.*

Mortar Joints

The mortar joints also are designated according to their position in the wall. The horizontal layer of mortar on which a unit is laid is a *bed joint.* The vertical mortar joint between the

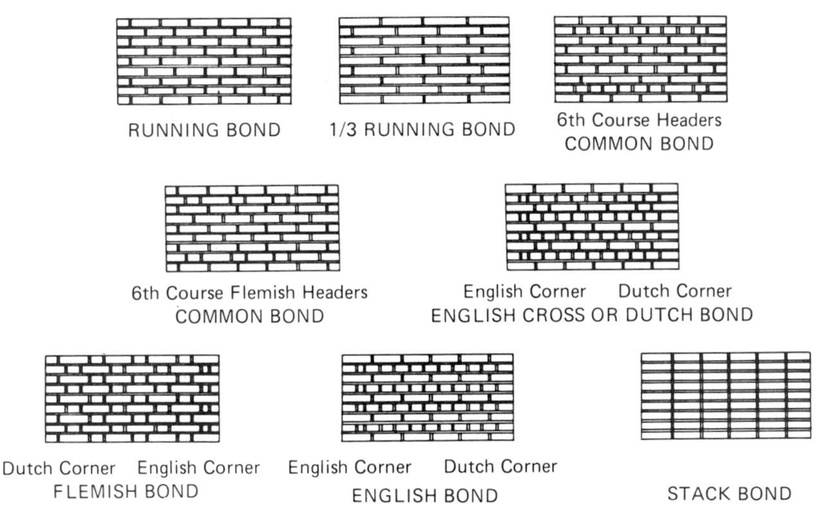

Fig. 14-3. Typical conventional patterns in brick.

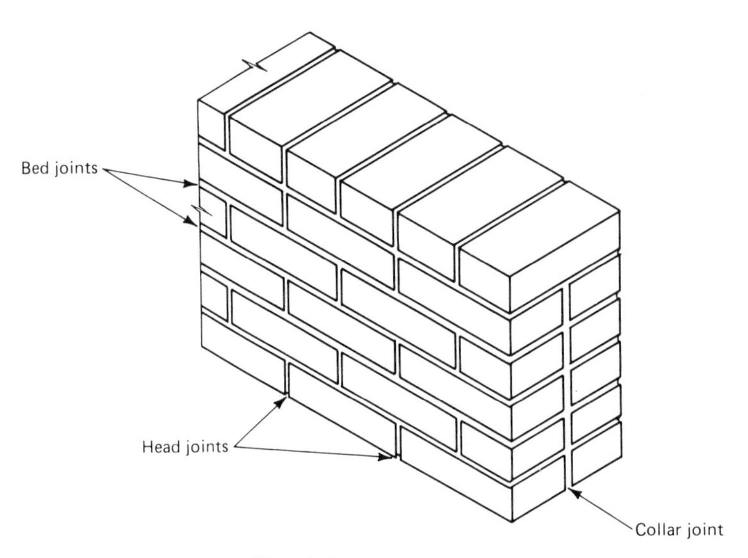

Fig. 14-4. Mortar joints.

ends of units or between units in a soldier course is a *head joint,* while the interior, vertical joint between wythes in a multiwythe wall is a *collar joint.* (See Fig. 14-4.) The actual application of mortar to a brick is known as *buttering.*

The mortar joints may be finished in several different ways, as illustrated in Figure 14-5. Concave, beaded, and v-joints are normally quite small, are formed by a *jointing tool* and, because they have been compacted, quite effectively resist rain penetration. The weathered joint requires careful treatment but is a fairly well compacted and weather-resistant joint. The struck joint, on the other hand, although equally well compacted, provides a ledge for the accumulation of dust and moisture. The flush joint is the easiest to make, but since it is not compacted and therefore not weather-resistant, it is usually used only for backup walls. The raked joint may be compacted but is not recommended where high resistance to moisture penetration is required. In an extruded joint, the mortar is not touched

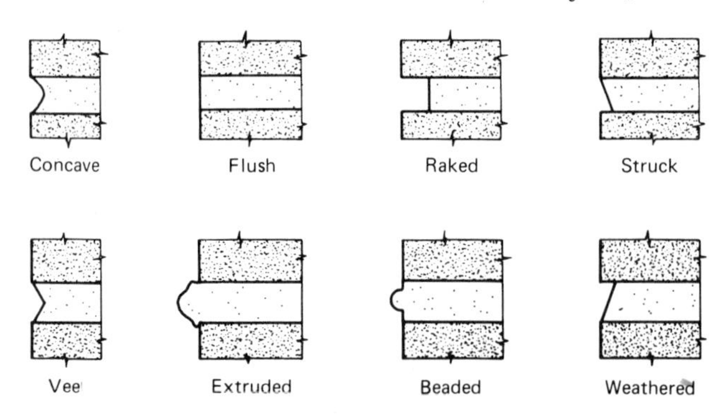

Fig. 14-5. Styles of mortar joints.

but is forced out beyond the face of the wall by the pressure applied during laying of the brick to the line.

MORTAR APPLICATION

In order to obtain good results with masonry construction, it is first of all essential that the proper mortar be used. It should be one of those specified in Table 5-1, which will contribute to good mortar bond and work well on a trowel. It is also important that the brick being used have the proper *initial rate of absorption* (see Chapter 8) in order to obtain good adhesion of mortar to brick. Figure 14-6 illustrates good mortar adhesion, which, when the mortar has set and hardened, should result in a good bond between mortar and brick.

All mortar joints must be completely filled as the bricks are being laid. Failure to do so will result in voids being left in the joint through which water may penetrate. Mortar for the bed joint should be spread thickly. A shallow furrow may be made down the center of the bed (see Fig. 14-7), which will be filled when the brick is bedded, except in the case of load-bearing walls, when no furrow should be made.

Head joints must also be completely filled. For stretcher courses, enough mortar should be applied to the end of the brick to be placed so that, when it is set, some mortar will be forced out at the top of the head joint. (See Fig. 14-8.) In header courses, mortar may be spread over the entire edge of the header to be placed; or one long edge and one cross-edge of both the brick already in place and the one to be placed (see Fig. 14-9) is buttered. Then, as the brick is set to the line, mortar should extrude from the face and the top of the joint. (See Fig. 14-10.)

Closures in both stretcher and header courses need to be carefully laid. Mortar should be spotted on the sides or ends of *both* bricks

Fig. 14-7. Well-laid bed joint mortar.

Fig. 14-8. Mortar appears at top of head joint.

Fig. 14-6. Good mortar adhesion.

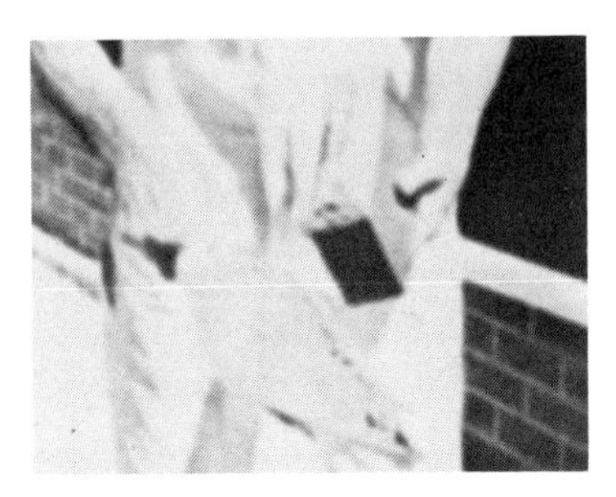

Fig. 14-9. One long edge and one cross-edge buttered.

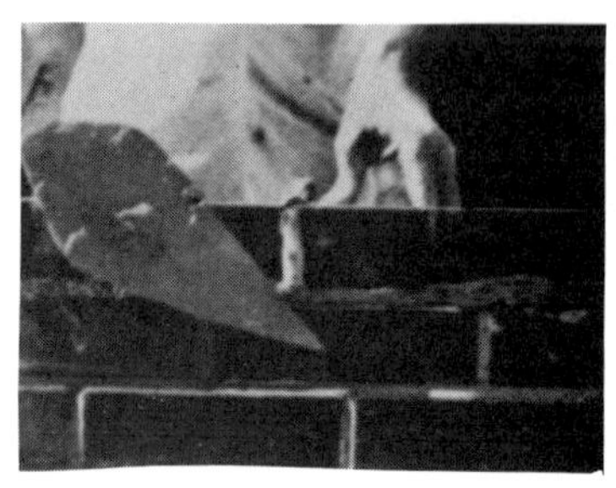

Fig. 14-10. Mortar extrudes from face and top of header joint.

already in place, and both sides or ends of the brick to be placed should be well buttered. Then the closure brick is set in without disturbing those already in place. (See Fig. 14-11.)

After the mortar has partially set, the joints are *tooled* (that is, smoothed or shaped to some pattern, a number of which are shown in Figure 14-5.) Tooling compacts the mortar, making it more dense, and helps to seal any fine apertures between brick and mortar. (See Fig. 14-12.)

Fig. 14-11. Closure brick being set.

Fig. 14-12. Mortar joint before and after tooling.

Basic Procedures for Laying Brick

As indicated earlier, bricks are laid in horizontal courses, using a line as a guide. The line helps the mason build a straight, level and plumb wall and is strung tightly from one corner to the next, sometimes on *corner poles* but more often from *corner leads.*

The first step is to establish lines representing the outside of the wall to be built. In some cases, the outside edge of the foundation will be that line, but in others, a chalk line is snapped on the surface of the footing or foundation to indicate the outside line of the wall. (See Fig. 14-13.)

It is important at this point to check to see that the foundation does not deviate from "level" to the degree that the first bed joint measures less than 6 mm (¼ in) or more than 19 mm (¾ in) thick at any point along its length. If a check reveals that such would be the case, then the surface of the foundation must be levelled with a layer of load-bearing grout, or the lower surface of the units in the first course will have to be cut to accommodate the unevenness.

Leads are usually built at each corner to establish the proper course height and to provide a means of attaching the mason's line. (See Fig. 14-14.) The height of each course is established by a *storey pole* with the course heights carefully marked on it or by a *mason's rule.*

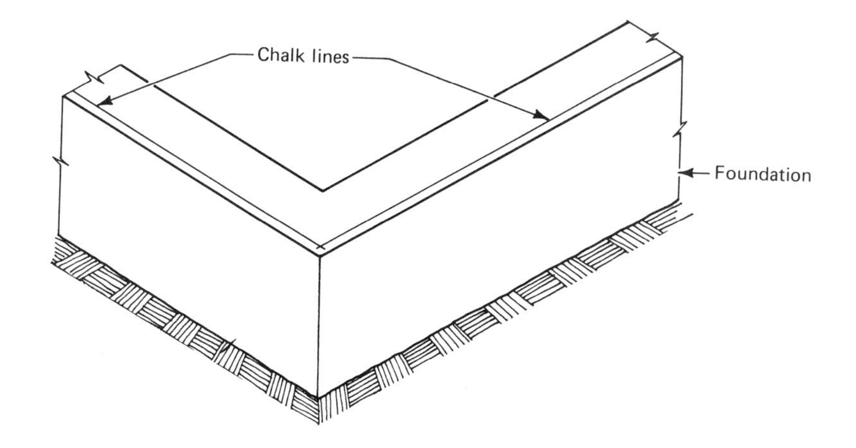

Fig. 14-13. Chalk line on foundation wall.

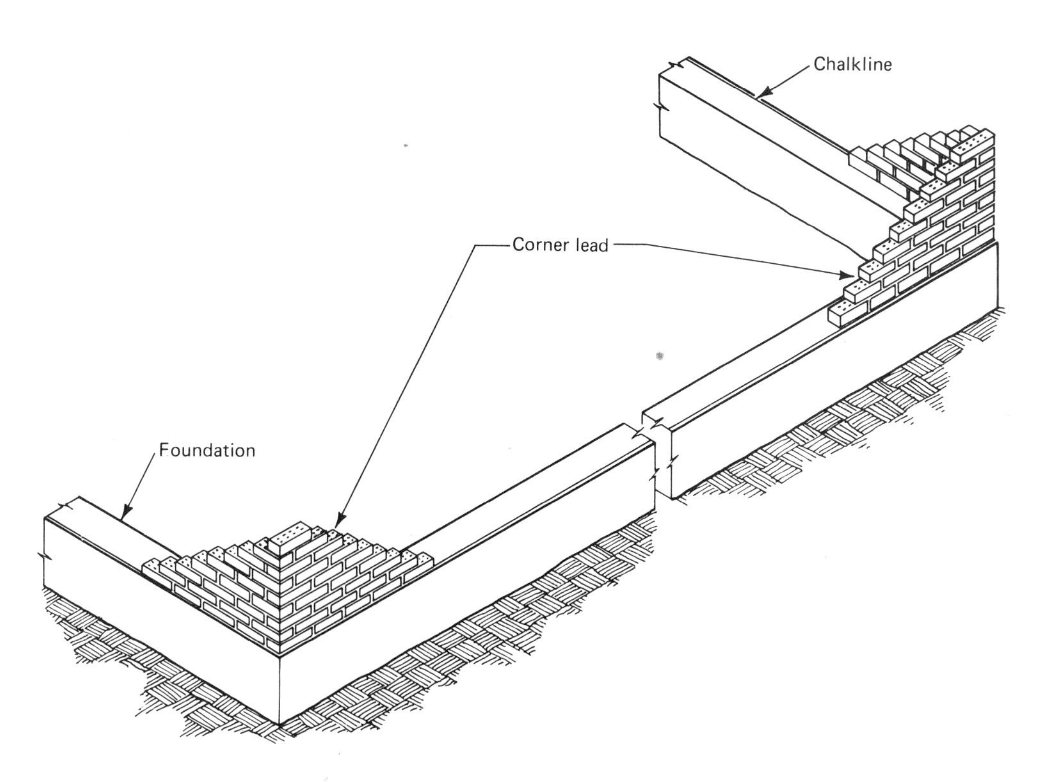

Fig. 14-14. Brick corner leads.

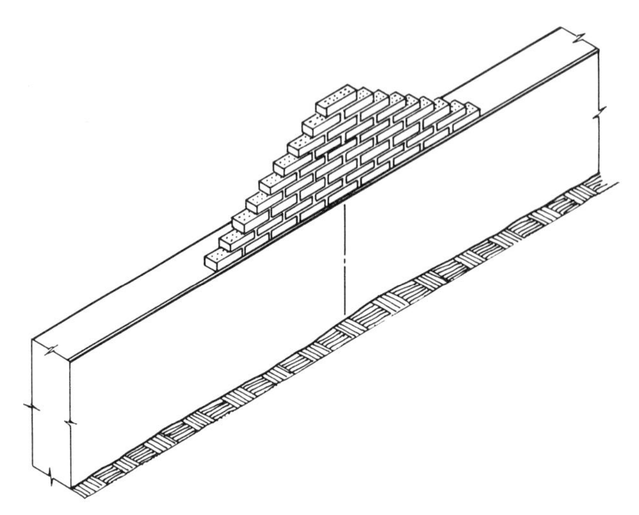

Fig. 14-15. Rack-back lead at center of wall.

Lines are strung from special *line blocks* made to attach to the lead corners or from *pins* driven into head joints in the corner leads at the proper levels. The line should be supported away from the face of the corner leads by about 1.5 mm ($^1/_{16}$ in) so that bricks do not actually touch it as they are being laid.

In the case of long walls, it may be necessary to build a *rack-back lead* in the center of the wall to support the line. (See Fig. 14-15.) It is of prime importance that all of these leads be laid up very accurately to establish proper course levels and to ensure that the wall will be straight and level and that mortar beds are maintained at the correct thickness.

As the walls are built up, new corner leads are

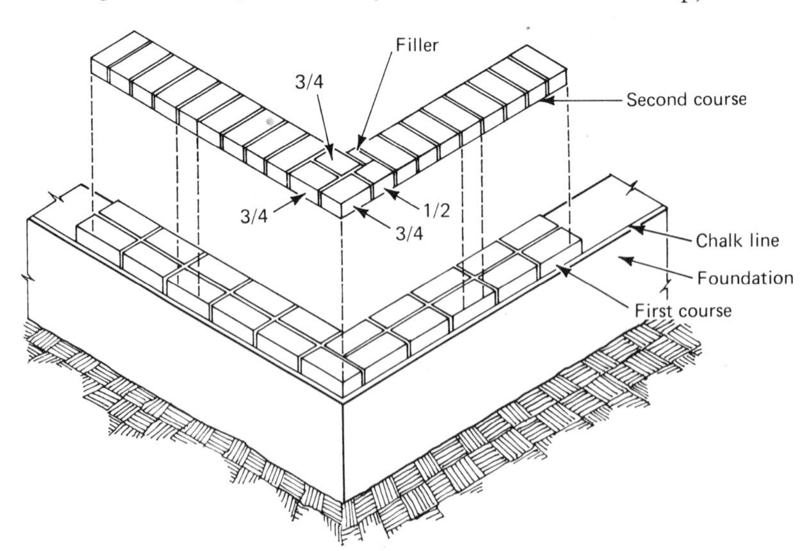

Fig. 14-16. First and second courses of corner lead for common bond pattern with sixth course headers.

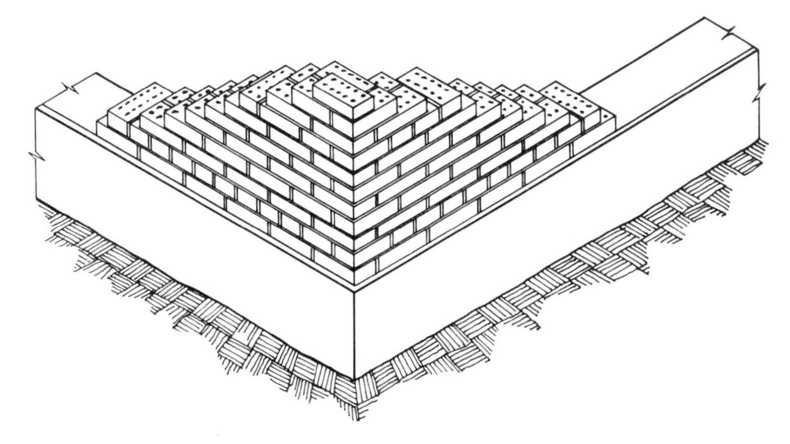

Fig. 14-17. Completed corner lead for common bond with sixth course headers.

made, and laying continues until the full height of the wall is reached.

The corner leads for a two-wythe wall will be somewhat more complicated and may require the cutting of some units. For example, the first two courses of a corner lead for a 200 mm (8 in) wall laid in common bond with sixth course headers may be arranged as illustrated in Figure 14-16. The complete eleven-course lead for the above wall is illustrated in Figure 14-17. The corner leads for single-wythe or multiwythe walls using other patterns (see Fig. 14-3) should be planned in similar detail.

Structural Bonding With Brick

When the individual units in a wall are interlocked or tied together so that the entire mass acts as a structural unit, *structural bonding* occurs. This is accomplished when the units are *overlapped* on one another (see Fig. 14-15), when a *header course* is used or when *metal ties* are introduced into the wall.

Overlapping occurs when the head joints of successive stretcher courses are offset from those in the course below, often by half the length of the stretcher. (See Fig. 14-14.) A header course is laid *across* a multiwythe wall, as illustrated in Figure 14-18.

Metal ties are an additional means used to hold the units in a wall together. Walls may be tied longitudinally, and multiwythe walls must also be tied laterally. Longitudinal ties are galvanized sections of wire that are referred to as *joint reinforcement*. (See Fig. 14-19.) Lateral ties may be continuous or they may be individual pieces of galvanized metal (Fig. 14-19).

Ties are placed in the bed joints in a full bed of mortar. When used in cavity walls, lateral ties should be of sufficient width to overlap at least 40 mm ($1^5/_8$ in) onto each wythe. They

Fig. 14-18. Brick headers with block backup. (*Courtesy Portland Cement Assn.*)

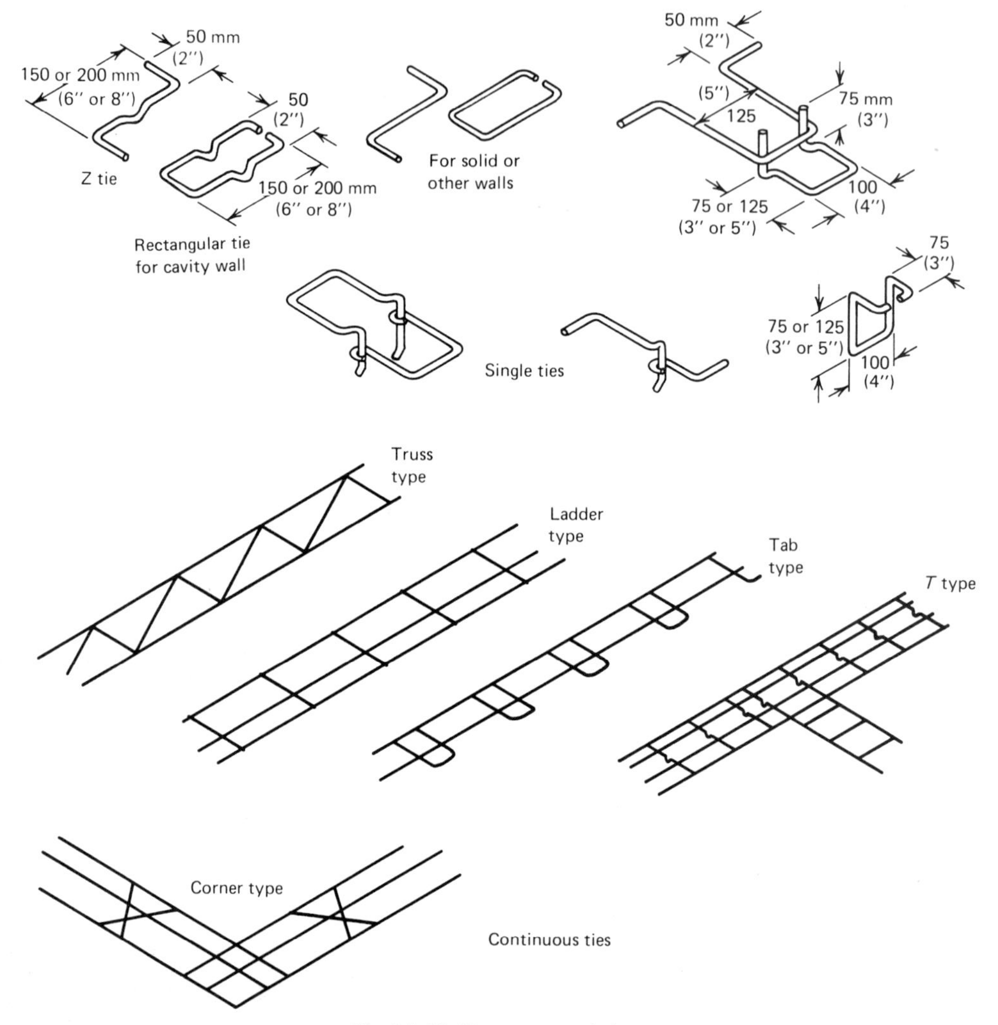

Fig. 14-19. Masonry metal ties.

are usually about 5 mm (0.2 in) in diameter and are crimped to allow moisture to drip, although crimping may lead to a significant reduction in strength. In grouted walls, the crimp is not required. (See Fig. 14-20.)

Types of Wall Ties

Ties are either *unit* or *continuous.* Unit ties may be *rectangular* or *Z-shaped* (see Fig. 14-19) and are either *adjustable* for joints of different heights or *nonadjustable* for joints of equal heights. The rectangular ties should have a minimum width of 50 mm (2 in) and have welded joints if less than 75 mm (3 in) wide. Z-ties should have 50 mm. 90° legs at either end. Unit ties are usually spaced not more than 1000 mm (40 in) o.c. horizontally and 500 mm (20 in) vertically. Extra ties should be placed within 300 mm (12 in) of the edge around the perimeter of an opening.

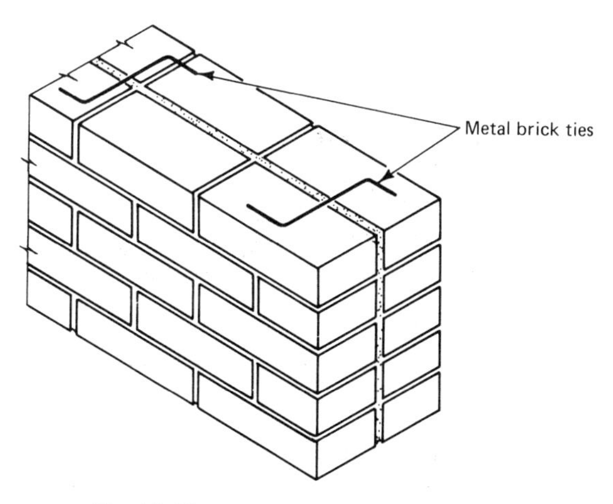

Fig. 14-20. Wythes tied with metal ties.

Corrugated ties are commonly used to tie masonry veneer to a wood backup wall, using at least one tie for each 0.185 m^2 (2 ft^2) of wall area. The nails for such ties should be placed near the bend in the tie to diminish movement. (See Fig. 14-21.)

Continuous metal ties that provide both longitudinal and lateral reinforcement are generally of the *truss, ladder, looped ladder* or *X type.* (See Fig. 14-19.) The width of the tie should be 50 mm (2 in) smaller than the width of the wall. The number of cross-wires is a function of the wire size and the type of wall, though there will usually be one for each 0.185 m^2 (2 ft^2) of wall area.

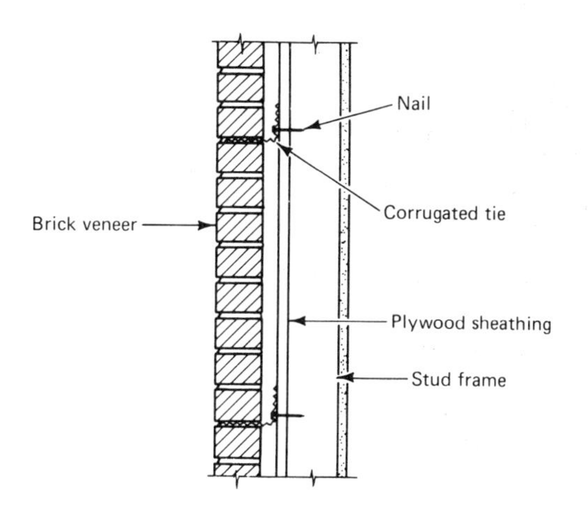

Fig. 14-21. Corrugated ties for wood backup wall.

Flashing

The purpose of flashing is to exclude moisture from entering a wall or to direct any moisture that may penetrate and return it to the outside. Impermeable products such as *plastic sheet, bituminous membrane materials,* or *sheet metal* are used for the purpose, with 3000 to 6000 g/m^2 (10 to 20 oz/ft^2) sheet copper being the generally preferred material for the best work.

Two types of flashing are in general use—*external* and *internal.* External flashing prevents the entrance of water into a wall where it meets a relatively flat surface, such as a roof. Usually external flashing consists of two members—the *base flashing* at the bottom and the *counter* flashing that is built into the vertical surface and turned down over the base flashing. (See Fig. 14-22.)

Internal flashing is built into and usually concealed within the wall to control the travel of moisture and direct it to the exterior surface. It will consist very often of only one piece. (See Fig. 14-23.)

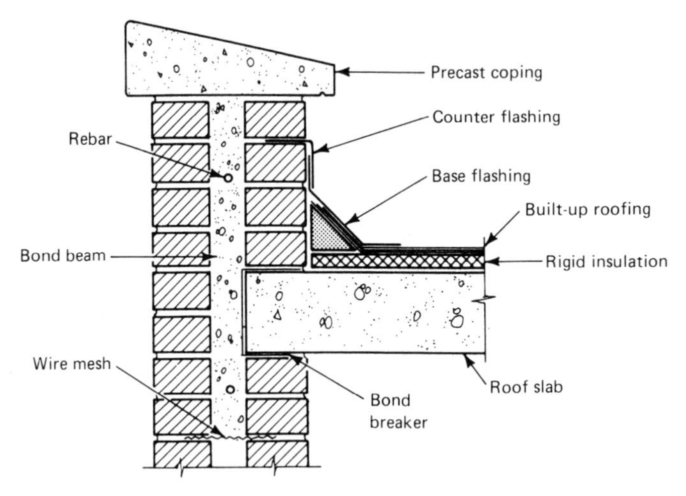

Fig. 14-22. External flashing for brick wall.

There are a number of vulnerable points in masonry wall assemblies that require flashing, including the *junction of wall and flat roof*, the *junction of parapet wall and flat roof* (Fig. 14-22), *projections and recesses* in a wall, *bases of masonry walls* at spandrel beams (Fig. 14-23) and at the head and sills of openings in the wall. (See Fig. 14-24.)

Weep Holes

Although internal flashing will direct moisture toward the exterior surface, it must be given a passage through which to escape, and this is provided by the use of *weep holes.* They are openings located in the head joints of the outer wythe immediately above the

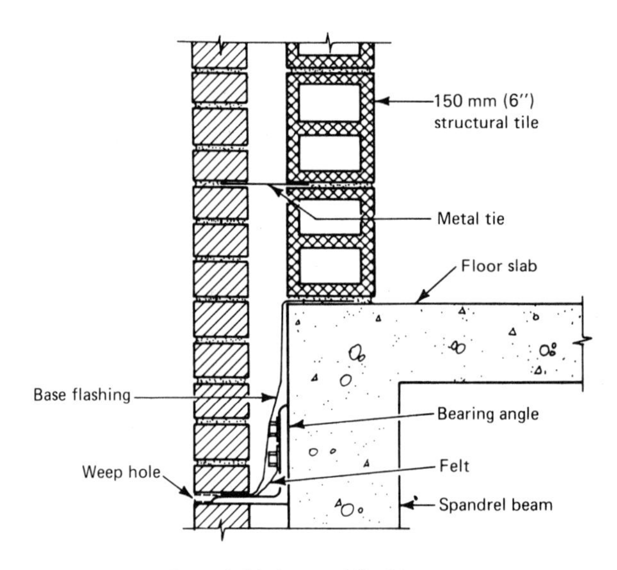

Fig. 14-23. Internal flashing.

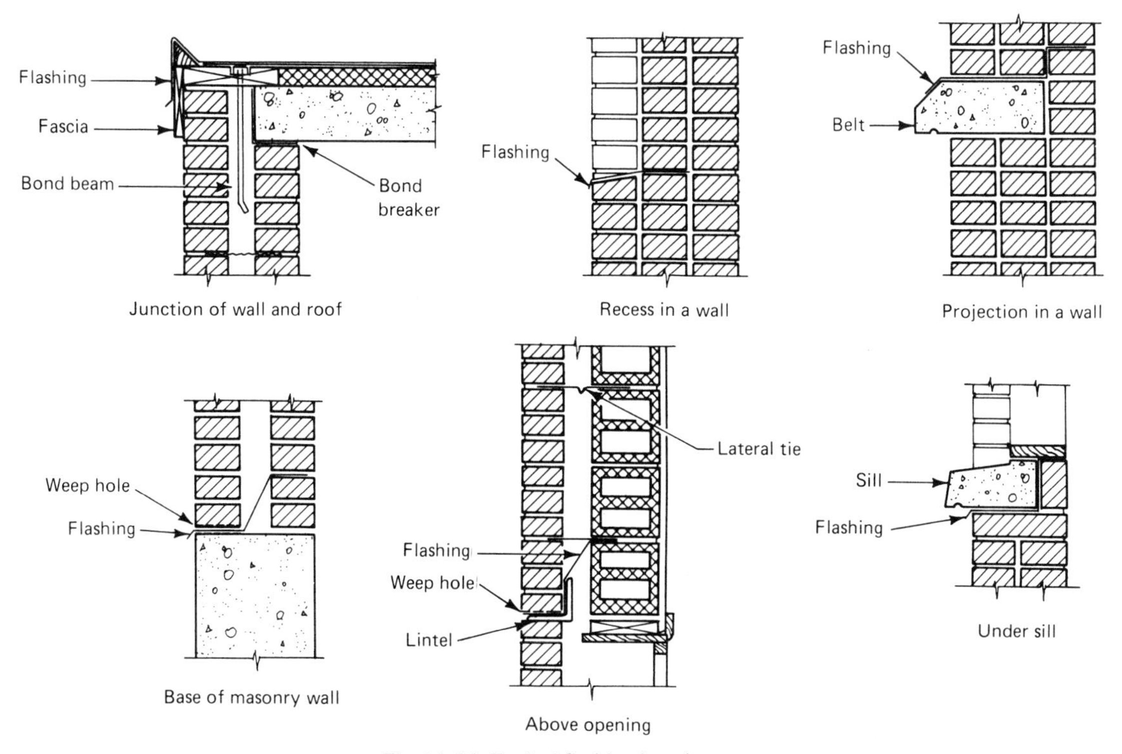

Fig. 14-24. Typical flashing locations.

flashing and are normally spaced at 600 mm (12 in) o.c. (See Fig. 14-25.)

Weep holes are formed by omitting the mortar from all or part of the joint, by inserting oiled rods into the joints, to be removed later, or by inserting such materials as sash cord, small plastic tubes or fiberglass pads into the joints and allowing them to remain. Such a material as sash cord will act as a wick, drawing moisture out of the cavity. Weep holes are required at all flashing locations except at the tops of walls.

Control Joints in Brick Walls

As indicated in Chapter 12, control joints are breaks in the continuity of a wall, designed to relieve excessive stress resulting from movement in the wall. Though most control joints will be vertical, consideration should be given to horizontal control joints in masonry that is carried on a structural frame, as there is differential movement in such a structure, particularly with a reinforced concrete frame.

Fig. 14-25. Weep holes in brick wall.

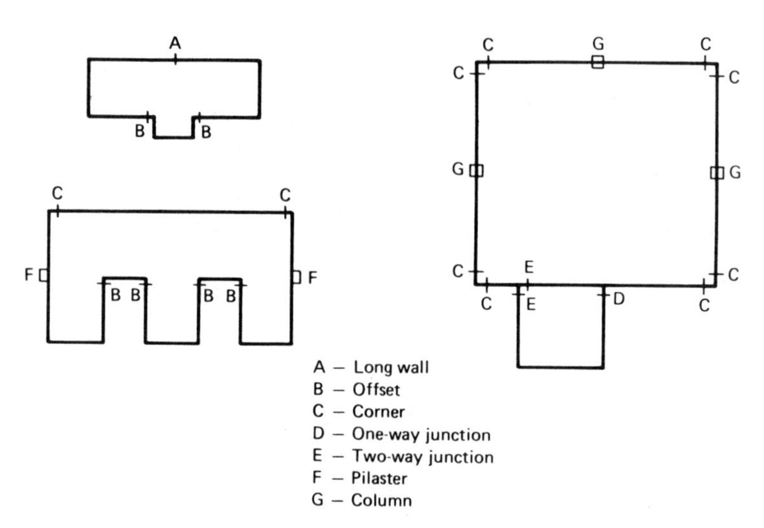

Fig. 14-26. Control joint locations.

This joint will occur at the bearing angles and can be constructed by placing roofing felt above and below the angle and sealing the joint with an elastic sealant. (See Fig. 14-23.)

Typical vertical control joint locations are indicated in Figure 14-26.

Very careful attention must be paid to the construction of a vertical control joint. All units must be butted to the joint, which means that some of them may have to be cut. The joint itself must be kept plumb and be maintained at the same width throughout its length.

A bond breaker, often building felt, is used to ensure that the meeting ends of the units at the joint will be free to move. Lateral ties on either side of the joint keep the wall in alignment.

The joint opening may be stopped with a compressible filler such as *preformed copper water stop, premolded foam rubber padding,* a *formed neoprene gasket* or an *extruded plastic stop.* (See Fig. 14-27.) The outer portion of the joint opening is caulked with an elastic joint sealant.

Figure 14-28 illustrates the construction of some typical control joints for various locations in a wall assembly.

Intersecting Walls

Intersecting walls must be securely tied together to ensure that building loads can be safely transmitted from one to the other. This may be done by mortar bonding the end of one wall into a *chase* left in the other, by *butting* one to the other and tying the two together with metal ties or by *brick bonding.* (See Fig. 14-29.)

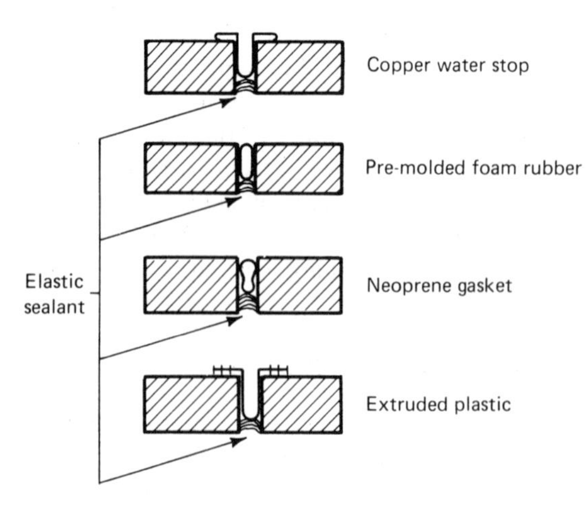

Fig. 14-27. Compressible control joint fillers.

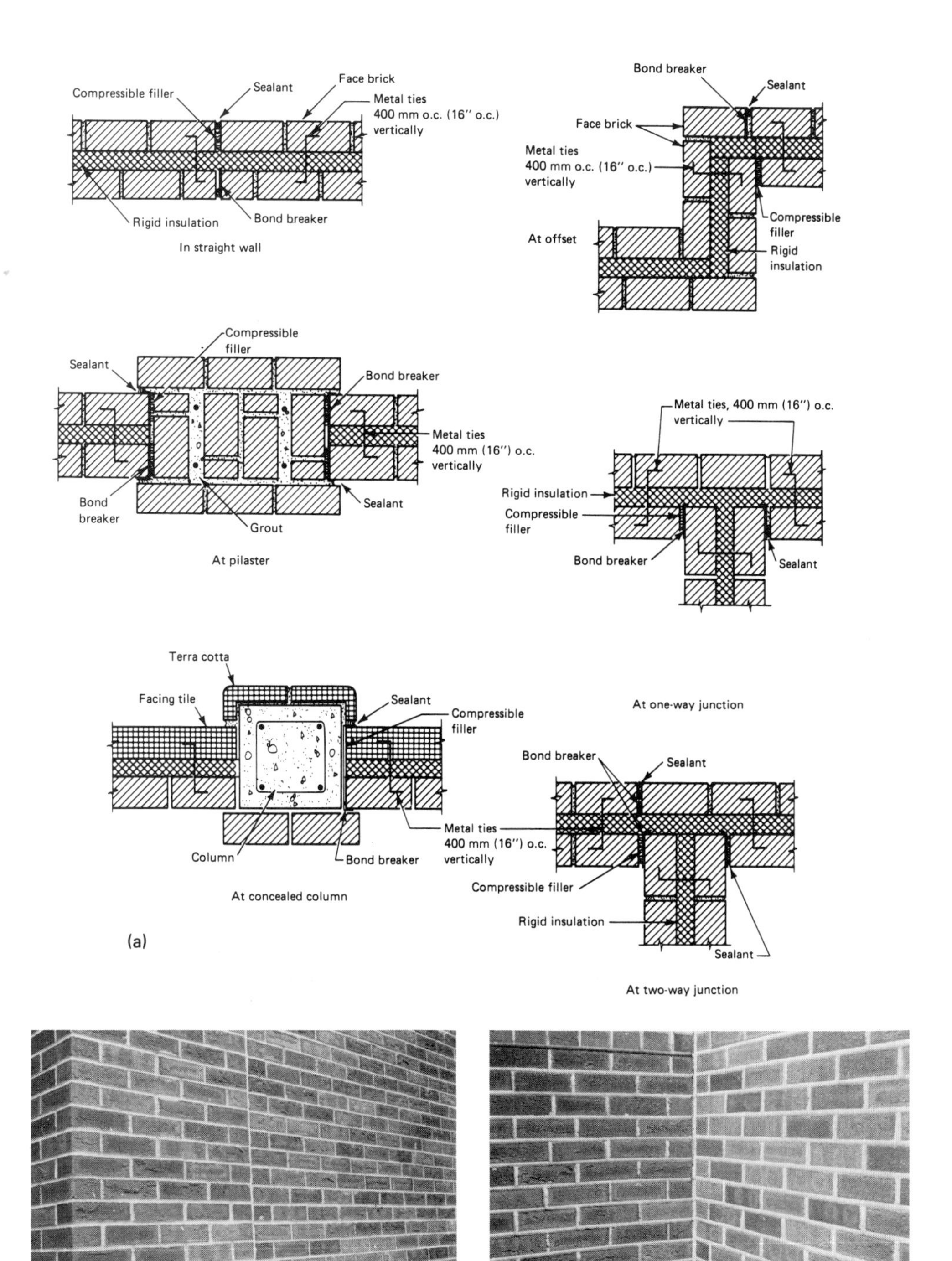

Fig. 14-28. (a) Control joints at various locations; (b) control joint in straight wall; (c) control joint at offset.

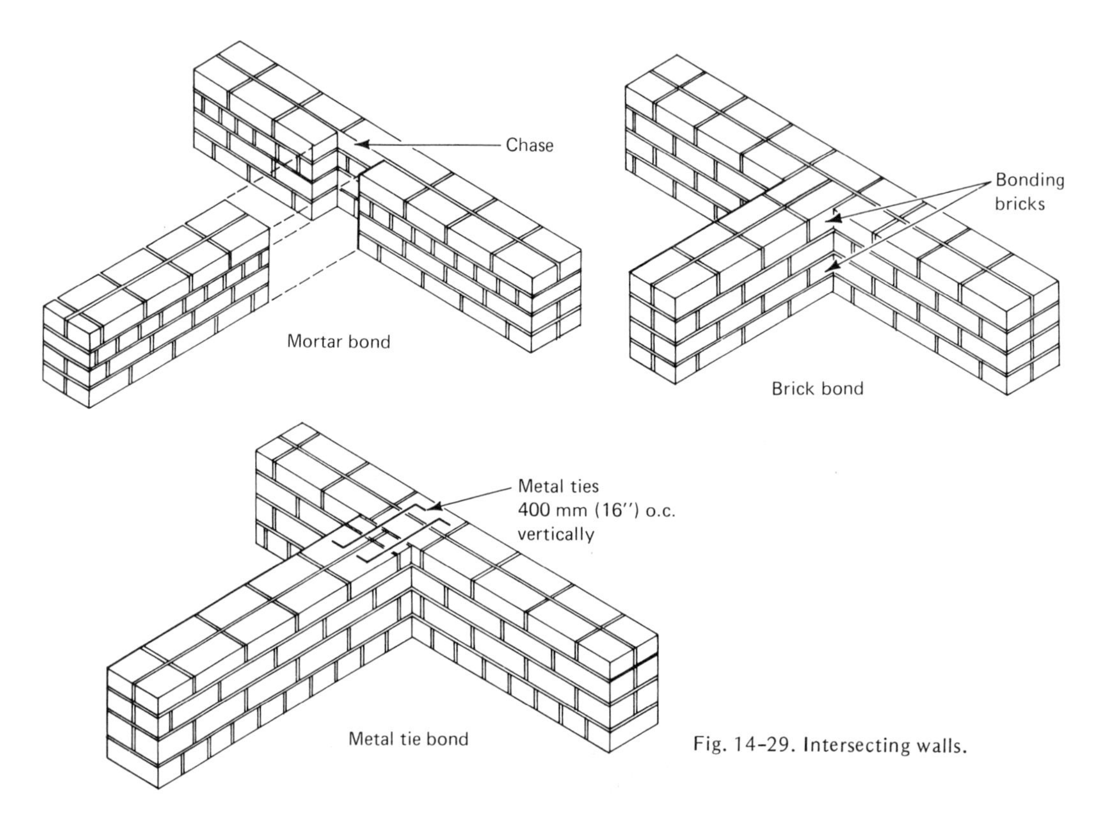

Fig. 14-29. Intersecting walls.

The location of the intersection in the building is important since that may determine whether the connection is to be a *solid tie* or a *hinged control joint.* If it is likely that there will be movement between the two walls, the control joint type of intersection will be required. (See Fig. 14-28, one-way junction.)

Corners

Wall corners in brick masonry structures usually require some special consideration, depending on the wall pattern and the thickness of the units. If the units being laid have nominal thicknesses of 100 or 200 mm (4 or 8 in) and are laid in running bond, there should be no problems. But with other patterns or other widths the construction may have to involve the inclusion of extra pieces or the cutting of a unit into a shorter length. (See Fig. 14-16.)

Figure 14-30 illustrates some typical corner construction with brick masonry.

Piers, Pilasters and Buttresses

Piers are isolated, free-standing unit masonry elements of a structure, intended to support vertical concentrated loads. The number required and their cross-sectional area will depend on the magnitude of the load that they must support. The height of a pier is restricted to not more than three times the least lateral dimension. Typical brick piers are illustrated in Figure 14-31.

Pilasters and buttresses, on the other hand, are built into and form integral parts of masonry walls. In addition to supporting vertical concentrated loads, they are used to provide additional stiffness and lateral support for walls.

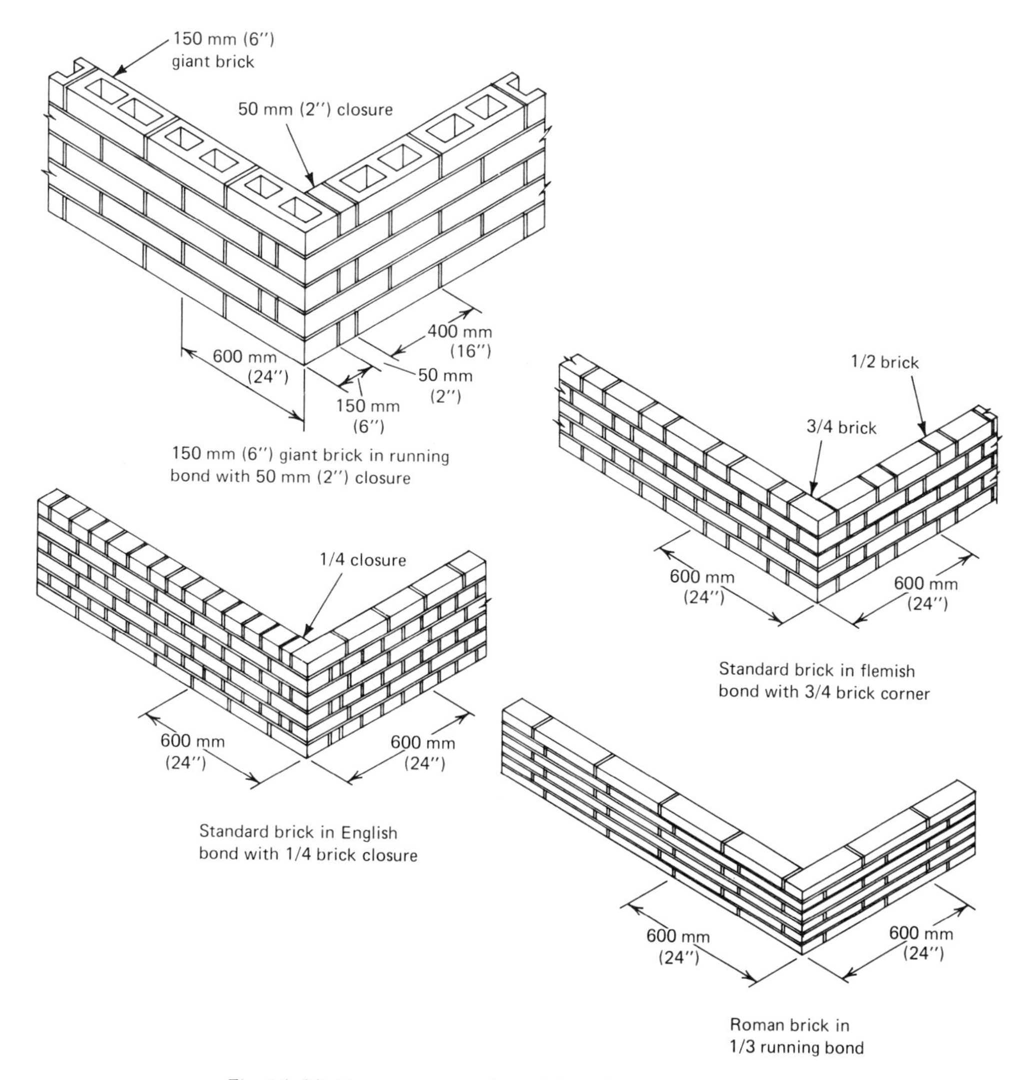

Fig. 14-30. Corner construction with various types of brick.

Pilasters may project beyond the wall plane on one or both sides, with the outer face or faces parallel to the wall. Some typical pilaster layouts are shown in Figure 14-32.

The primary purpose of a buttress is to provide lateral support for a wall; consequently, to do this most efficiently, its outer face is tapered from bottom to top, as illustrated in Figure 14-33. The taper is achieved by setting back each course a sufficient amount to keep the top edge of that course in line with the designed slope, as shown in the enlarged portion of Figure 14-33.

In addition to its lateral support role, a buttress will normally also carry concentrated vertical loads.

Fig. 14-31. Brick piers.

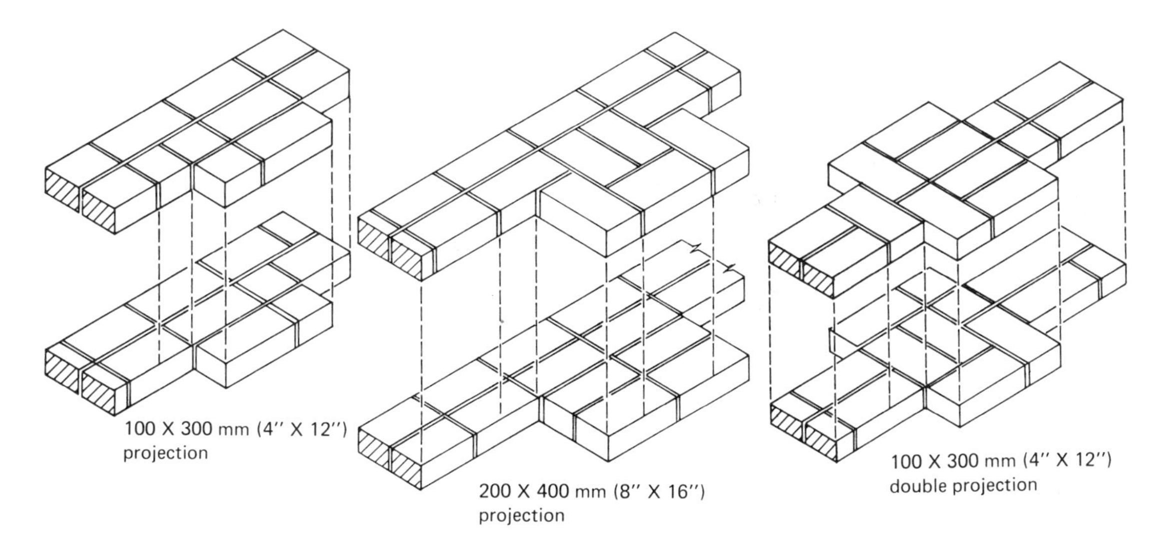

Fig. 14-32. Brick pilaster layouts.

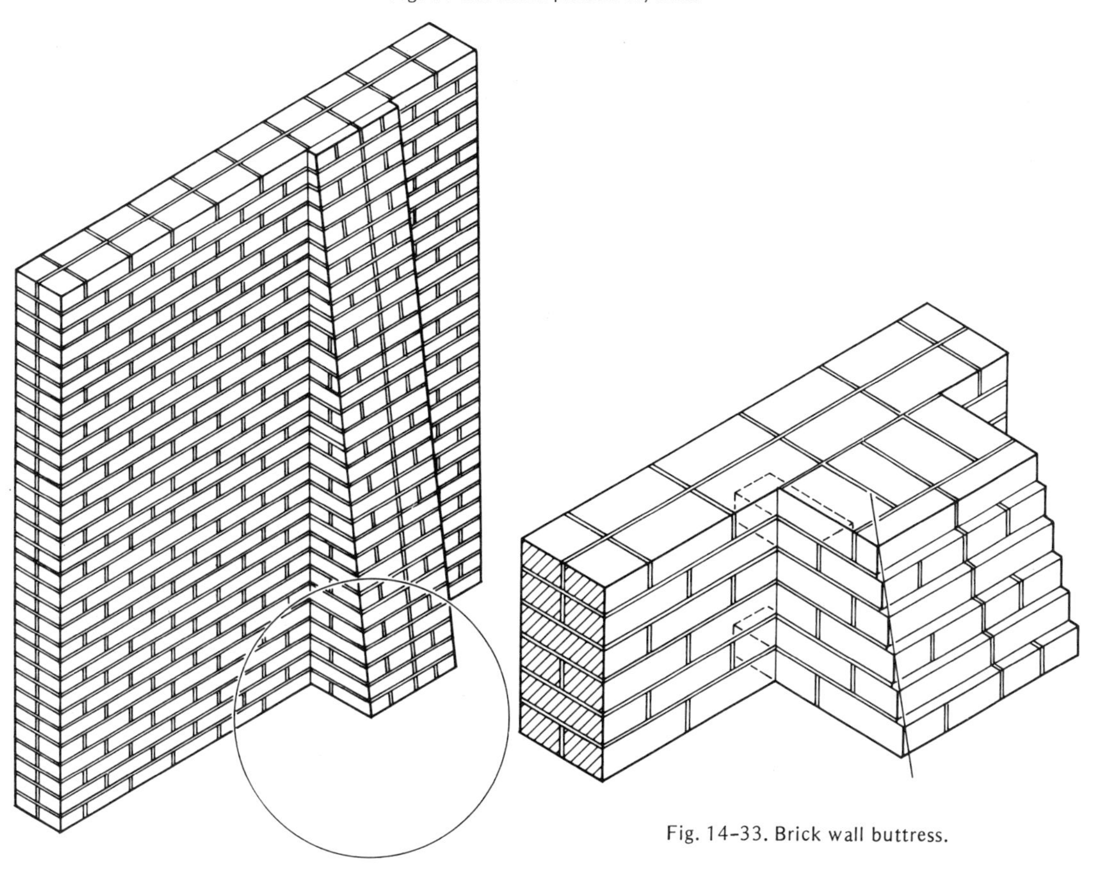
Fig. 14-33. Brick wall buttress.

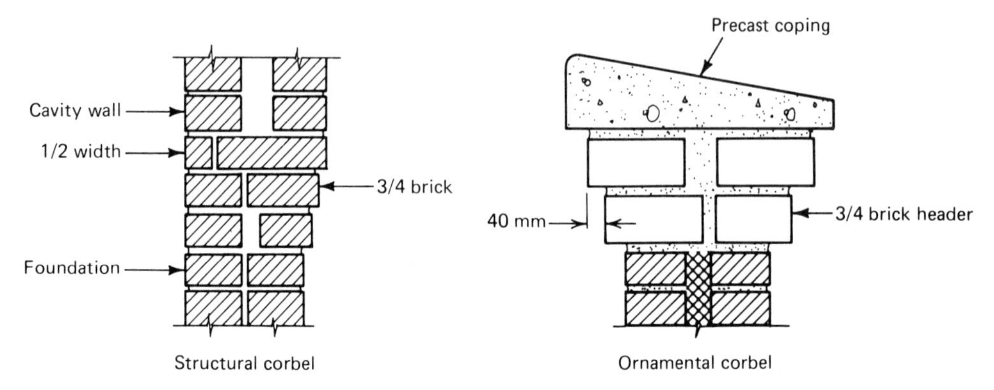

Fig. 14-34. Brick corbels.

Corbels

A *corbel* is a projection of masonry units beyond the regular vertical plane of the wall, in order to form a shelf or ledge. It may be strictly decorative or it may be used to thicken a wall to suit a particular situation.

Brick courses used to make a corbel may be either header or stretcher courses, although headers will provide greater strength by tying further into the wall than they extend beyond it. In any case, the maximum projection for one course should not exceed one-half the unit's height and, as a rule, the total projection of the

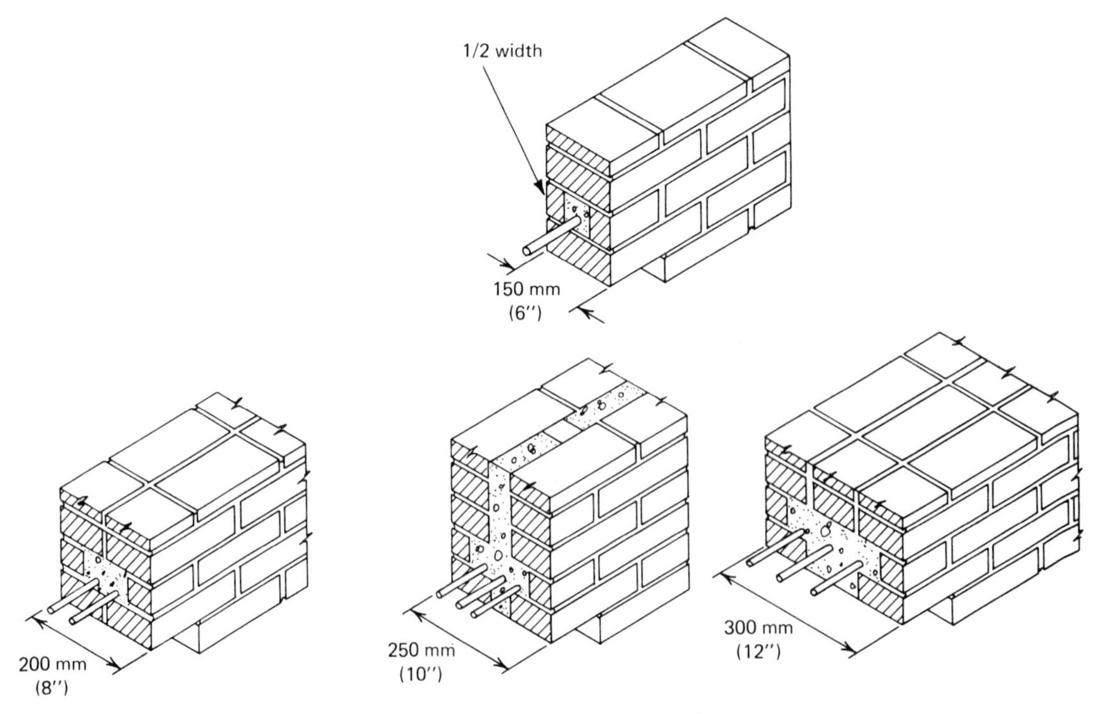

Fig. 14-35. Reinforced brick lintels.

corbel should not exceed half the thickness of the wall. (See Fig. 14-34.) It is particularly important when building a corbel to use well-filled mortar joints and to ensure the best possible bond.

Lintels

A *lintel* is a horizontal member placed over a wall opening to carry the superimposed loads from above. In brick masonry, it may be a reinforced brick member, a steel member covered with brick, or a reinforced concrete member.

The size of the lintel should always be governed by an analysis of the loads involved, in order to avoid cracking due to excessive deflection resulting from inadequate design.

A reinforced brick lintel is formed by introducing reinforcing bars into the masonry. In some cases the design may simply require bars that can be laid in the bed joint between the *soffit course* and the one above it. In other cases a cavity is built, reinforced with the required number of bars and filled with grout. (See Fig. 14-35.)

The usual procedure is to provide temporary shoring upon which the soffit course is placed, with mortar in the head and collar joints only. (See Fig. 14-36.) The remainder of the lintel courses and the reinforcing required are then built on that soffit. After the mortar or grout is sufficiently cured, the shoring is removed.

Steel lintels may consist of one or more steel angles to span the opening or, for wider spans and heavier loads, a rolled shape with a suspended soffit plate may be used. (See Fig. 14-37.) In any case, the lintel should have a minimum bearing of 100 mm (4 in) on the masonry at the sides of the opening.

Brick Arches

Masonry *arches* are an alternative to lintels as a means of spanning openings in masonry walls and supporting the superimposed loads.

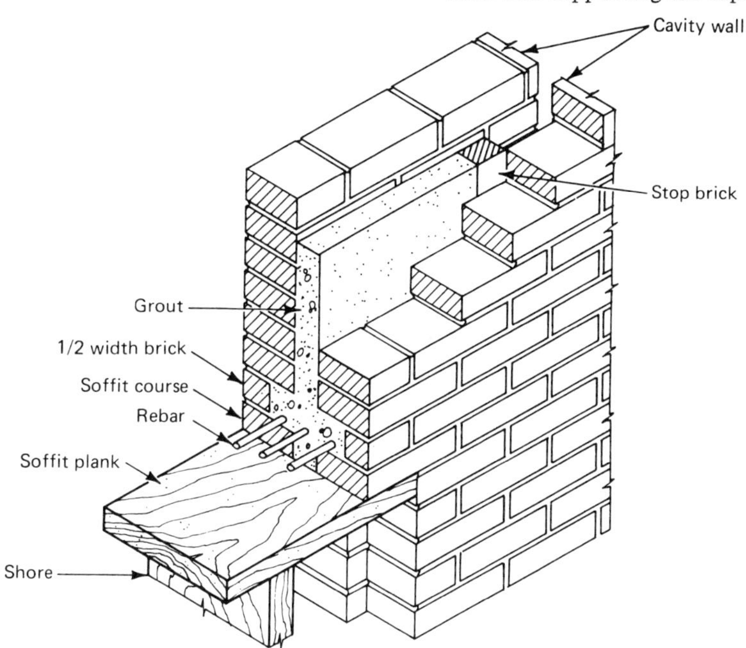

Fig. 14-36. Shoring for reinforced brick lintel.

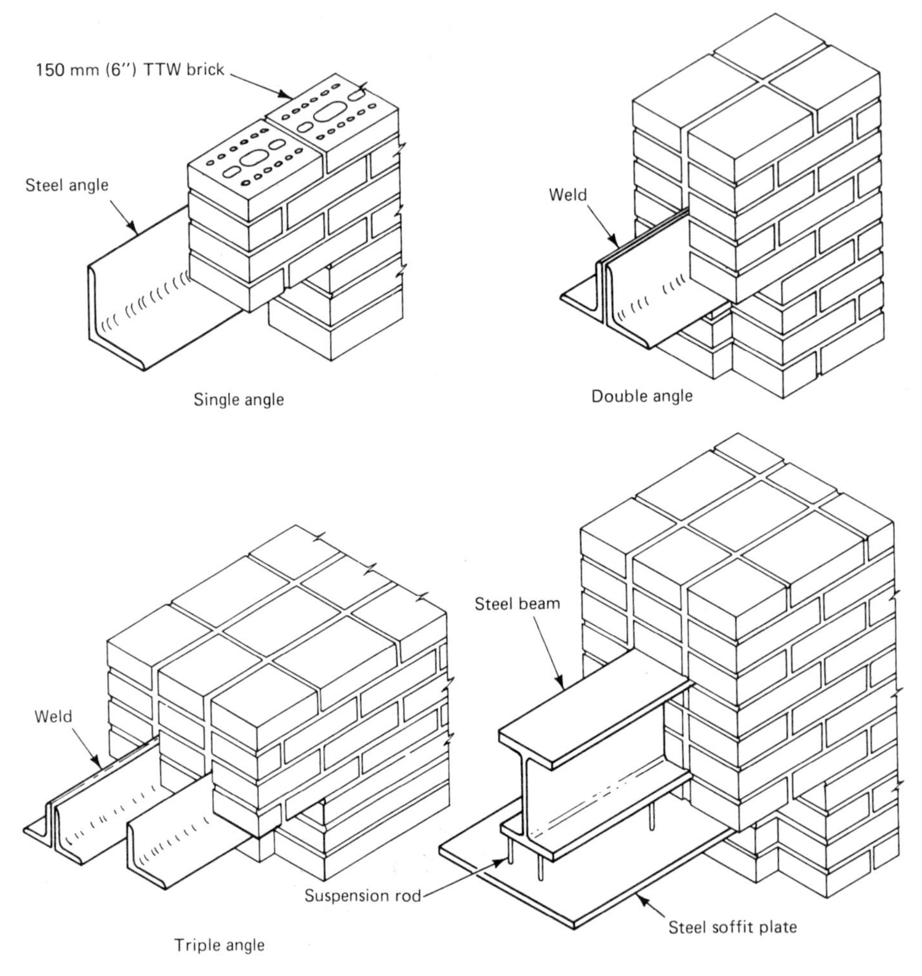

Fig. 14-37. Steel lintels for brick structures.

Arches were developed in ancient times and their use has continued to the present, with a number of well-known arch shapes having evolved over the years. Those shapes include: the *jack arch,* a flat arch commonly having a camber (curvature) of 3 mm (1/8 in) per 300 mm (12 in) of span; the *segmental arch,* with a curve that is an arc of a circle less than a semicircle; the *semicircular arch,* whose curve is a semicircle; the *multicentered arch,* whose curve consists of arcs of two or more circles, normally tangent at their intersections; the *tudor arch,* a pointed, four-center arch, with medium rise-to-span ratio; the *gothic arch,* with relatively high rise-to-span ratio, whose sides consist of arcs of circles of the same radius and whose centers are located on the springline; and the *parabolic arch,* whose curve is a parabola. (See Fig. 14-38.)

Arches are normally divided into two classes,

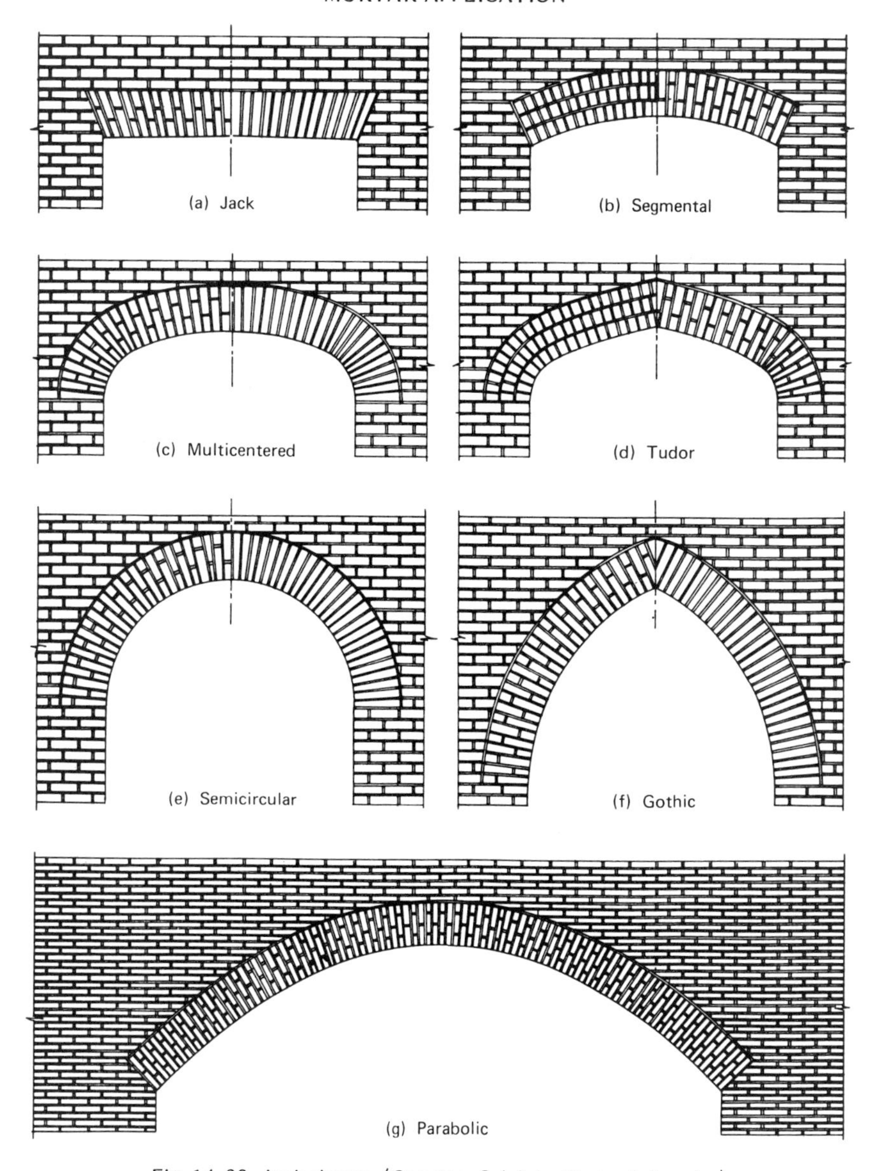

Fig. 14-38. Arch shapes. (*Courtesy Brick Institute of America*)

minor and *major,* depending on span, rise and loading. Minor arches are those with a maximum span of 1830 mm (72 in), maximum rise-to-span ratios of 0.15 and maximum equivalent uniform load of approximately 16 kN/m (1100 lbs/ft). Major arches are those with *spans, rise-to-span-ratios* or *loadings* exceeding the maximum for minor arches.

Generally, segmental, multicentered and jack arches are used in minor arch construction, while the parabolic arch is often the most efficient for a major arch from a structural standpoint, when subjected primarily to uniformly distributed loads.

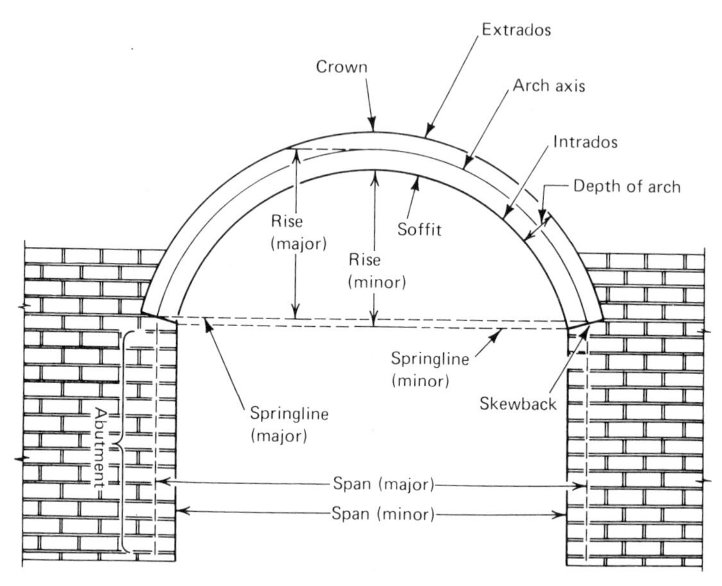

Fig. 14-39. Arch terminology.

Arch Terminology

A number of specific terms are used in connection with the layout and construction of masonry arches, as indicated in Figure 14-39.

The ends of an arch are supported by *abutments,* a portion of the wall that consists of a *skewback* and the masonry that supports it.

A skewback is the inclined surface at which the arch joins the supporting abutment. It may consist of a precast or stone shape set into the wall (see Fig. 14-40) or it may simply consist of the ends of the units at the top of the abutment cut to the proper angle. (See Fig. 14-39.)

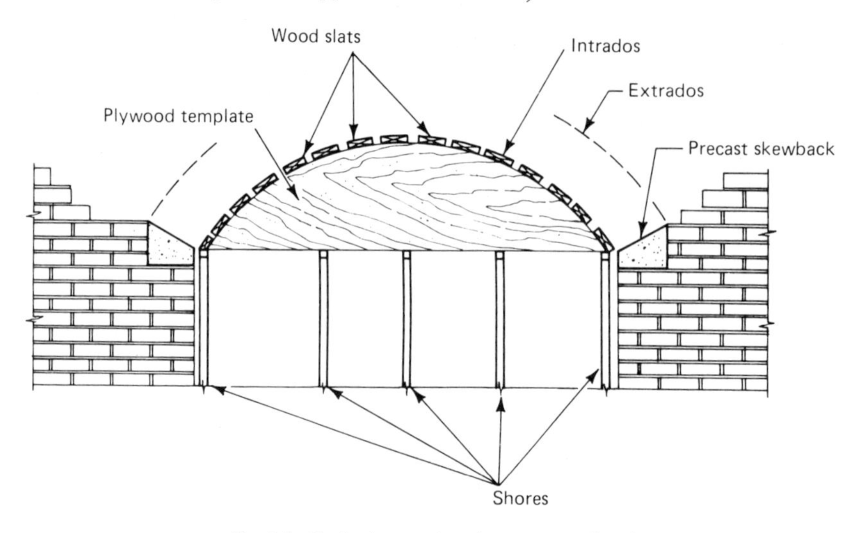

Fig. 14-40. Arch template for segmental arch.

The *springline* of an arch is a horizontal line that joins the intersection of the skewbacks and the soffit (the under-surface of the arch) in minor arches. In major arches, it joins the intersection of the *arch axis* and the skewbacks. The arch axis is the median line between the *intrados* and the *extrados* of the arch. The intrados is the concave-curved line that bounds the lower surface of the arch, while the extrados is the convex curve which outlines the upper extremities of the arch. The apex of the extrados is the arch *crown*.

The *span* of an arch varies somewhat, depending on the class to which it belongs. For a minor arch, the span is the clear span of the opening. For a major arch, the span is the distance between the ends of the arch axis at the skewback.

The *rise* of a minor arch is the maximum height of the intrados above the level of the springline, while the rise of a major arch is the height of the apex of the arch axis above its springline.

The *depth* of an arch is measured on a line that is perpendicular to a tangent of the arch axis.

Arch Construction

Arches are built with the aid of wooden templates, made to conform to the shape of the arch intrados. (See Fig. 14-40.) The template is supported by shores, and the whole structure must be strong enough to carry the dead load of the arch and other loads that may be imposed before the arch has gained sufficient strength to support itself and the imposed loads. It is recommended that, for plain masonry arches, under normal curing conditions, the shoring remain in place for seven days after the arch is completed. In cases where the majority of the load will not be applied until later on in the construction of the building, that time may be reduced somewhat.

There are two methods commonly used to build masonry arches. In one, wedge-shaped masonry units, known as *voussoirs,* are used, so that mortar joints of constant thickness are obtained. (See Fig. 14-38, e.) In many cases, those special shapes may be obtained by cutting standard rectangular units at the jobsite. In the other, units of uniform thickness are used, and the joint thickness is varied to obtain the desired curvature. The method to be used in a particular case will be determined by the arch dimensions, the appearance desired, the availability of special shapes or the availability of machinery to cut standard units to shape.

Brick masonry arches are usually built in such a way that the units will be laid in soldier or rolok header bond. (See Fig. 14-38.) It is very important that all mortar joints be completely filled. This may be difficult to do with an arch that has a short radius and in which mortar joints of varying thicknesses are to be used. In such a situation, it is better to use two or more rings of rolok headers, rather than soldier courses. In addition to making it easier to achieve full mortar joints, the headers provide a bond through the wall, thus strengthening the arch. (See Fig. 14-38, b.)

Type **M**, **S** or **N** mortar is recommended for use in arch construction. (See Table 5-1 and Table 5-3).

Bond Beams

As indicated in Chapter 12, a *bond beam* is a horizontal, reinforced member that is introduced into a masonry wall to provide added resistance to cracking resulting from movement. It is normally incorporated into the structure at the base of the wall and at floor and roof levels. In addition to aiding in tying the wall together, bond beams also tend to distribute vertical concentrated loads that may be imposed upon the wall.

There are a number of ways by which a bond

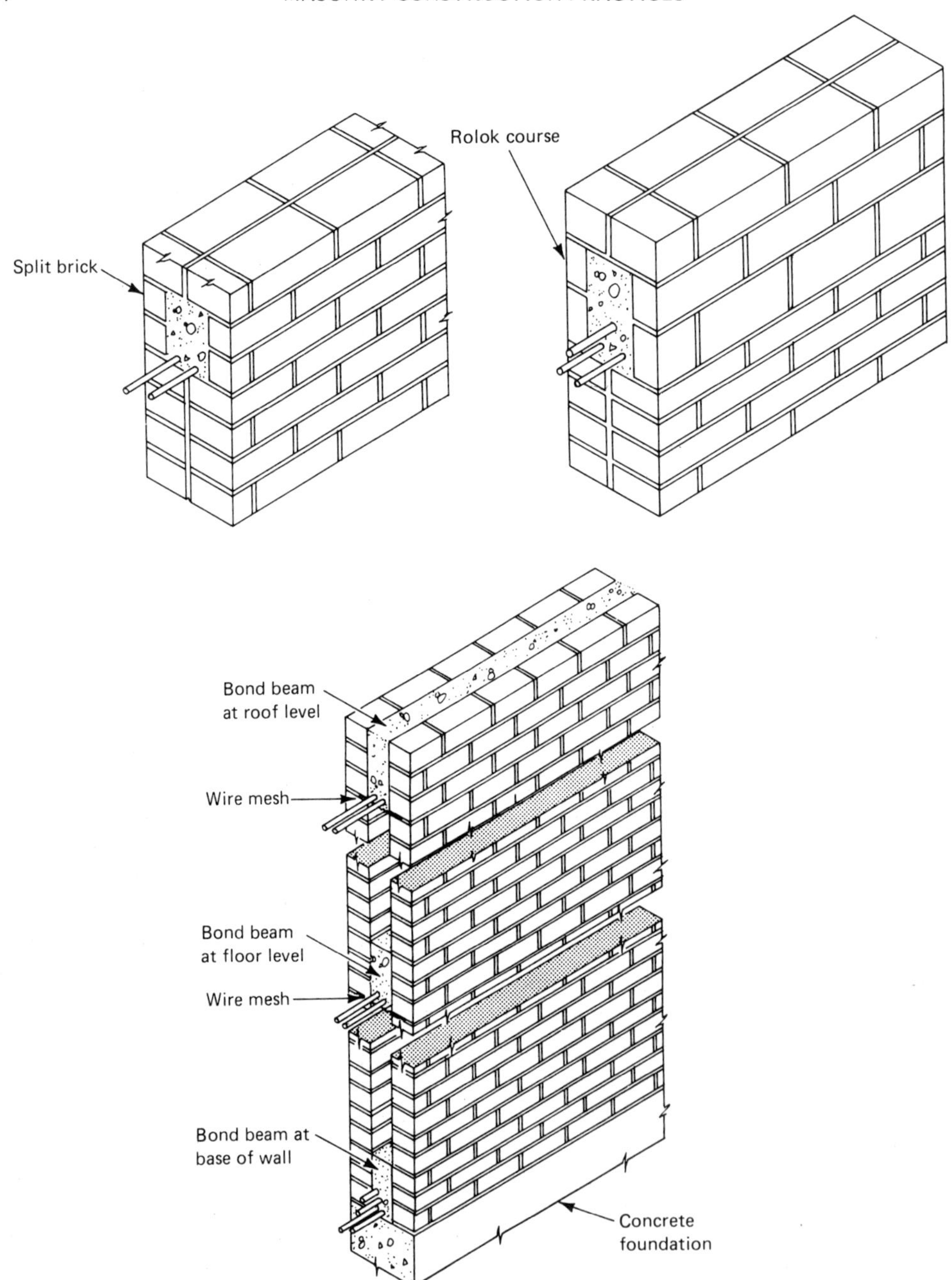

Fig. 14-41. Bond beams in brick walls.

beam may be introduced into a brick wall. One method is to use the same technique that is employed in the construction of a reinforced lintel in a brick wall. In that case, brick are split lengthwise in order to produce units that, when laid, will leave a cavity that may be filled with reinforced concrete. A similar method involves the use of one or more rolok courses to provide the cavity.

When a cavity wall is being built, wire mesh may be laid into the bed joints across the cavity at the appropriate level and reinforced concrete cast into the cavity to the required depth. (See Fig. 14-41.)

STRUCTURAL TILE CONSTRUCTION

Because brick and structural tile are similar products, made from the same materials, fired by the same methods and possessing many of the same physical and surface characteristics, techniques for building with them are also similar, including mortar types, mortar joints and basic procedures for laying. (See Fig. 14-42.) Tile are normally laid in running bond and most commonly used for backup walls (see Fig. 14-43), although facing tile (see Fig. 3-3) are sometimes used for the outer wythe.

Since tile are *hollow* units, the techniques used to construct a bond beam or reinforced lintel with them are different from those used with brick. With tile, the space into which reinforcing steel and grout are to be introduced can be made by cutting the top lip and the center web from a structural tile and placing the reinforced grout in the resulting channel. (See Fig. 14-44.)

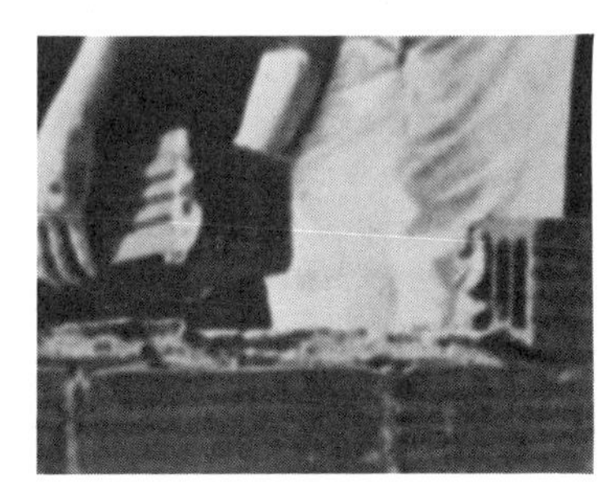

Fig. 14-42. Laying structural tile.

Fig. 14-43. Cavity wall with brick face and tile backup.

CONCRETE MASONRY CONSTRUCTION

Since concrete masonry is a relatively new product compared to brick, tile or stone and used in the same general field, the methods and techniques for building with concrete masonry have, for the most part, followed quite closely after those used with the older materials.

Concrete Masonry Terms

Some of the terms used in brick masonry are also applied to concrete masonry. Units are laid in *courses,* normally in a *single-wythe* wall and almost always with the units as *stretchers* in a course. The solid material in the long faces of

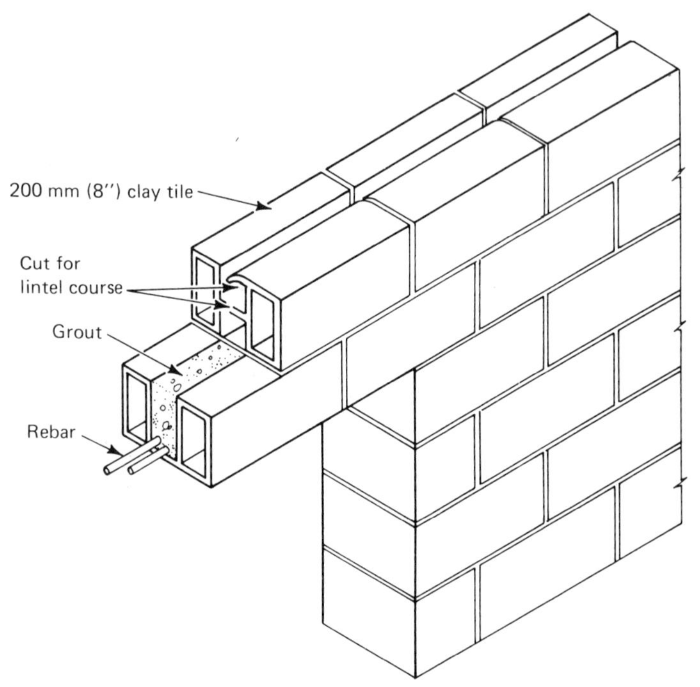

Fig. 14-44. Reinforced lintel with structural tile.

the units are *face shells,* while the cross-members are *webs.* The hollow portions of the units are the *cores,* while the units themselves are commonly referred to as *concrete blocks.*

Wall Patterns

A great variety of wall patterns are possible with concrete block, due to the fact that units are made in *full* and *half* lengths and heights and that *concrete brick* are also available for use with the block. Figure 14-45 illustrates a number of wall pattern possibilities with concrete masonry.

Mortar Joints

Bed joints and *head joints* are used with concrete block as with brick and tile and *collar joints* in composite walls involving concrete block. Jointing tools are used to finish mortar joints, in much the same way as with brick masonry. (See Fig. 14-46.)

Two types of bed joint are used with concrete block—*full mortar bedding* and *face-shell bedding.* With the former, the webs, as well as the face shells, are bedded in mortar, while with the latter, only the face shells are buttered.

Full mortar bedding is used when laying the *starter course* of block on a footing or foundation, when laying *solid units* such as concrete brick or solid block, when building concrete masonry *piers or pilasters* or for the *webs on either side of grouted cores* in partially grouted walls. For all other concrete masonry work with hollow units, it is common practice to use face-shell bedding only. (See Fig. 14-47.)

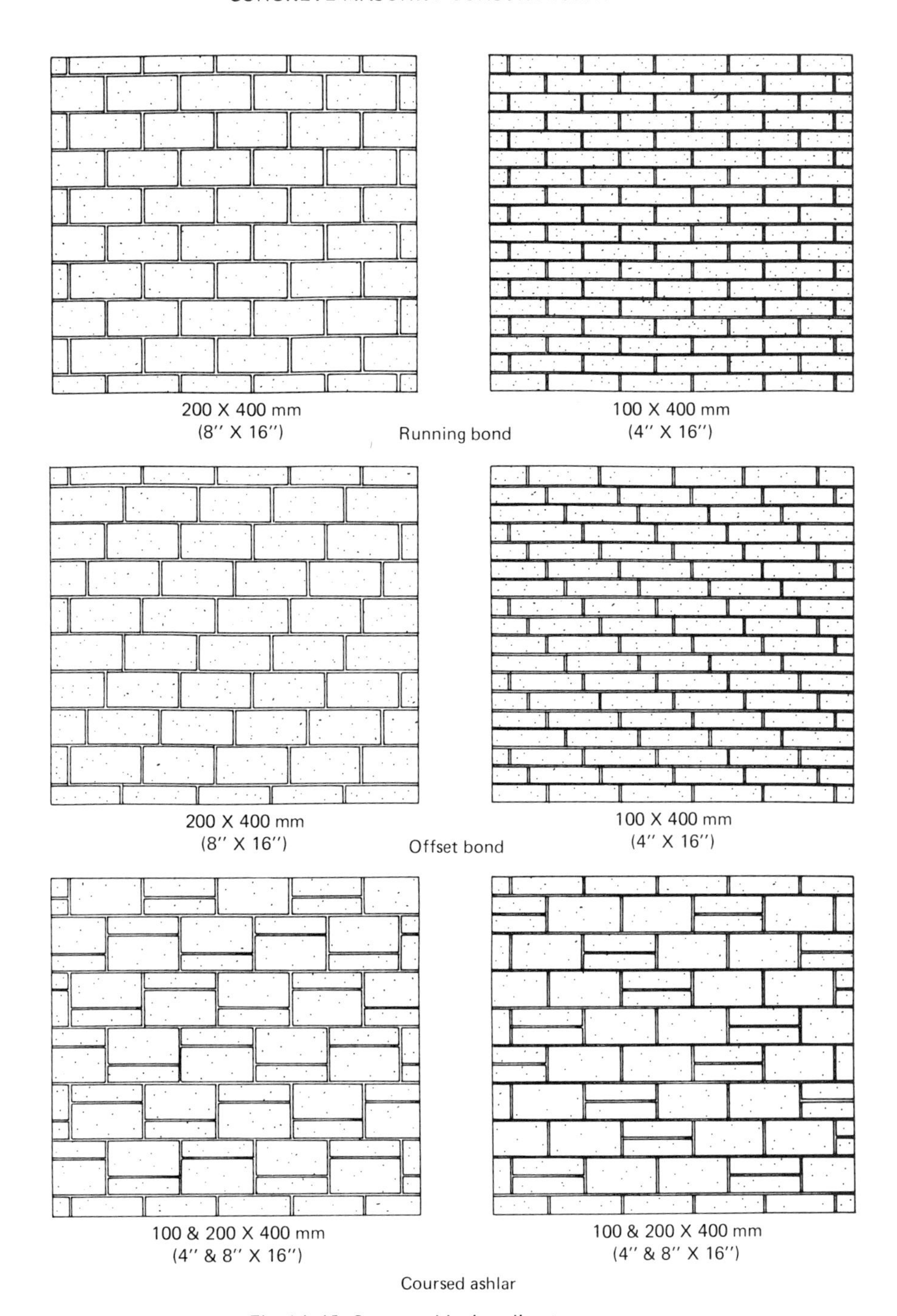

Fig. 14-45. Concrete block wall patterns.

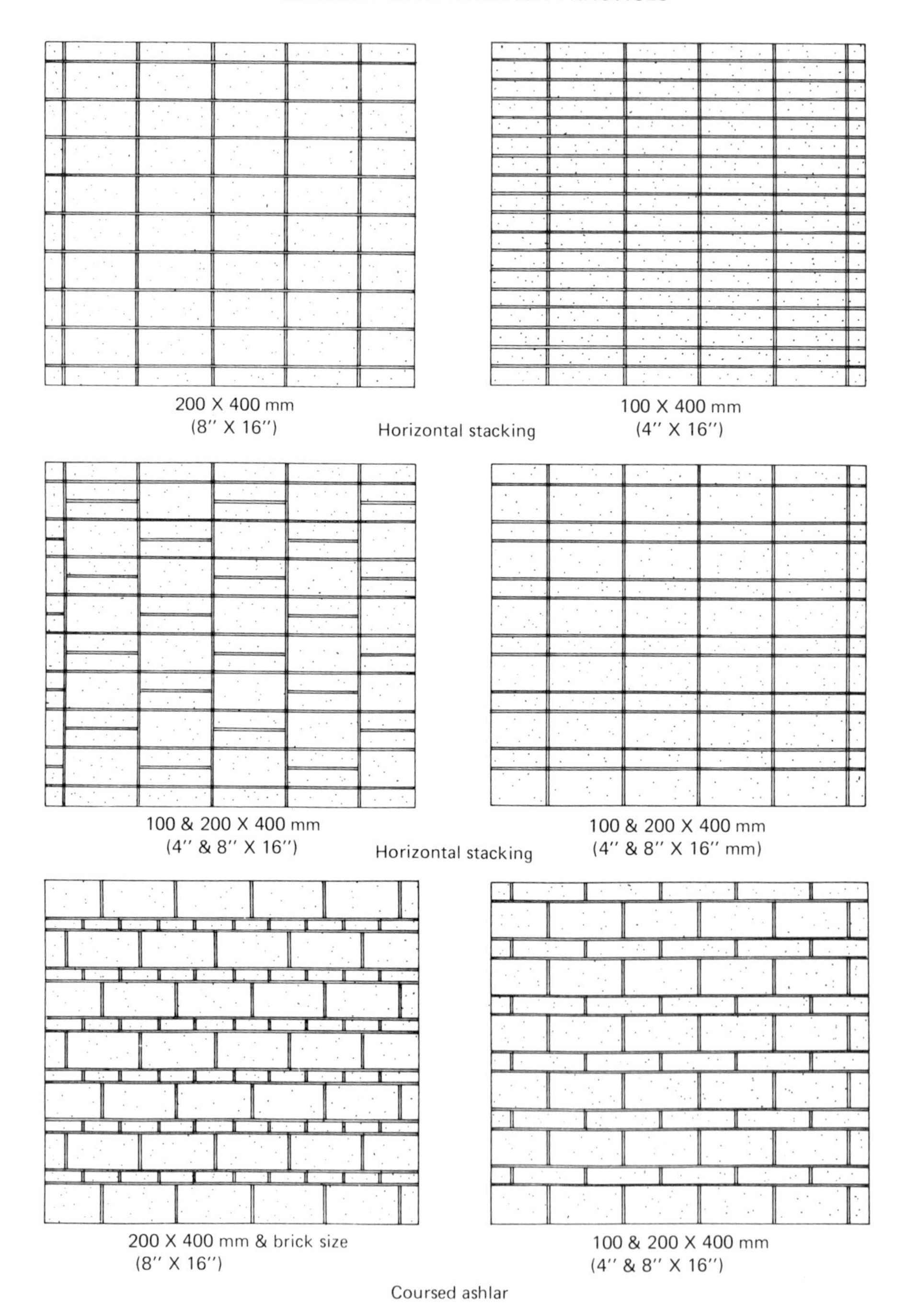

Fig. 14-45 (cont'd)

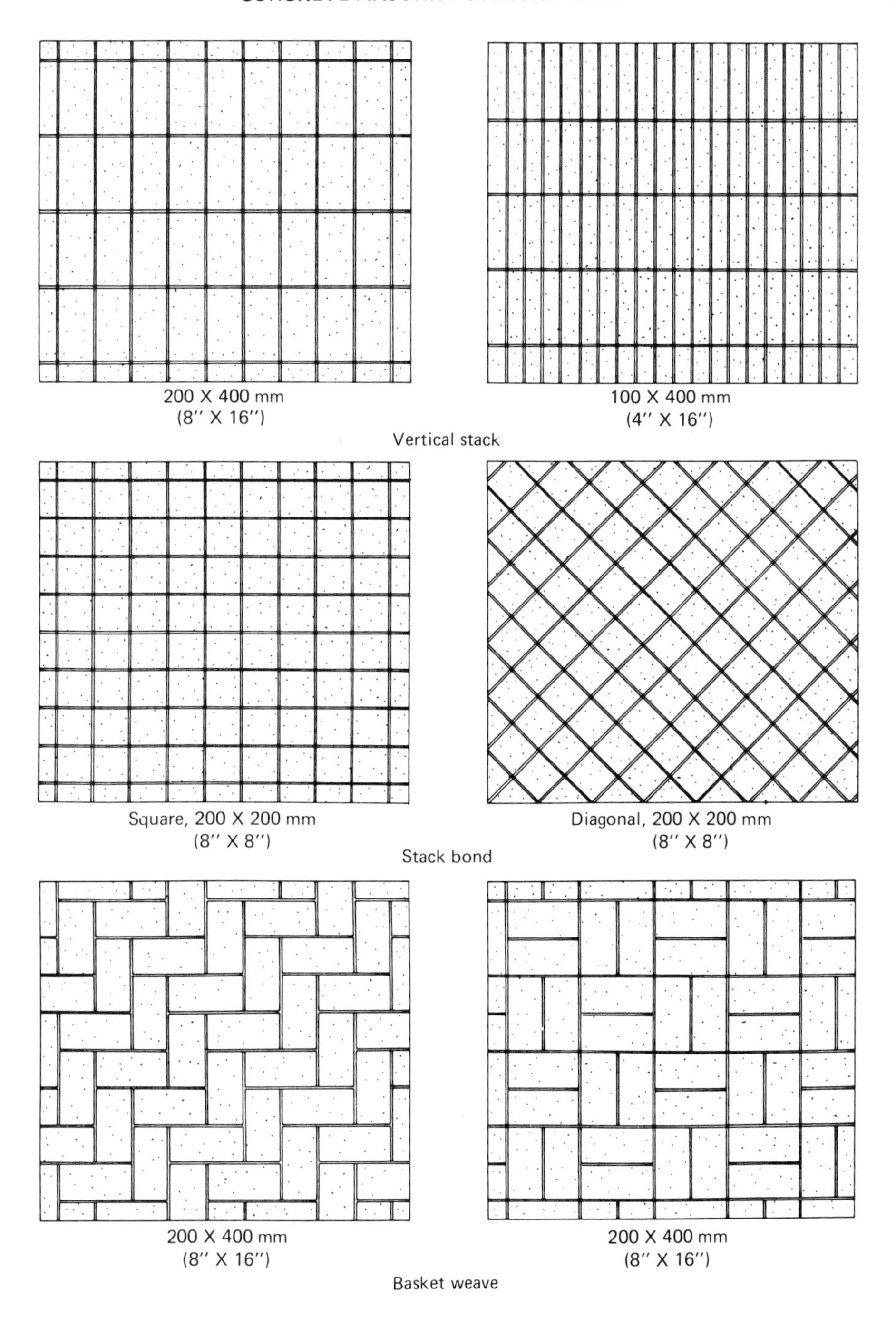

Fig. 14-45 (cont'd)

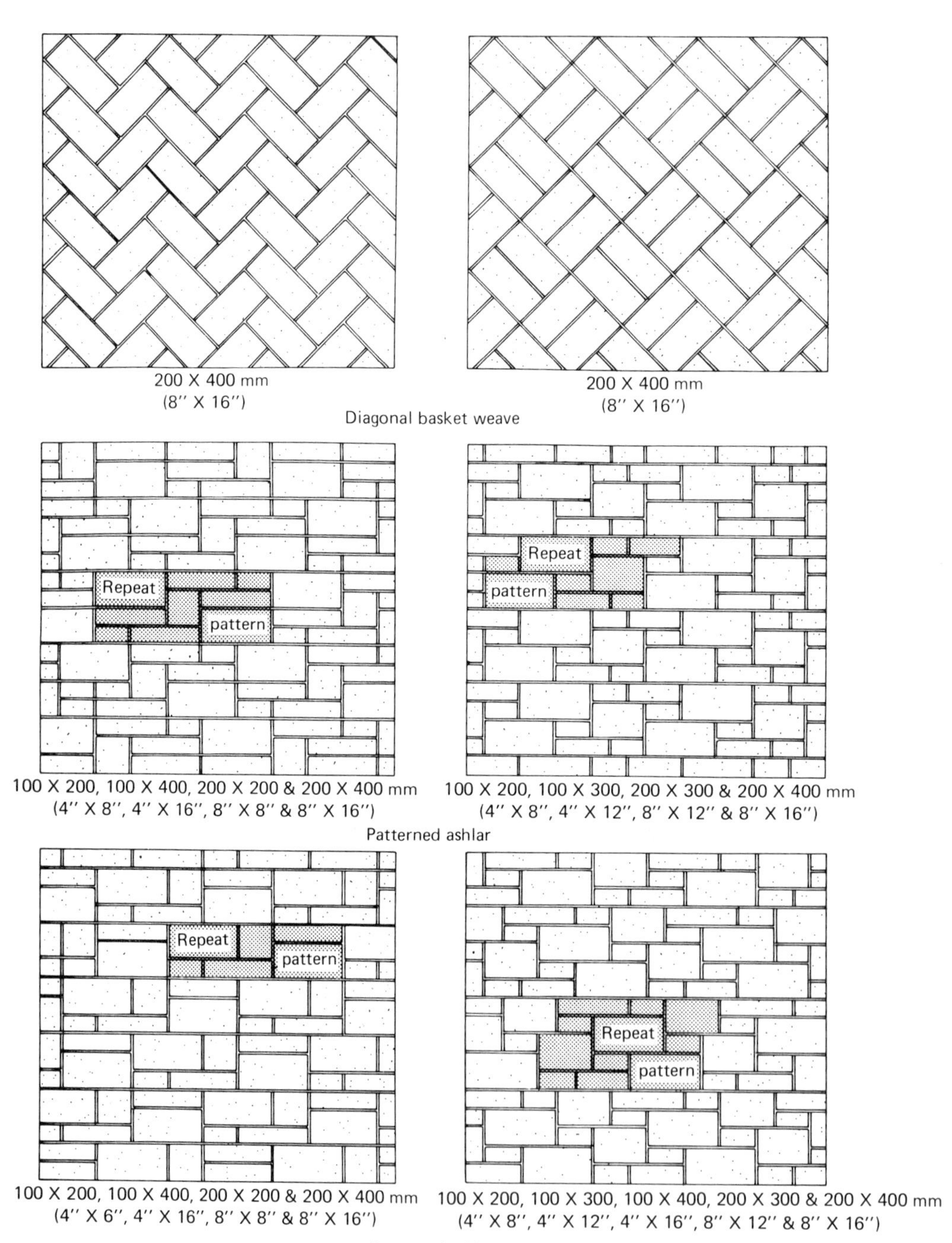

Fig. 14-45 (cont'd)

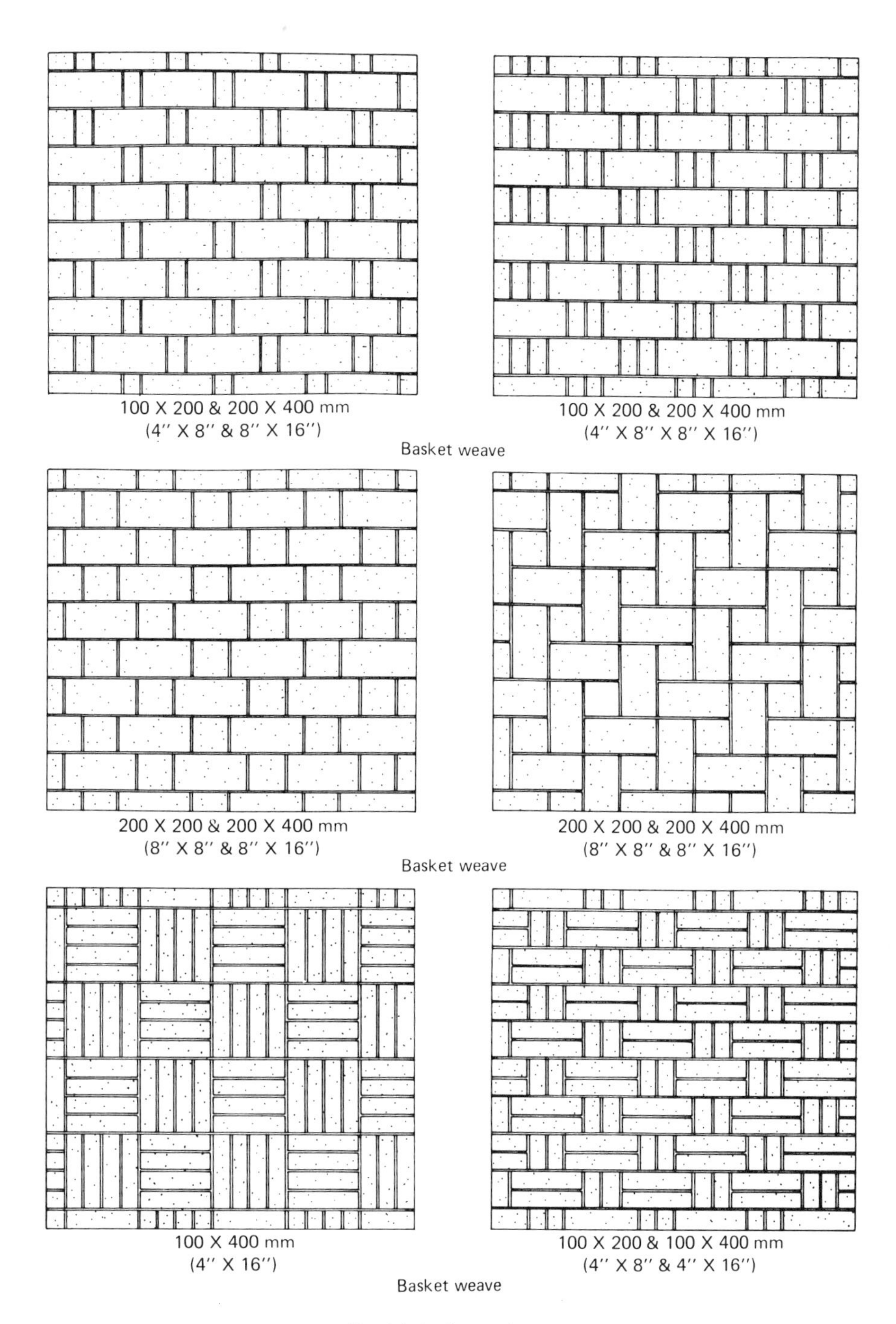

Fig. 14-45 (cont'd)

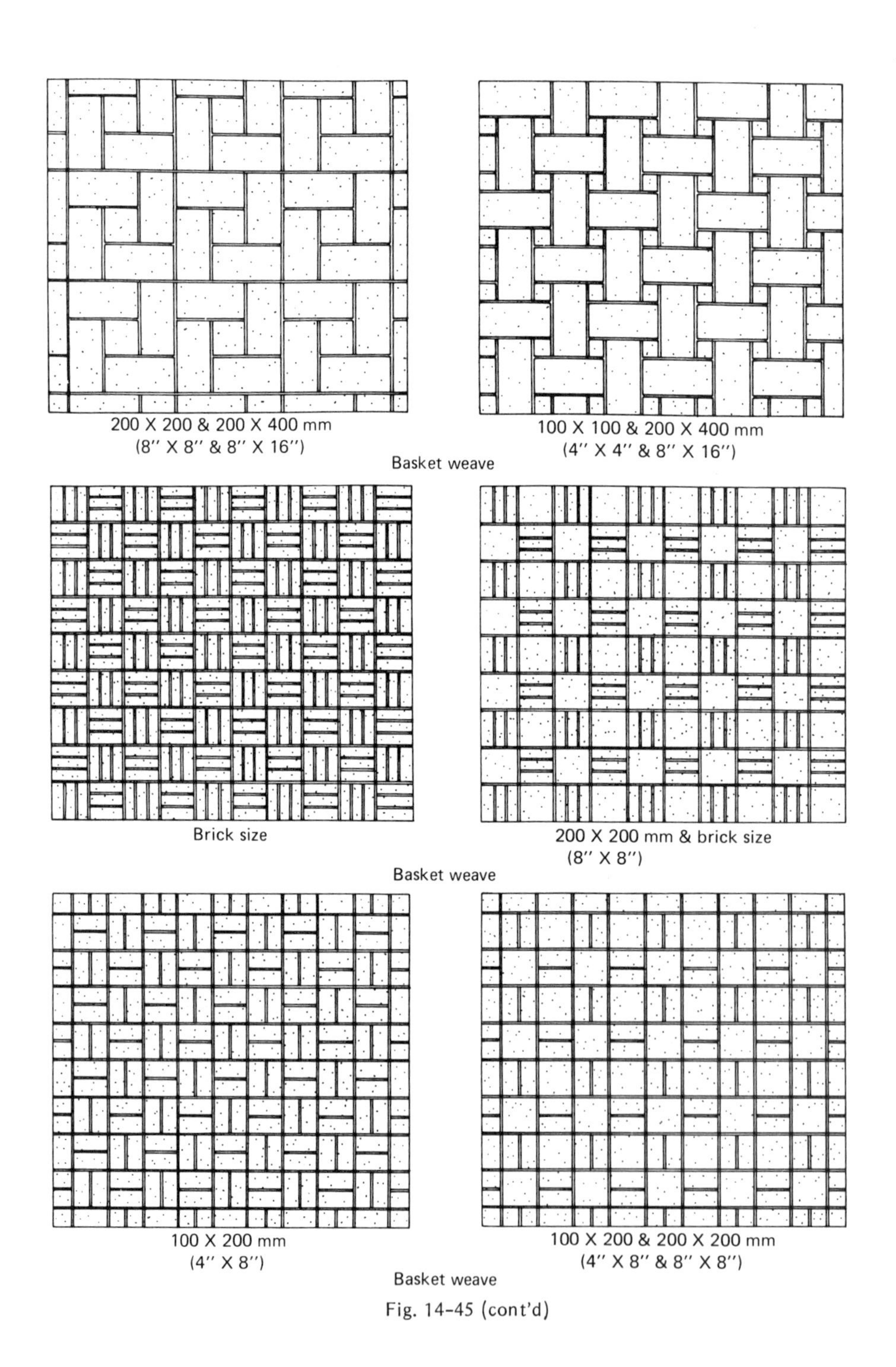

Fig. 14-45 (cont'd)

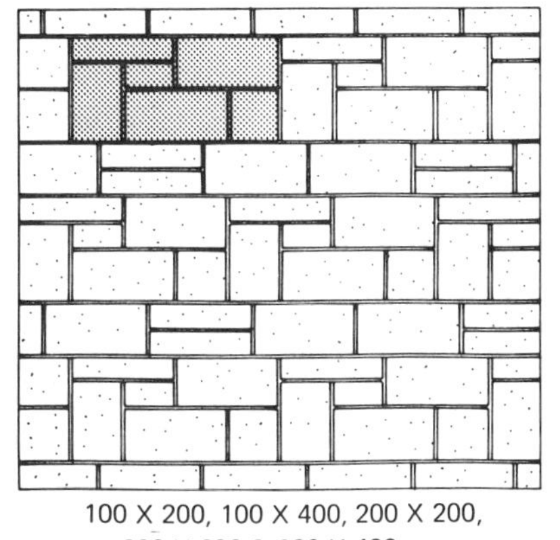

100 X 200, 100 X 400, 200 X 200,
200 X 300 & 200 X 400 mm
(4″ X 8″, 4″ X 16″, 8″ X 8″, 8″ X 12″ & 8″ X 16″)

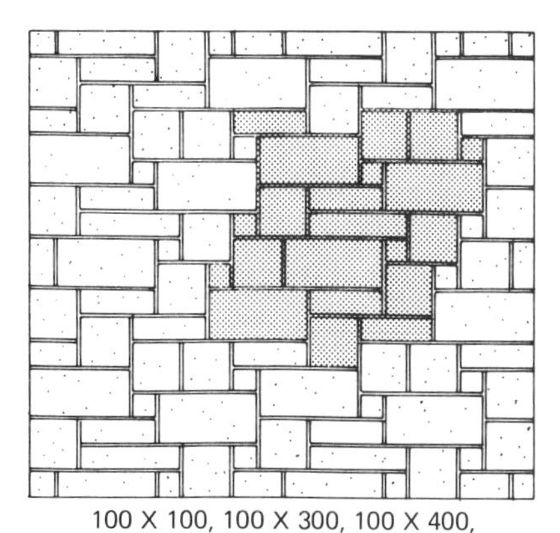

100 X 100, 100 X 300, 100 X 400,
200 X 200 & 200 X 400 mm
(4″ X 4″, 4″ X 12″, 4″ X 16″, 8″ X 8″ & 8″ X 16″)

Patterned ashlar

Diagonal bond
200 X 400 mm
(8″ X 16″)

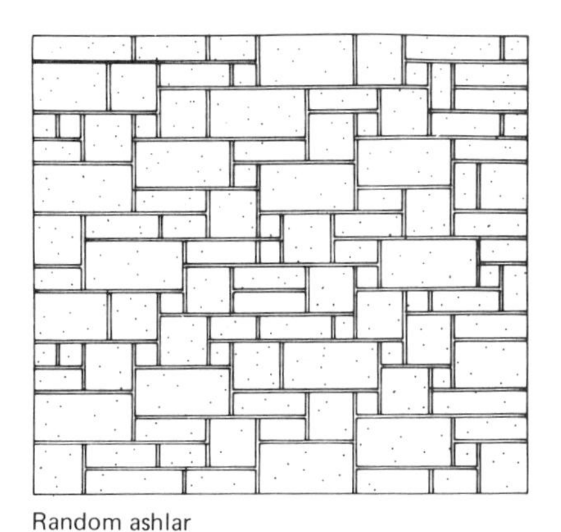

Random ashlar
100 X 100, 100 X 200, 100 X 300,
200 X 200 & 200 X 400 mm
(4″ X 4″, 4″ X 8″, 4″ X 12″, 8″ X 8″ & 8″ X 16″)

Fig. 14-45 (cont'd)

Fig. 14-46. (left) Jointing tool for concave joints (*Courtesy Portland Cement Assn.*); (right) tooling a struck joint (*Courtesy Portland Cement Assn.*)

Fig. 14-47. (left) Full mortar bedding (*Courtesy Portland Cement Assn.*); (right) face shell bedding (*Courtesy Portland Cement Assn.*)

Basic Procedures for Laying Block

Brick and tile units are often wetted before laying if the initial rate of absorption is high, but concrete masonry units should *never be wetted* immediately before and during placement. Moist concrete blocks shrink with loss of moisture and, if the shrinkage is restrained, as it very often is, stresses will develop that may cause cracking in the wall.

Concrete block should be laid with the thicker part of the face shell *up* (see Fig. 2-6), in order to provide a larger area for the bedding joint and to provide a better grip on the block for the mason. For head joints, mortar is applied only on the face ends of the block. It is common practice to butter the ends of several blocks before they are laid (see Fig. 14-48), which provides some time for adhesion to develop between mortar and block. Some masons also butter the ends of the block already in place.

Fig. 14-48. Buttering ends of several block before laying. (*Courtesy Portland Cement Assn.*)

Fig. 14-49. Corner lead with concrete block. (*Courtesy Portland Cement Assn.*)

(a)

(b)

(c)

Fig. 14-50. (a) Plumbing starter course (*Courtesy Portland Cement Assn.*); (b) Levelling starter course (*Courtesy Portland Cement Assn.*); (c) Squaring corner in starter course (*Courtesy Portland Cement Assn.*)

As in bricklaying, blocklaying is begun by building up *corner leads* (see Fig. 14-49), and in so doing, great care must be taken to ensure that the starter course is *plumb, level* and the corner *square.* (See Fig. 14-50.)

With the corner leads built, blocks are laid *to a line* strung from one corner lead to the other. (See Fig. 14-51.) The *closure block* in each course must be laid with great care. Both ends of the closure block are buttered, as well as the ends of the two blocks on either side. Then the block is carefully set in place, as illustrated in Figure 14-52.

In many situations in concrete block construction, it is necessary to cut a unit. Cutting may be done either by hand, using a mason's hammer and broad chisel, or by machine, using a power-driven abrasive wheel (see Fig. 14-53).

Fig. 14-51. Blocks laid to a line.

Fig. 14-52. Closure block set in place. (*Courtesy Portland Cement Assn.*)

Hand cutting, in most cases, is confined to cutting *across* the block–shortening its length–whereas blocks may be cut in any direction with the abrasive wheel.

For some types of construction, it will be necessary to attach a wood plate or *sill* to the top of a concrete masonry wall. It will be held in place by bolts that are anchored into grouted cores in the wall. The first step is to place a piece of wire mesh in the course below, directly under the core that is to be filled. That core is filled with grout that is allowed to stiffen enough to hold the bolt upright in position when it is embedded in the grout. (See Fig. 14-54.) An alternative is to suspend the bolt through a small piece of board placed across the core, in which case the bolt can be set as soon as the grout is placed.

Structural Bonding with Concrete Masonry

Structural bonding in a single-wythe concrete masonry wall is achieved by overlapping the blocks and by using continuous metal ties that are laid in the bed joints (see Fig. 14-19).

In a cavity wall, structural bonding is achieved by metal ties, either individual or continuous, from wythe to wythe across the cavity. In a composite wall, headers in the outer wythe, backed up by concrete brick or header block (see Fig. 2-7), may be used to provide structural bond. Header block may be laid with the cut-away portion either up or down. (See Fig. 14-55.)

Fig. 14-53. (left) Cutting concrete block with hand tools (*Courtesy Portland Cement Assn.*); (right) cutting concrete block with abrasive wheel (*Courtesy Portland Cement Assn.*)

Fig. 14-54. Anchor bolt embedded in grout. (*Courtesy Portland Cement Assn.*)

Control Joints in Concrete Masonry Walls

Control joints are constructed in concrete masonry walls in several ways, some of which are illustrated in Figure 14-56. In (a), a strip of building felt is set on one side of the joint to act as a bond breaker and so ensure freedom of movement at the joint. In (b), sash blocks (see Fig. 2-7) are used at the joint, with a premolded neoprene water stop fitted into the notches in the ends of the units. The blocks in (c) are specially made units, in full and half lengths, used to form a *tongue-and-groove* type of joint. Here mortar is placed in the head joints and some of it is allowed to remain as backing for the caulking that will be used to finish and seal the joint. (See discussion of sealants, Chapter 16.) In (d) standard blocks are used, with a Z-type tie placed across the joint to provide lateral stability. Again, a portion of the mortar is removed from the head joint to provide space for the elastic caulking. In (e), *jamb blocks* are used at the joint, with a Z-type tie across the joint to provide shear strength.

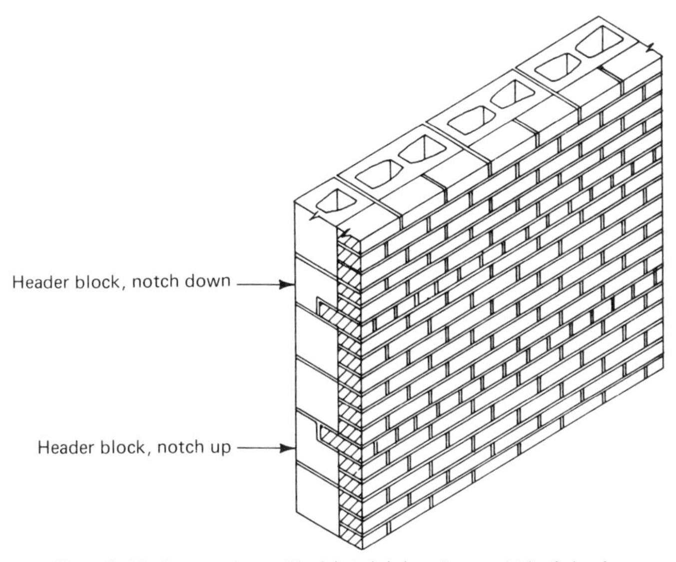

Fig. 14-55. Composite wall with brick headers and block backup.

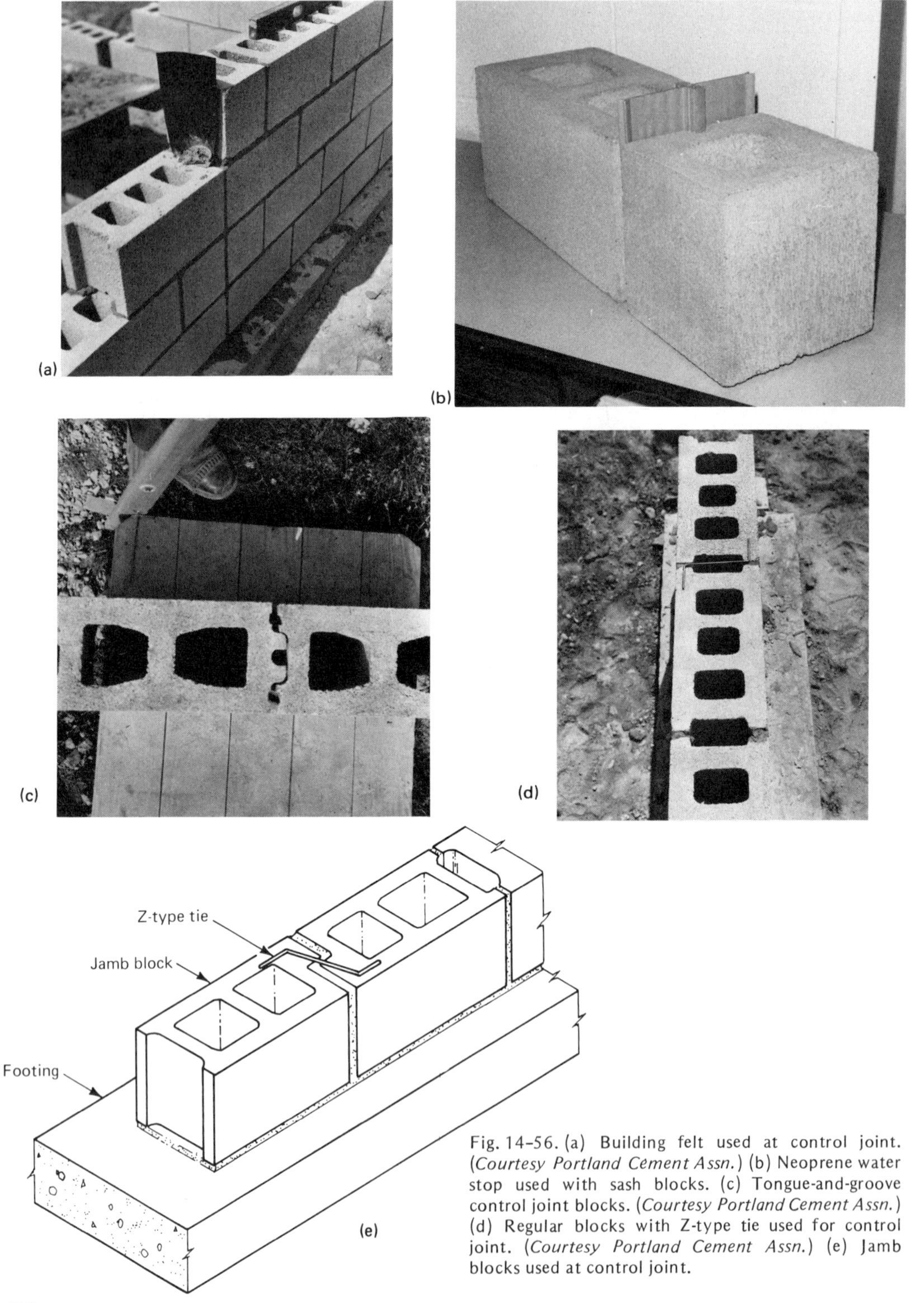

Fig. 14-56. (a) Building felt used at control joint. (*Courtesy Portland Cement Assn.*) (b) Neoprene water stop used with sash blocks. (c) Tongue-and-groove control joint blocks. (*Courtesy Portland Cement Assn.*) (d) Regular blocks with Z-type tie used for control joint. (*Courtesy Portland Cement Assn.*) (e) Jamb blocks used at control joint.

Intersecting Concrete Masonry Walls

Walls intersect one another under various conditions, that must be taken into account in determining the kind of connection to be used between them. If intersecting load-bearing walls depend on one another for continuity and lateral support, they must be rigidly bonded together to resist the forces that might tend to separate them. If, on the other hand, they should be allowed to move in relation to one another, then a flexible, control joint type of connection must be used.

The best type of rigid connection for intersecting walls is that produced by half-lapping the units of one over those of the other, as illustrated in Figure 14-57(a), and bonding them with mortar. One alternative method is to use a steel bar [Fig. 14-57(b)], with its ends embedded in grouted cores in the two walls and the end core between the walls filled with grout. Notice the wire mesh used to close off the bottom end of a core for grouting. Another alternative method is to use wire mesh laid in the bed joints of the corresponding courses of the intersecting walls to tie the two together. [See Fig. 14-57(c).] Such ties should be spaced not more than 400 mm (16 in) o.c. vertically.

Still another method is to use *looped ties* embedded in grouted cores, as illustrated in Figure 14-57(d). The end core is grouted in both cases, and when the open-ended loop is used, vertical bars are set into the grouted cores to improve the strength of the tie.

For a control joint type of connection, the two walls must be *tied* together laterally but *not bonded.* One method of doing this is to use the same type of bar connector as in Figure 14-57(b) but with no grout in the end core between the walls. (See Fig. 14-58.)

Corners

The arrangement of units at corners is important if the wall pattern is to be maintained properly. Units with a 200 mm (8 in) thickness dimension present no problems, but the arrangement of thicker or thinner units requires some attention in order to preserve the pattern and adhere to the 600 mm (12 in) planning module. Figure 14-59 illustrates some corner layouts with units of various thicknesses.

Piers and Pilasters

Concrete masonry piers may be constructed of *standard-plain-ended units* filled with grout, *solid masonry units,* or from special *pier blocks,* as illustrated in Figure 14-60.

Pilaster blocks (see Fig. 2-11) are commonly used to construct concrete masonry pilasters, as illustrated in Figure 14-61(a), using two different styles of block. An alternative method is illustrated in Figure 14-61(b), where standard blocks are used to form the pilaster.

Lintels

Reinforced lintels over openings in concrete masonry walls are constructed by using full or half-lintel blocks (see Fig. 2-11), reinforced and filled with grout or concrete. Temporary shoring (see Fig. 14-36) is required to support the structure until the grout has hardened. (See Fig. 14-62.)

Bond Beams

Bond beams are used in concrete masonry construction for the same purposes and in the same general locations in the structure as with brick and tile walls.

In addition to the *single-* and *double-channel* and *knockout style* bond beam blocks shown in Figure 2-11, bond beams may be constructed using *open-end* bond beam blocks, H-blocks or low-web bond beam blocks. (See Fig. 14-63) Typical bond beam construction is illustrated in Figure 12-9.

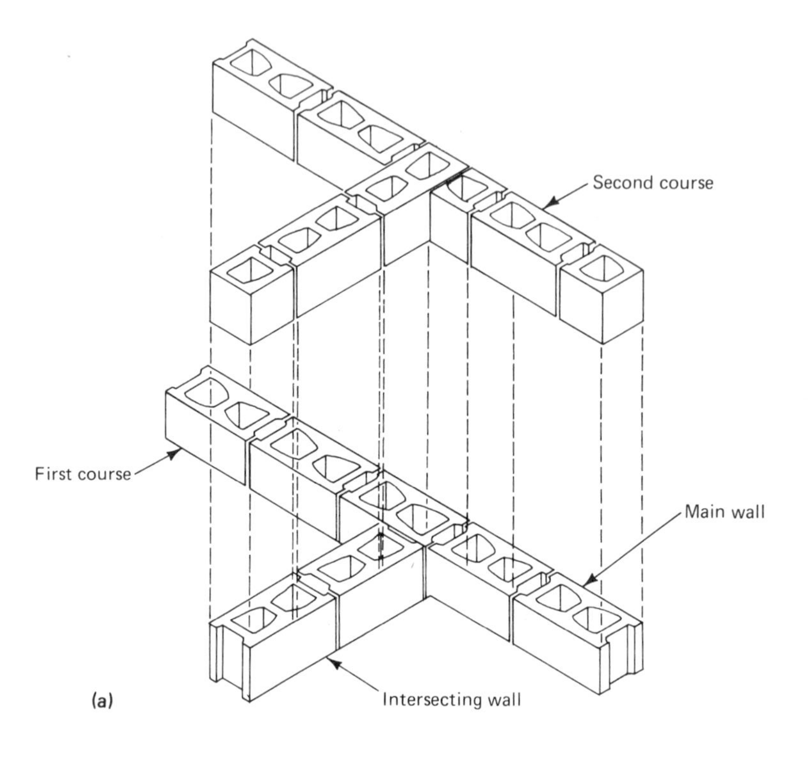

Fig. 14-57. (a) Intersecting walls bonded by overlapping blocks. (b) Steel bar tie for intersecting walls. (*Courtesy Portland Cement Assn.*)

(c)

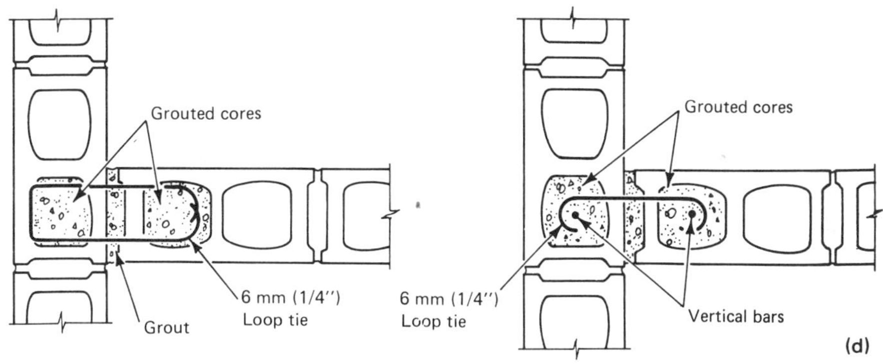

Fig. 14-57. (c) Wire mesh ties for intersecting walls (*Courtesy Portland Cement Assn.*) (d) Looped ties for intersecting walls.

Fig. 14-58. Intersecting walls tied but not bonded. (*Courtesy Portland Cement Assn.*)

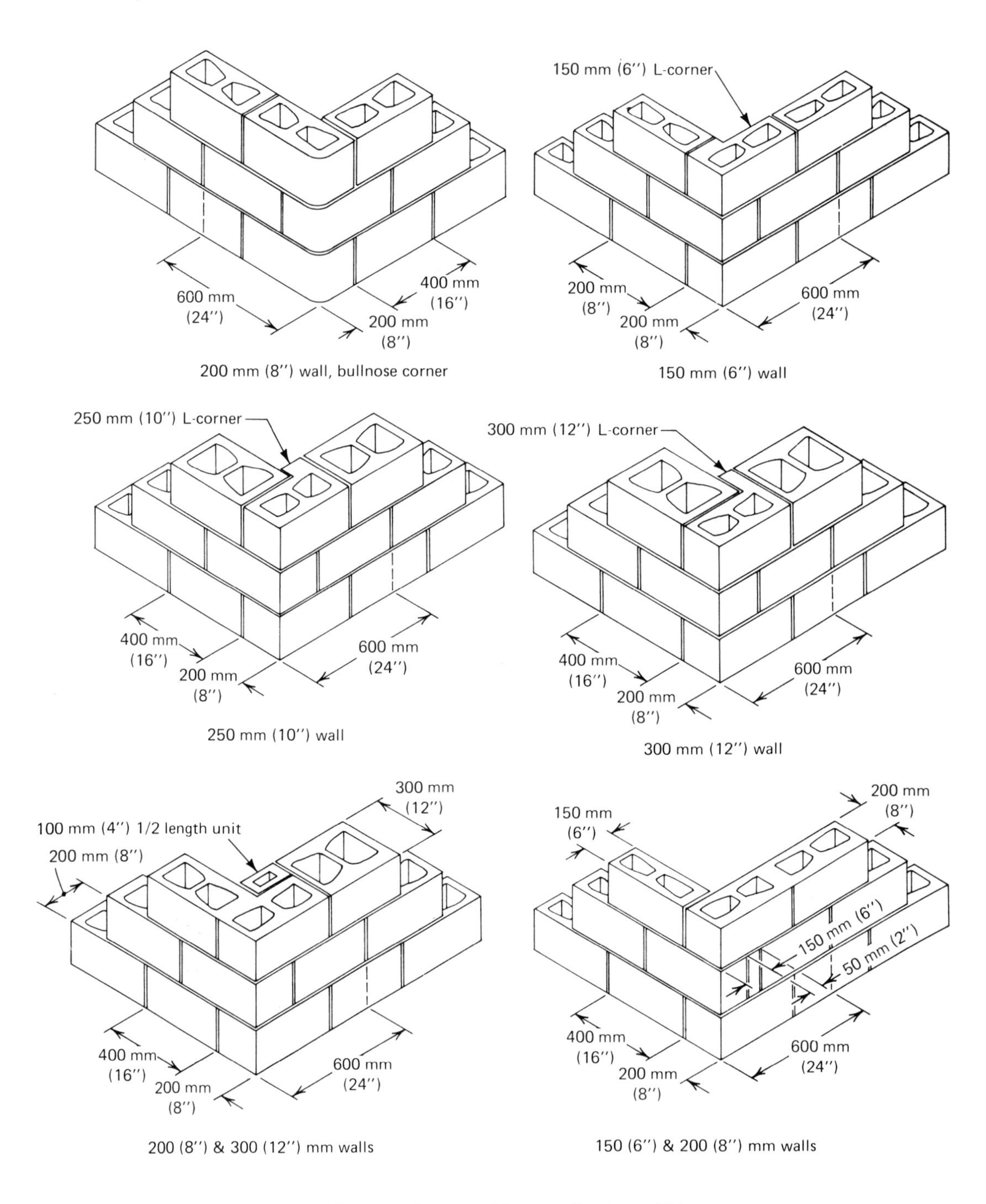

Fig. 14-59. Corner layouts with units of various widths.

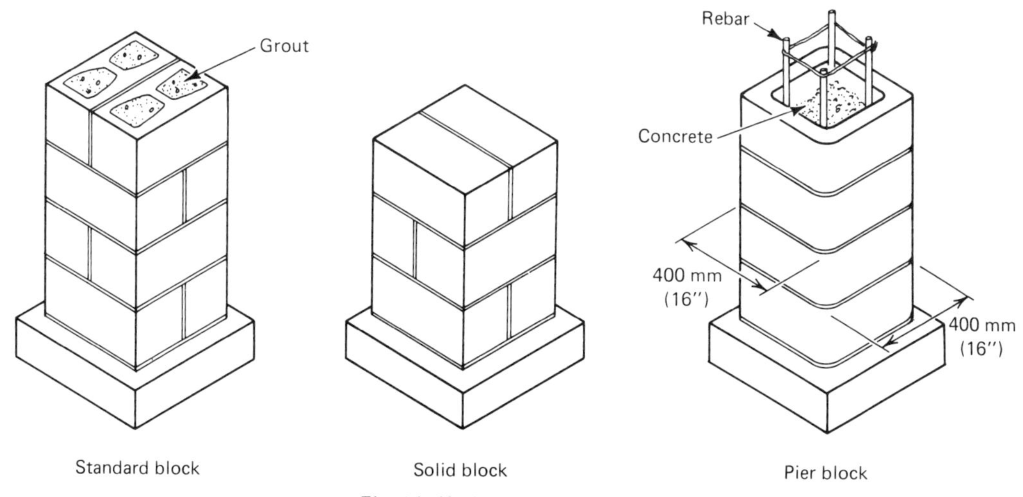

Fig. 14-60. Concrete masonry piers.

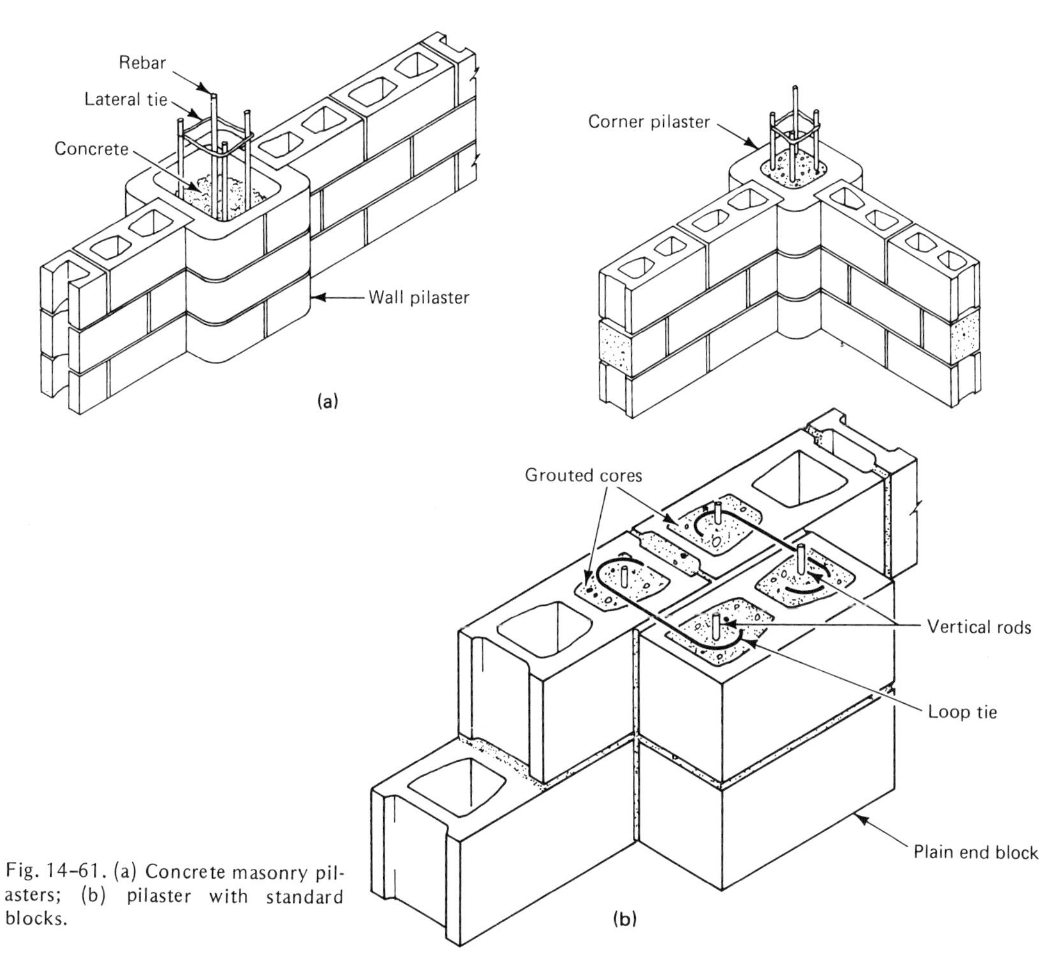

Fig. 14-61. (a) Concrete masonry pilasters; (b) pilaster with standard blocks.

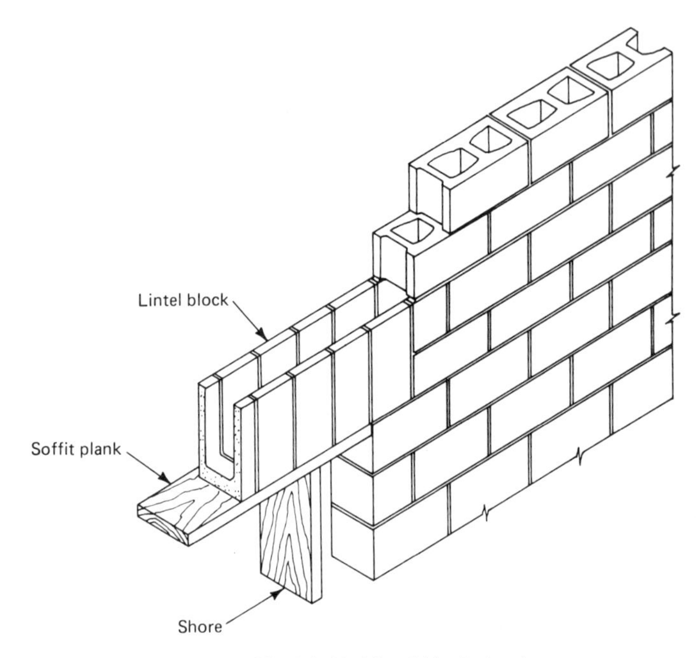

Fig. 14-62. Lintel blocks in place.

Flashing and Weep Holes

The method of securing the upper edge of counter-flashing used on concrete masonry parapet walls may differ from that shown in Figure 14-22 (review earlier discussion of flashing and weep holes). Instead of being bonded into a mortar joint, the upper edge of the counter-flashing may be retained in a *metal reglet* that is inserted into the bed joint between two courses of block. In such a case, the webs of the units should be bedded in mortar to compensate for the loss of mortar bond resulting from the presence of the reglet in the face-shell joint. (See Fig. 14-64.)

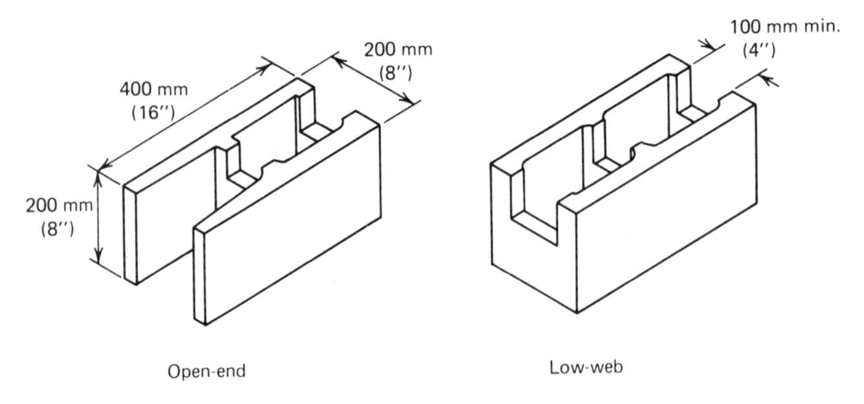

Fig. 14-63. Special bond beam units.

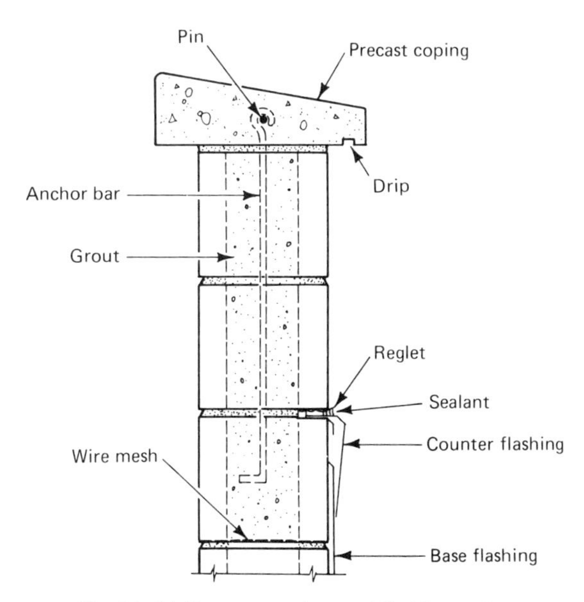

Fig. 14-64. Parapet, coping, and flashing reglet.

Copings

All masonry walls having an upper edge exposed to the weather must be protected against water penetration. A *coping*, which is a cap that rests on top of the wall, extends beyond both faces and normally slopes inward toward the roof, serves this purpose. (See Fig. 14-22.) It may be made of *stone* or *precast concrete* or, in the case of concrete masonry construction, *coping blocks* (see Fig. 2-7) may be used.

Stone and precast concrete copings will usually be provided with metal *anchors* at the joints to secure them in place (see Fig. 14-64). Where concrete masonry coping blocks are used, they will normally be anchored by the mortar bond between them and the wall on which they rest.

PREFAB CONSTRUCTION

Prefabricated Panel Anchors

As indicated in Chapter 12, a number of systems have been developed for tying prefabricated masonry wall panels to one another and to the floors of structural frame buildings into which they are introduced.

Wall-to-Floor (Vertical) Ties

The *spliced reinforcement* system is much the same as that used to tie precast concrete panels to one another and to the floor. A coupling nut is attached to the bottom ends of reinforcing bars that have been grouted in place at the plant. When the panel is in place, a short, straight bar, threaded at one end is attached to the upper bar by the coupling nut and to the one in the panel below by a welded joint. At the same time, an L-shaped bar is laid with one end in a pocket left in the floor and the other end down in the block core of the lower panel, to tie wall and floor together. All pockets will be grouted after the connections are complete. (See Fig. 14-65.)

The *tension connection* system provides a method of tying the panels together, through the cores, from top to bottom of the building and, at the same time, tying each panel to the floor on which it rests.

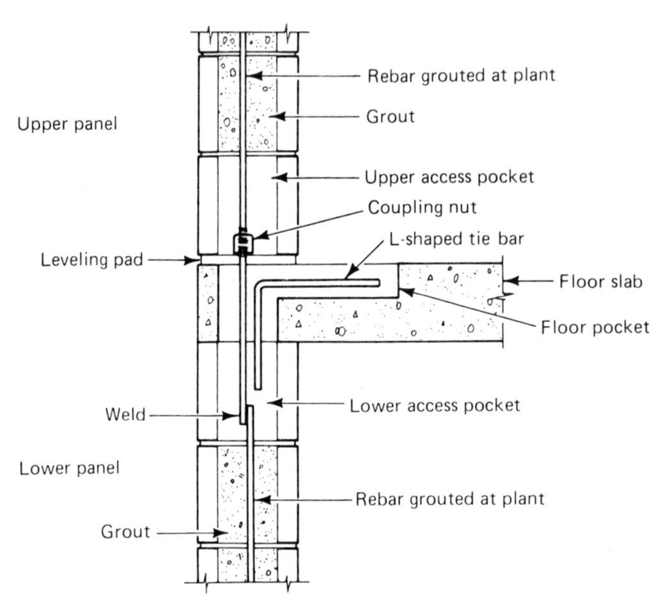

Fig. 14-65. Spliced reinforcement connection. (*Courtesy National Concrete Masonry Assn.*)

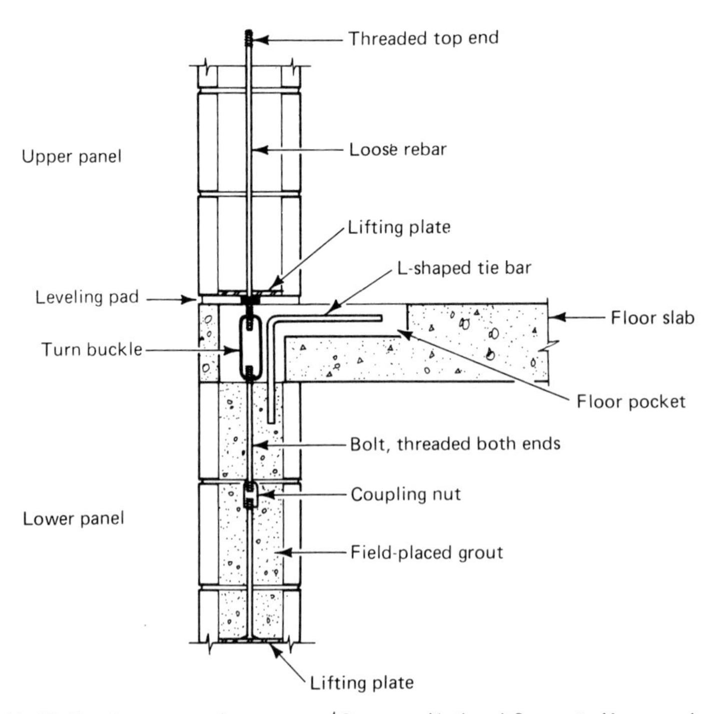

Fig. 14-66. Tension connection system. (*Courtesy National Concrete Masonry Assn.*)

Panels for such a system have *loose vertical bars* threaded at both ends in some of the cores. Each bar has a bolt, threaded at both ends and attached to the top. also a *steel-lifting plate and nut* attached at the bottom end, to act as a lifting device.

After a panel has been set in place, a *turnbuckle* is used to join the corresponding top and bottom bars together. The cores in the bottom panel are then grouted, but the ones in the top panel are left empty until the next panel above is installed. The last panel in the tier must have a plate at the top to complete the tension connection. Finally, the turnbuckles are tightened, the L-shaped bars are placed to tie wall and floors together, and the pockets are grouted. (See Fig. 14-66.)

In the *bolted-tension* connection, vertical reinforcing bars are grouted into the panels in the plant, with threaded eyebolts attached at both top and bottom ends. When the bottom panel is in position in the wall, a connecting device consisting of a *turnbuckle with a yoke at each end* is attached to the end of each bar to be tightened. When the next panel above is installed, the connection is attached to the bars above and the turnbuckle tightened to join the two panels. Finally, after the floor anchors are placed, grout is placed in the cores and access pockets. (See Fig. 14-67.)

When bolted and welded connections are used, anchor plates are set into the top and bottom of panels during fabrication. These are bolted or welded to connections attached to or cast into the floor slabs. Access pockets are grouted after the connections are complete.

Wall-to-Wall (Horizontal) Ties

One common type of connection for this purpose is a steel bar with its ends bent at right angles. One end is preset in the top end block of one panel, and after the adjoining panel has been positioned (notice the cutout in the top end block of the adjoining panel), its end core is grouted to anchor the other end of the bar. (See Fig. 14-68.)

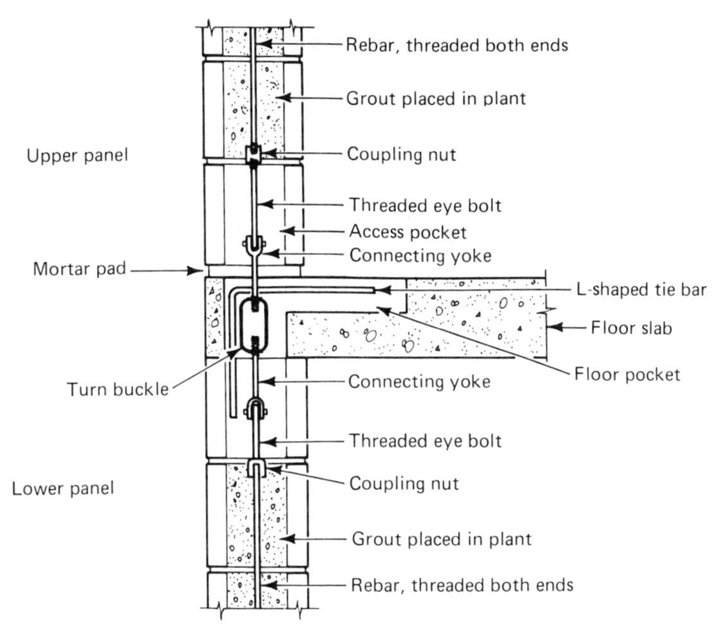

Fig. 14-67. Bolted-tension connection. (*Courtesy National Concrete Masonry Assn.*)

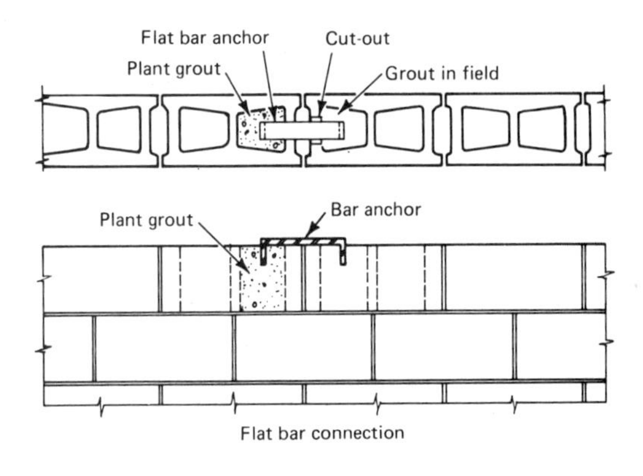

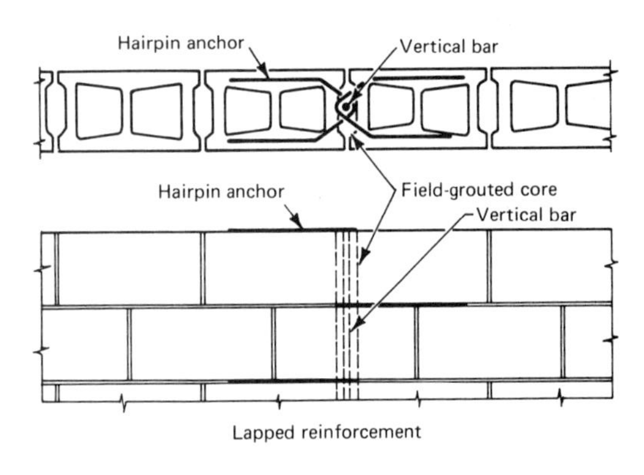

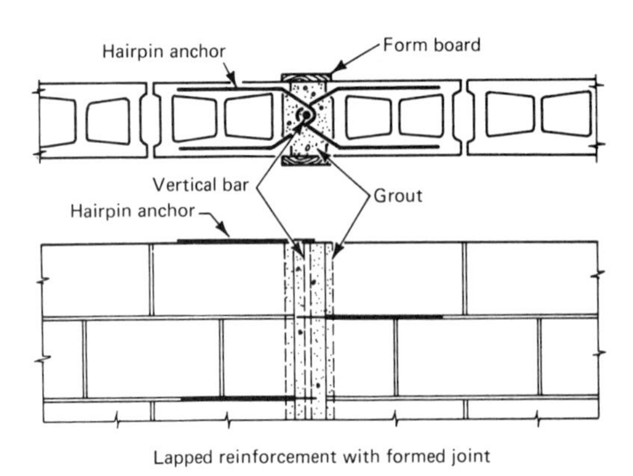

Fig. 14-68. Horizontal panel ties.

Lapped reinforcement is another type of horizontal tie. Hairpin-shaped bars are placed in alternate block courses, with their loops projecting about 50 mm (2 in) past the vertical edge of the panel. After two adjoining panels are brought together, a vertical bar is dropped through the loops, and the core formed by the ends of the two panels is grouted. (See Fig. 14-68.)

Lapped reinforcement with a formed joint is similar to the above except that the panels are spaced 50 to 75 mm (2 to 3 in) apart when in position. The vertical bar is inserted and a form board is placed on each side of the space. Then the space is filled with grout. (See Fig. 14-68.)

STONE CONSTRUCTION

Although stone is still used in a few cases as a structural material, it is used far more extensively as a *veneer*, a *curtain wall material*, for *interior facing* and as *trim*.

As a veneer, stone is laid over a backup wall of masonry or wood-frame construction in rubble or ashlar patterns. (See Figs. 4-1 and 4-2.) As a curtain wall material, larger stone panels are anchored to a structural frame or to backup panels in a structural frame. Stone interior facings and trim may be secured by mortar or by mechanical anchors, depending on the thickness of the material. Figure 14-69 illustrates a variety of commonly used stone anchors.

Stone Veneer

Two types of stone are used in stone veneering. One consists of thin sheets resulting from natural cleavage, usually of slate or argillite, from 6 to 25 mm (¼ to 1 in) in thickness and up to 0.185 m^2 (2 ft^2) in area, bonded to a backup wall with mortar. The other consists of cut or semicut pieces from various kinds of building stone, from 50 to 150 mm (2 to 6 in) in thickness, in a variety of face dimensions,

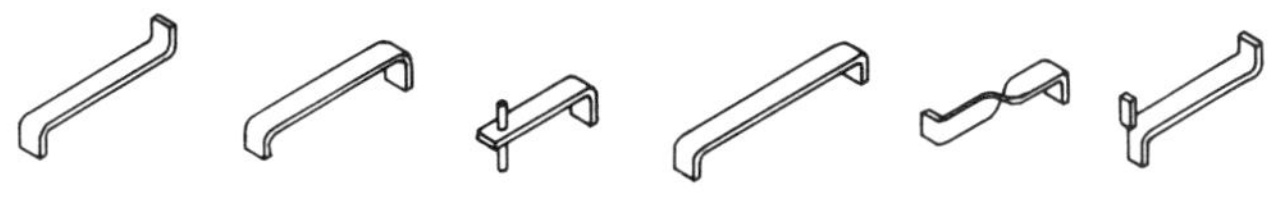

Strap anchors
(Recommended sizes–3 (1/8″) & 5 (3/16″) mm X 25 (1″) or 30 (1-1/4″) mm)

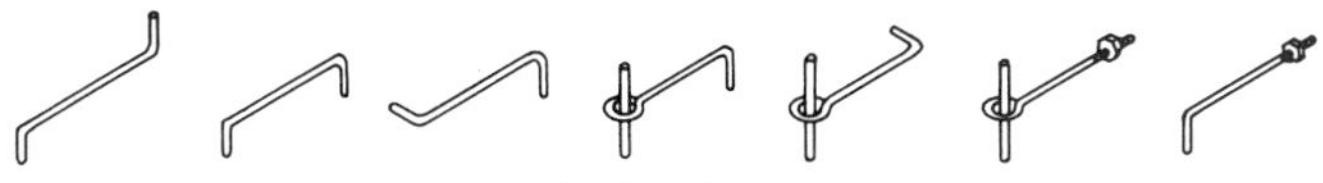

Rod anchors
(Recommended size–9.5 mm (3/8″) min. dia.)

Dovetail anchors
Recommended sizes–3 (1/8″) & 5 (3/16″) mm X 25 (1″) or 30 (1-1/4″) mm wide)

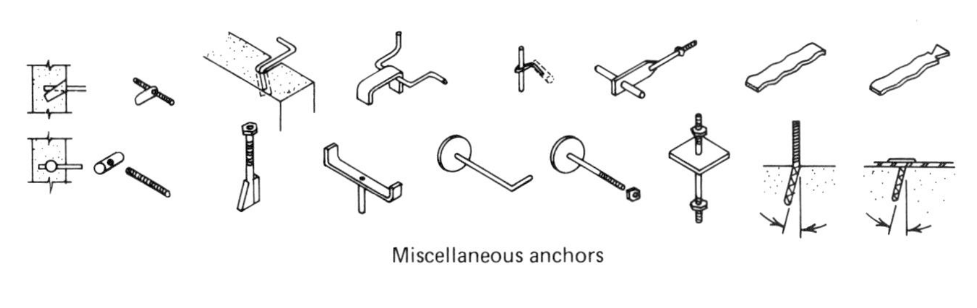

Miscellaneous anchors

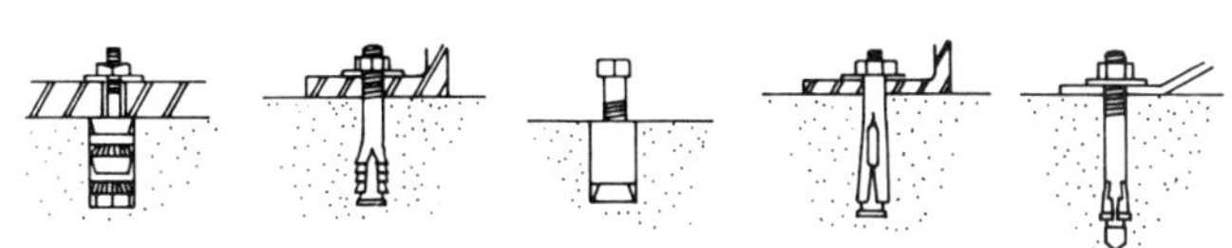

Expansion anchors

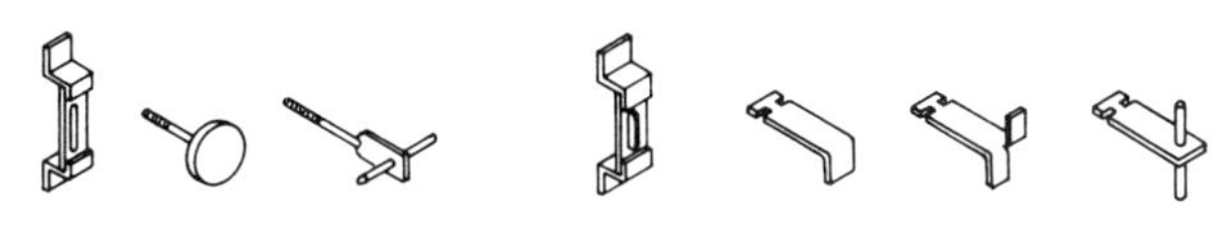

Special anchors
(Bracket mounted to masonry or steel)

Fig. 14-69. Stone anchors.

attached to a backup wall by some type of mechanical anchor.

When thin veneer is to be applied over a wood-frame wall, the sheathing is first covered with a good quality saturated felt paper, followed by stucco wire, securely nailed or stapled. A base coat of type M mortar is applied over the wire to a depth of about 10 mm (3/8 in), floated flat with a wood or cork float and damp-cured, preferably for at least 48 hours. At the end of that time the wall is dampened, one section at a time, and a type N setting mortar is applied to small areas that may be covered by one large or two small stones, to a

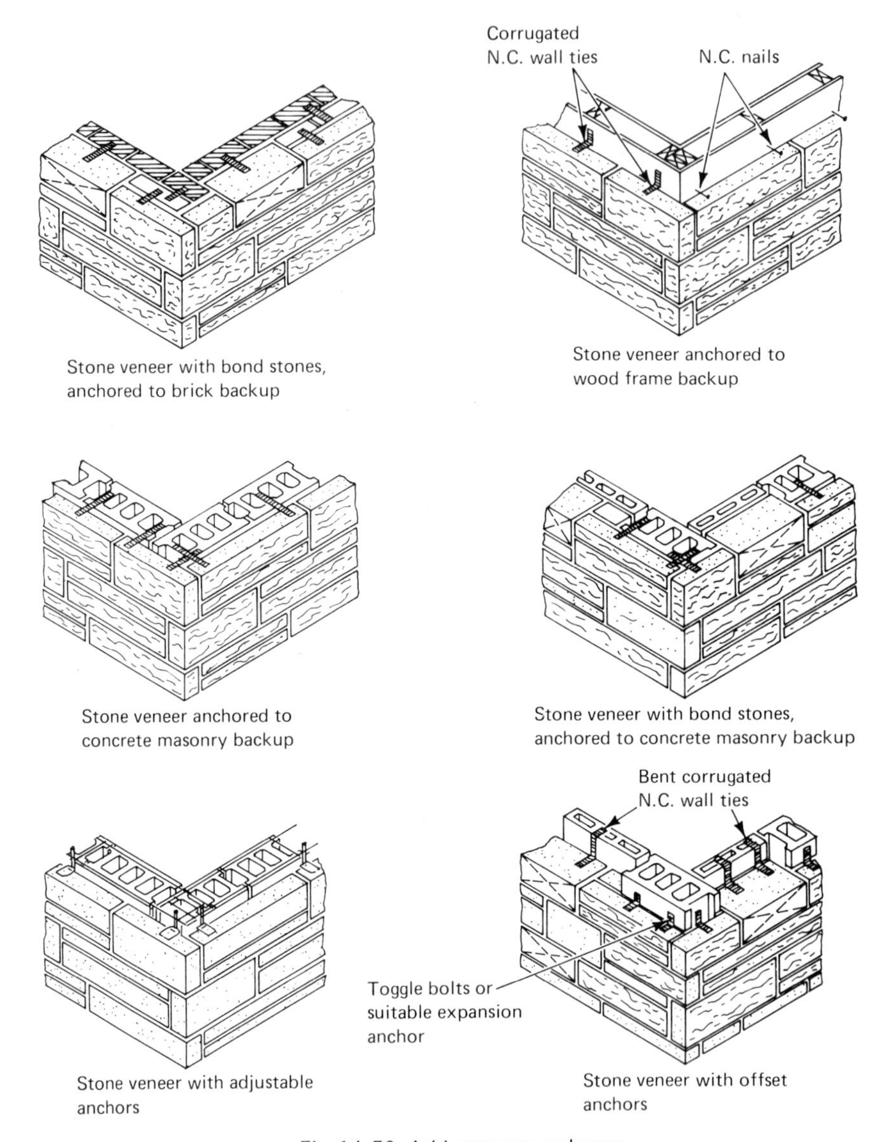

Fig. 14-70. Ashlar veneer anchorage.

depth of from 6 to 10 mm (¼ to ⅜ in). Each stone is then set firmly into the fresh mortar, and the excess that is forced out around its edges is removed. Care should be taken to keep the stone surface as clean as possible. After the mortar has set sufficiently so that it will not smear the stone, the joints can be rubbed and the entire surface cleaned with rough burlap and a rough fiber brush.

In the case of a masonry backup wall, the setting mortar may be applied directly over a dampened wall, following the same procedure as outlined above. Normally the pattern used will be random rubble in both cases.

Ashlar stone veneer is supported by the foundation or by shelf angles and anchored to a load-bearing backup wall (see Fig. 14-70) by a combination of *mortar, mechanical anchors* and *bond stones.*

Random and coursed ashlar are both made from strips of stone of various heights, ranging from 55 to 335 mm (2⅛ to 9¼ in) and cut to length as desired at the jobsite.

Stone Curtain Wall

Stone panels for curtain wall construction are produced in a number of thicknesses from 50 to 150 mm (2 to 6 in), depending on the kind of stone and the particular design requirements. Face dimensions also vary with the kind of stone, the manufacturer and the design but will range from 600 to 1500 mm (24 to 60 in) wide and from 1200 to 5400 mm high (48 to 216 in).

Stone Panel Support Systems

Stone facing panels may be carried directly *on the foundation and floors* of a building or *on shelf angles or plates* anchored to columns, spandrel beams or floors. There is little difference between supports and anchors for either a concrete or a steel frame; in either case, stone panels should be structurally supported at vertical intervals not greater than the distance between floors.

The most commonly used method of supporting stone panels is the shelf angle or plate method, in which a projecting angle or plate is attached to the building frame, as shown in Figures 14-71 and 14-72. This method allows all panels to be of the same thickness, and as a result, both fabrication and setting methods will be uniform.

With the direct bearing method, no structural support is required to be attached to the building frame. At each floor level a *bond stone* is introduced into the system, which bears directly on the foundation or floor slab and supports the weight of the stone above it. (See Fig. 14-73.) The main disadvantage of this system is that both stone and backup wall must be erected at the same time.

In both systems, lateral support is provided by anchors that may be located *above, below* or *in front of* the floors. (See Fig. 14-74.)

The details shown in Figures 14-72, 14-73 and 14-74 are representative of the many ways to support and anchor stone panels and are not meant to illustrate all of the methods possible or to indicate that all of the ones shown should be used on one building. For simplicity of construction, as few types of anchors as possible should be used for one building.

Mortar for Stone Panels

Cement for stone mortar should be white masonry cement and should meet the non-staining requirements of ASTM C91-70. The addition of hydrated lime or like amounts of ground limestone will further improve the working quality and the water retentivity, even though the initial shrinkage may be increased. The increased working quality should enable the mix to adjust to the extra shrinkage. If the setting mortar is also to be the pointing mortar, the sand used in the mix should also be white.

Setting Stone

Placing stone in position in a wall is known as *setting,* and the lifting into place or setting may be done by hand or by crane, depending on the weight of the units. Heavy stones are picked up by *grab hooks* or by *lewises*—pins or wedges—inserted into holes or slots in the back of the stone. (See Fig. 14-75)

Building stone should be stored clear of the ground and protected adequately from the weather. Before setting, the stone should be thoroughly cleaned and dampened to reduce the initial rate of absorption and so prevent

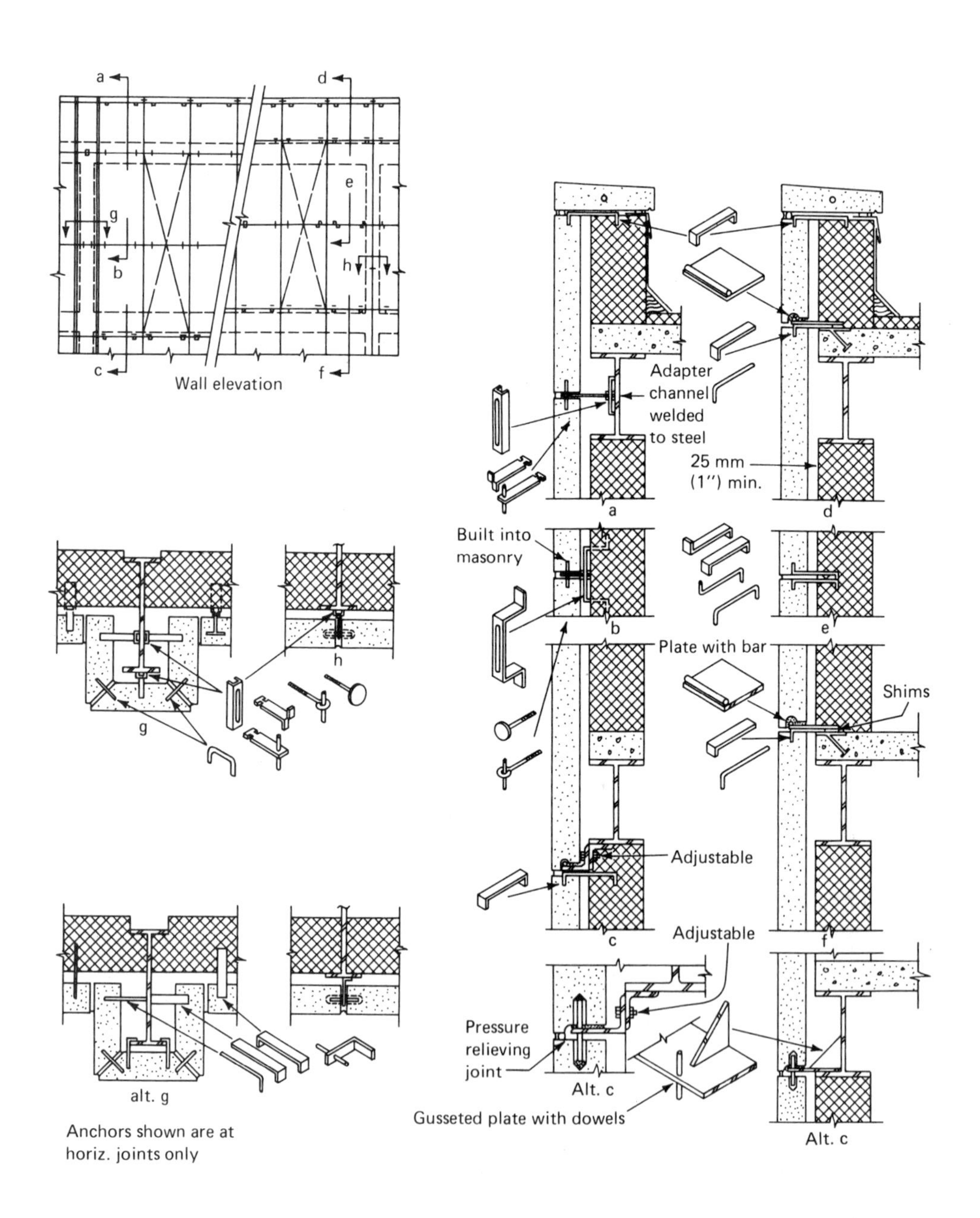

Fig. 14-71. Structural support by steel frame. (*Courtesy Indiana Limestone Co.*)

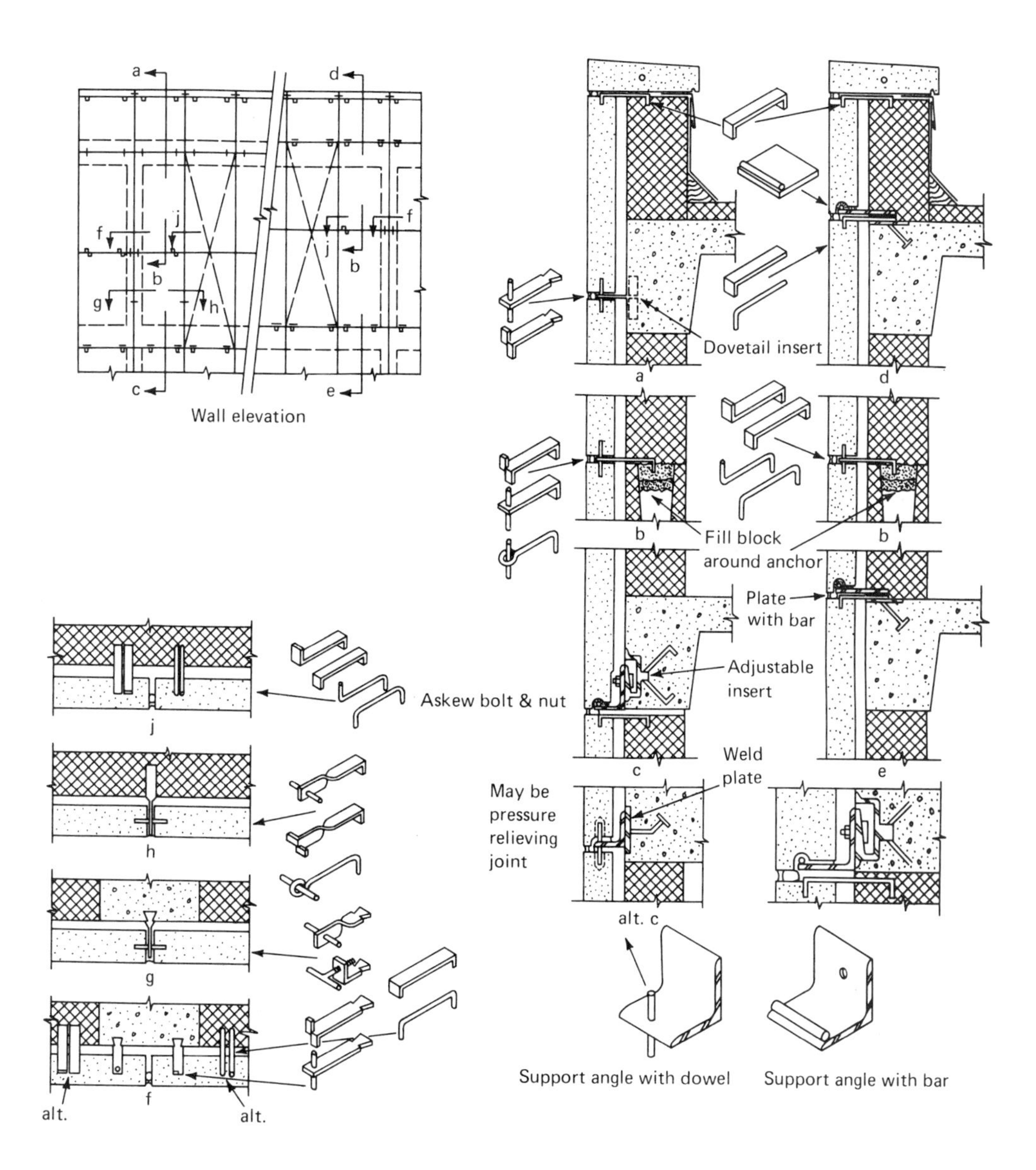

Fig. 14-72. Structural support by concrete frame. (*Courtesy Indiana Limestone Co.*)

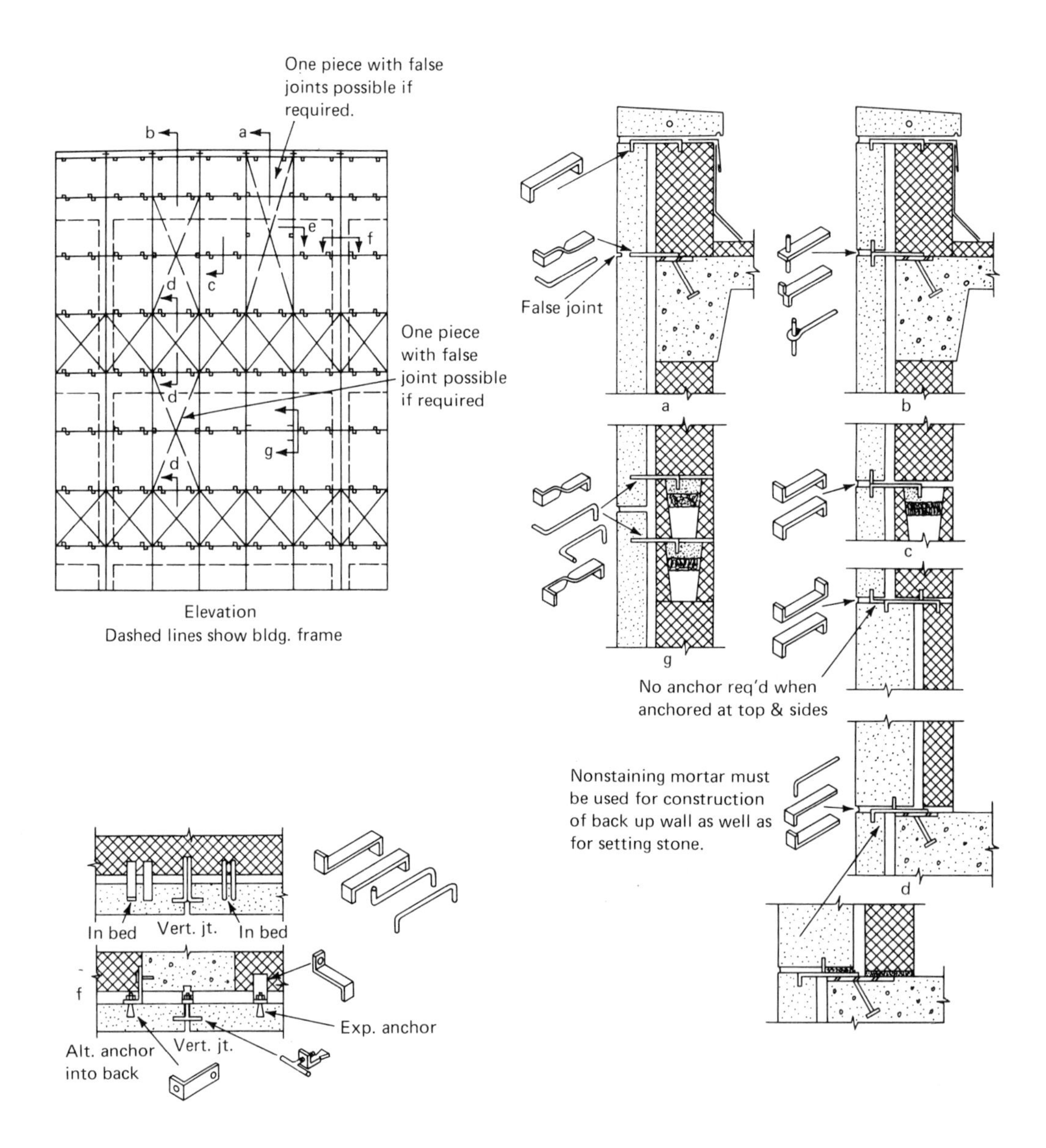

Fig. 14-73. Panels supported directly on floors. (*Courtesy Indiana Limestone Co.*)

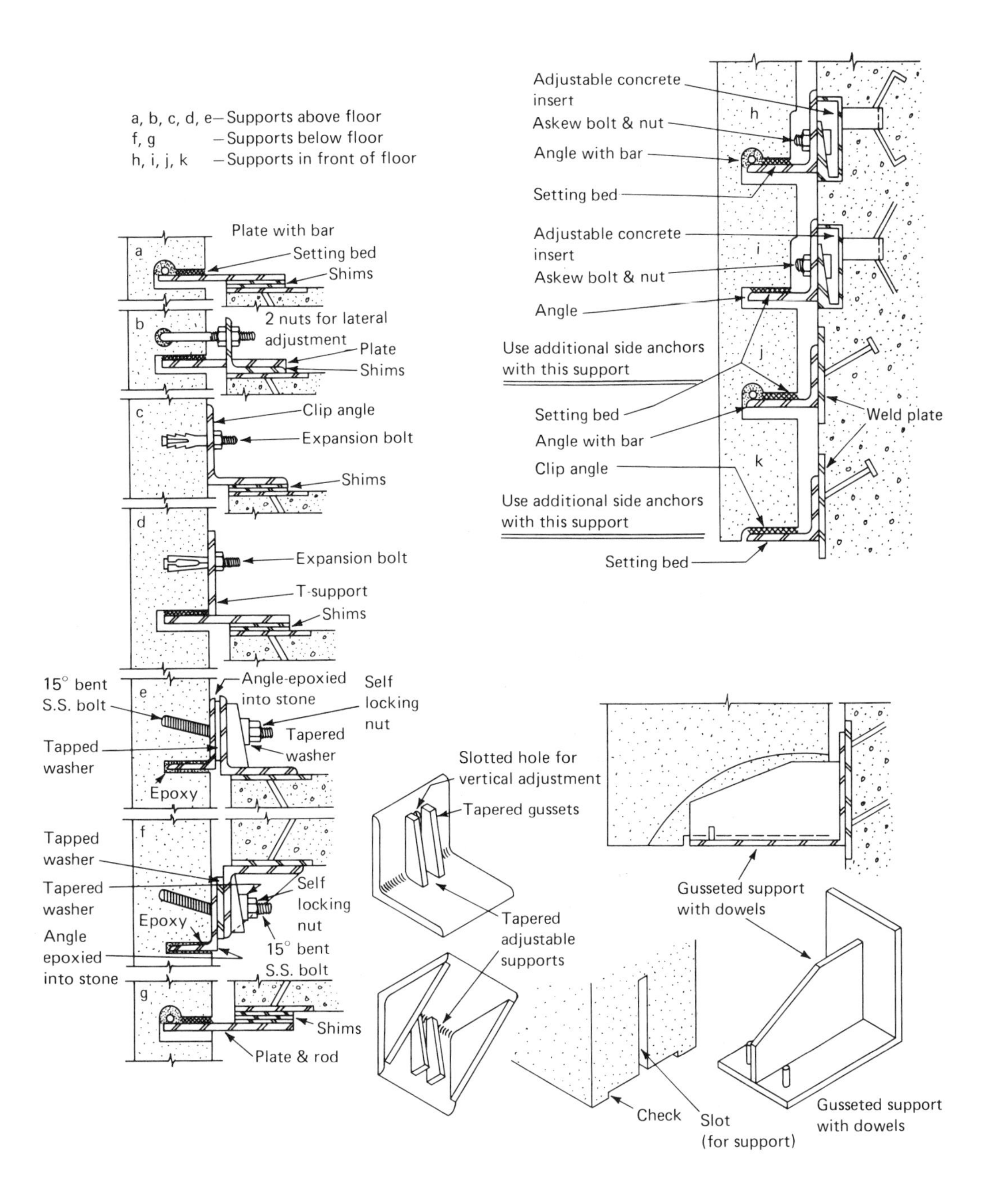

Fig. 14-74. Lateral support systems for stone panels. (*Courtesy Indiana Limestone Co.*)

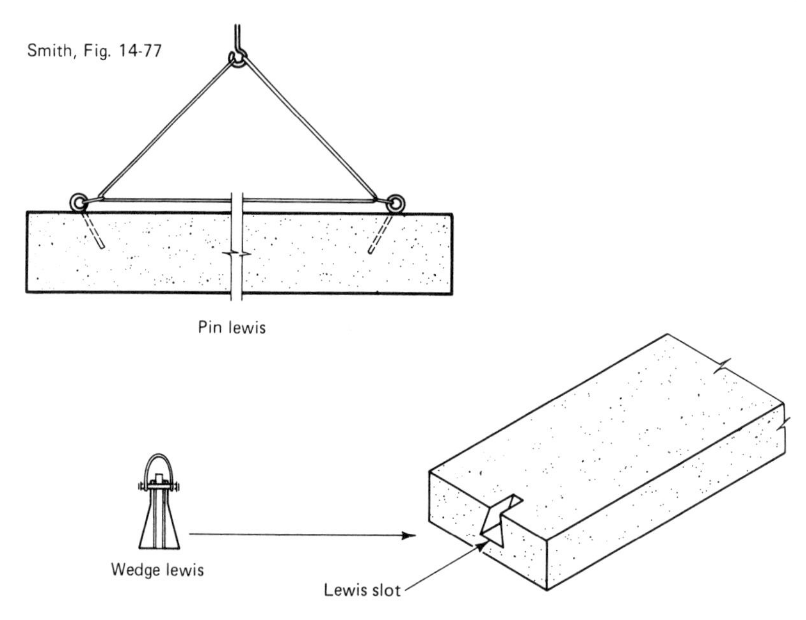

Fig. 14-75. Stone-lifting devices.

the extraction of moisture from the setting mortar.

Setting mortar should have as little slump as possible and still maintain good workability, and each stone should be set in a full mortar bed. All vertical joints, lewis holes, dowel holes and anchor holes should be filled with mortar to prevent the formation of water traps into which water might collect and freeze, thus splitting the stone.

Heavy stones may require supports in the horizontal joints to carry the weight until the mortar has set. These may be in the form of a number of *lead* or *plastic buttons* set into the joint, or they may be *soaked softwood wedges* inserted into the joint as indicated in Figure 14-76. When wood wedges are used, they are removed after the mortar has hardened sufficiently to carry the weight of the stone and after the wedges are dry.

Heavy stones should not be set until the mortar in the joints below is sufficiently hard to bear the load.

Cavity walls formed with a stone facing

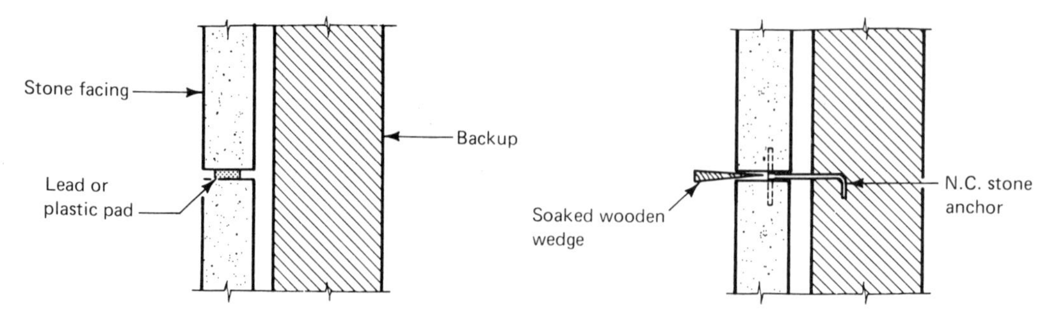

Fig. 14-76. Stone-setting supports.

should be kept clear of mortar droppings during construction, and the cavity should *not* be filled with grout made with normal portland cement, since alkali from grout or concrete can stain the stone.

Stone walls that are incomplete should always be covered at night and during rains to prevent future staining and efflorescence.

All ties and anchors used with stonework should be of the noncorrosive variety to prevent rusting and possible failure.

Pointing Stone Panel Joints

Pointing is the final filling and finishing of mortar joints that have been raked. Pointing cut stone after setting, rather than full bed setting and finishing in one operation, reduces a condition that tends to produce spalling and leakage.

Shrinkage of the mortar bed will allow some settling since the mortar bed hardens from the face inward. If the stone is set and the joints pointed in one operation, the settling, combined with the hardened mortar at the face, can set up stresses on the edge of the stone. For this reason, it is best to set the stone and rake out the mortar to a depth of 30 to 40 mm (1⅛ to 1⅝ in) for pointing with mortar or to a depth of 9 to 12 mm (⅜ to ½ in) for future application of sealant.

When joints are raked to a depth of 30 to 40 mm, pointing should be done in two or three stages, as shown in Figure 14-77. This allows each stage to seal shrinkage cracks in the preceding stage, and finally, the *concave-tooled* joint provides maximum protection against leakage.

Pointing Mortar

Pointing mortar should be composed of one part nonstaining cement, one part hydrated lime, six parts clean white sand that will pass a 1.25 mm (0.05 in) sieve and enough water to make the mix workable. An alternative mix would be one part nonstaining cement, 1½ parts lime putty and six parts clean, white sand.

To reduce shrinkage in pointing mortar, add just enough water to the mix at first to make a damp mixture; allow the mixture to sit from one to two hours. The mortar should then be remixed and enough water added to produce the desired plasticity and workability.

Stone Panel Joint Sealants

Joint sealants, which serve the same purpose as mortar and pointing, namely, the exclusion of moisture, are of two main types—*one-part* and *two-part* sealant systems. One-part sealants are called air-cure; that is, they depend on the presence of air to cure. The two-part systems depent on a catalyst for that purpose.

The materials in a sealant system consist of a *sealant backer* and the sealant itself. The backer is usually a plastic foam rope or rod, which is

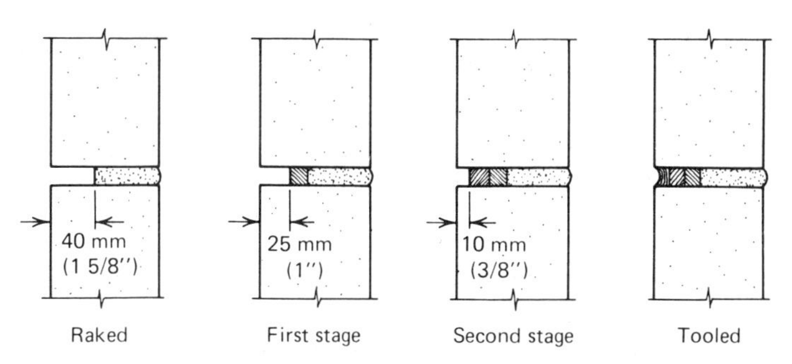

Fig. 14-77. Pointing steps.

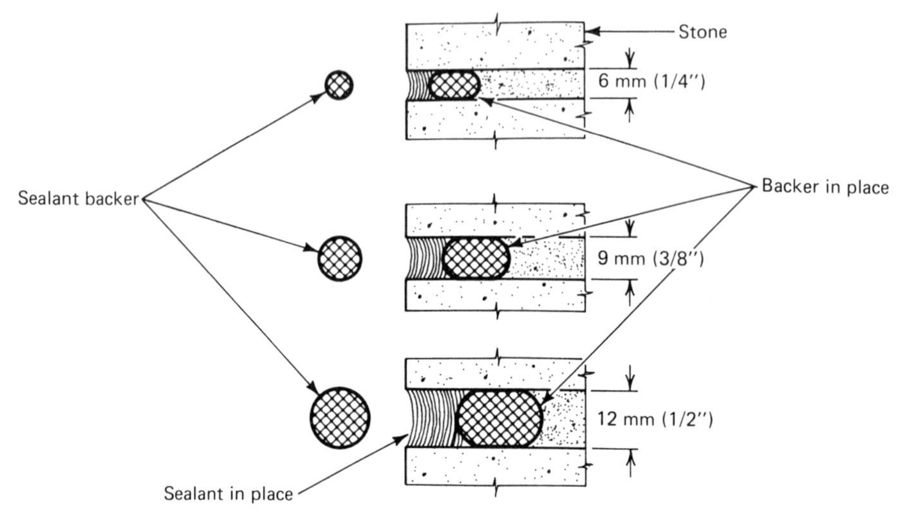

Fig. 14-78. Sealant and backer.

placed in the joint to a predetermined depth, after which the sealant is *gunned* into the joint against the stop. (See Fig. 14-78.) Some systems require a primer that must be applied to the joint's inner surface ahead of the sealant to ensure its adhesion.

Generally, the sealant should not adhere to the backer. Joints in which a sealant is used work best when the sealant is required to adhere only to parallel surfaces.

Expansion Joints

The coefficient of expansion of concrete and steel is much greater than that of most stones. As a result, differential movement between the building frame and the walls, due to thermal change, may cause damage to the stone facing. It is therefore important that expansion joints be provided in exterior stone curtain walls to reduce the effect of thermal expansion of the building frame.

In general, the location of expansion joints for stonework is the same as for other masonry materials. The joint itself normally consists of a premolded filler and a sealant compound. The filler should have sufficient compressibility to accommodate the amount of calculated expansion, with enough resilience to return to its original shape. The sealant compound should be completely elastic and provide lasting adhesion to the contact surfaces.

Pressure-Relieving Joints

In order to relieve the pressure between panel ends at shelf angle connections, a *pressure-relieving joint* may be placed under the shelf angle at each floor level. Figure 14-79 illustrates one method of designing such a joint. A compressible material—usually a semiplastic product—is placed in the bottom of the dowel holes in the ends of the meeting panels and under the shelf angle. In some cases, the material under the angle is omitted. The joint is then sealed with a backer rod and joint sealant.

Stone Interior Facings

Basically the same methods are used to apply stone to an interior facing as for exterior work. Thin stone veneer is set in a mortar bed (see Fig. 14-80), whereas ashlar is tied to the backup wall with strap or rod anchors.

Column facings are tied back with strap

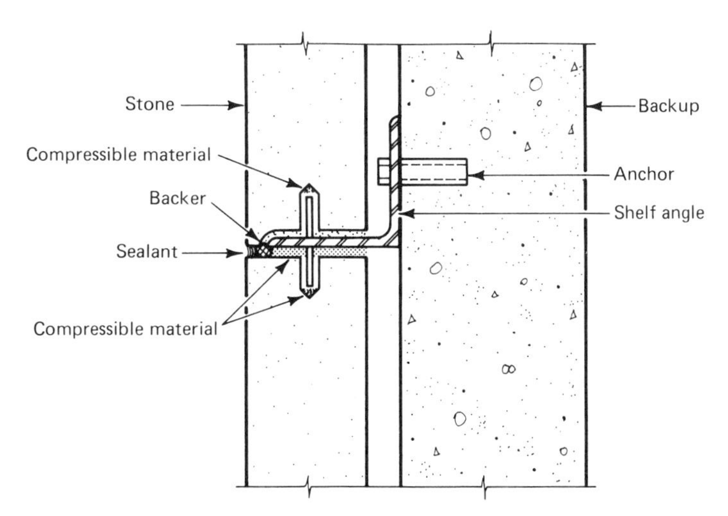

Fig. 14-79. Pressure-relieving joint.

anchors and dowels, with some variation in method between steel and reinforced concrete columns. In the case of concrete columns, one or more dovetailed anchor slots, made from sheet metal, are cast into the face, and a dovetailed strap anchor with dowel is used for lateral support. With a steel column, twisted and bent strap anchors are welded or pinned to the webs of the H-column.

Facing stones may meet at corners in a *butt* joint, or a *quirk* corner may be used, as shown in Figure 14-81.

Stone soffits are hung from noncorrosive rod hangers attached to concrete inserts or welded to a steel frame. Stone fascia are supported on shelf angles in the same way as exterior panels. (See Fig. 14-82.)

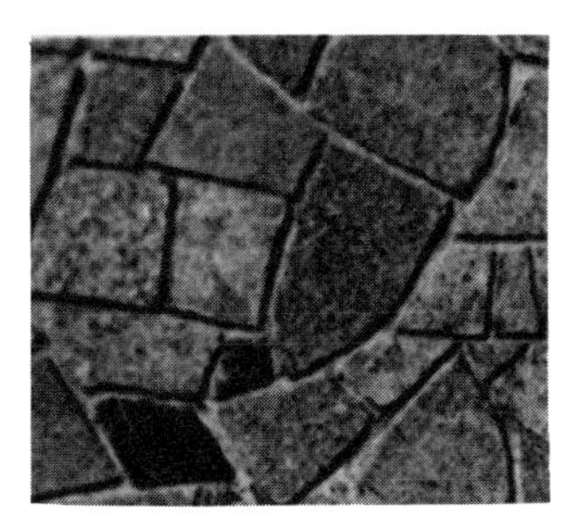

Fig. 14-80. Random rubble set in mortar bed.

Stone Trim

Stone trim includes such items as *base courses, watertable, quoins, belt courses, copings, sills, lintels, window and door jambs and heads* and *stair treads.*

A **base course** is the starting course for exterior stone paneling and normally will extend down to cover that portion of the foundation above grade. Flashing and weep holes must be provided at the bottom of the course. (See Fig. 14-83.) It is important that unprotected stone should never be covered with earth, and stone should be protected from being splashed with mud during construction.

A **watertable** is a stone course that is set above the base course to provide a *wash* surface at that point in the wall. (See Fig. 14-83.) Figure 14-84 illustrates some typical watertable designs.

Quoins are stones that are used to accentuate the corners of masonry walls and are usually used in conjunction with a brick structure. Sizes are such that they will coincide with a

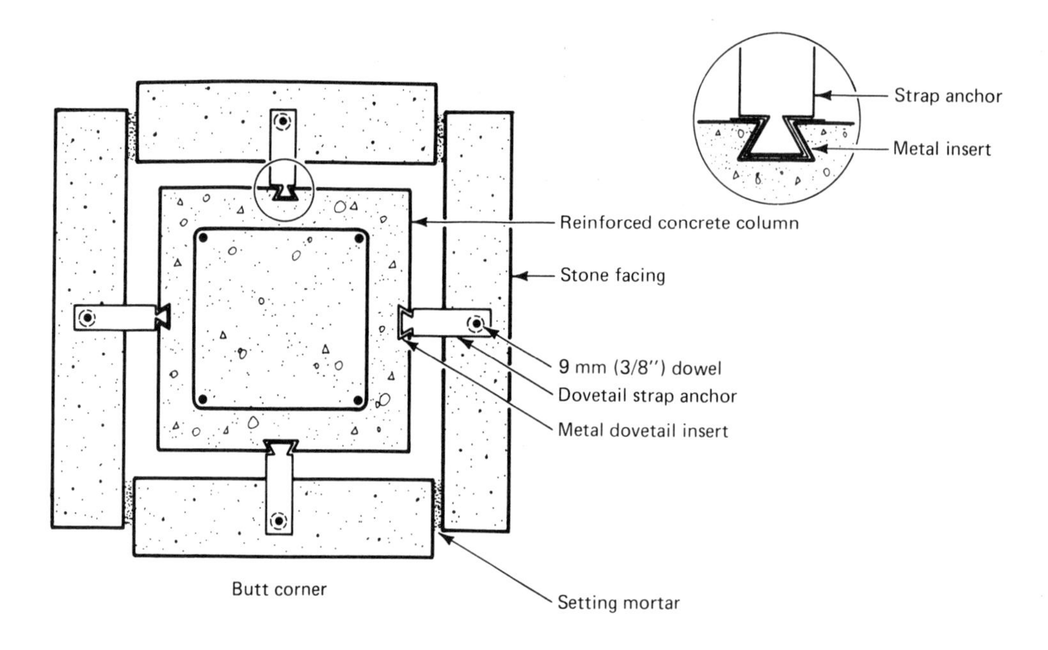

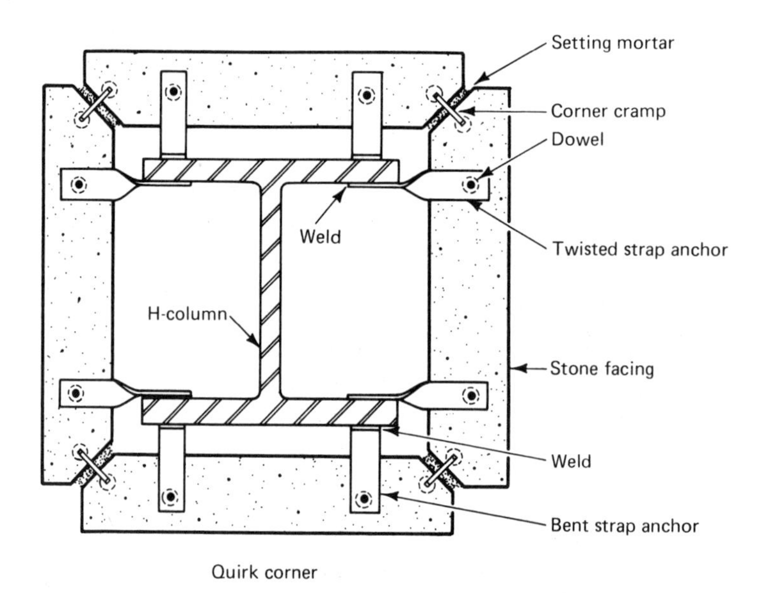

Fig. 14-81. Stone column facings.

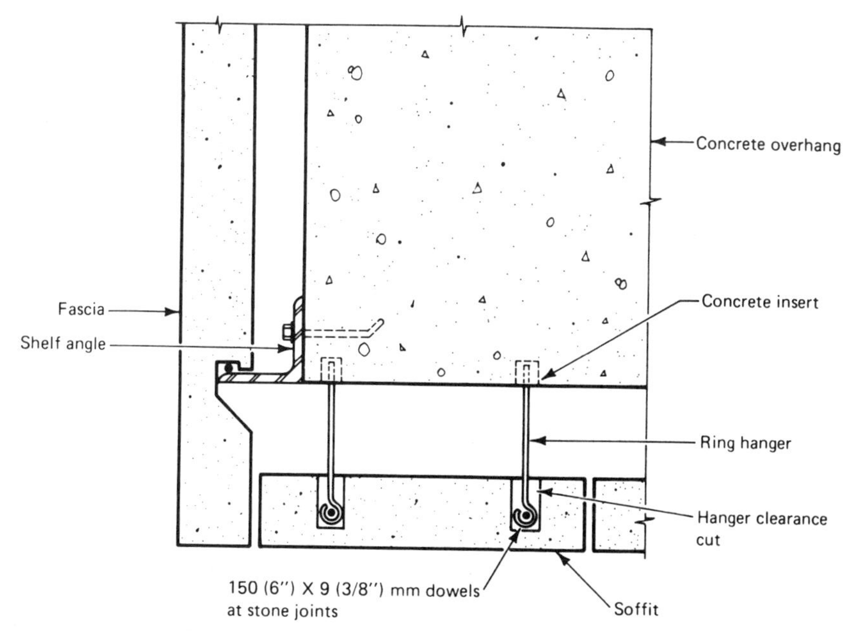

Fig. 14–82. Stone soffit and fascia supports.

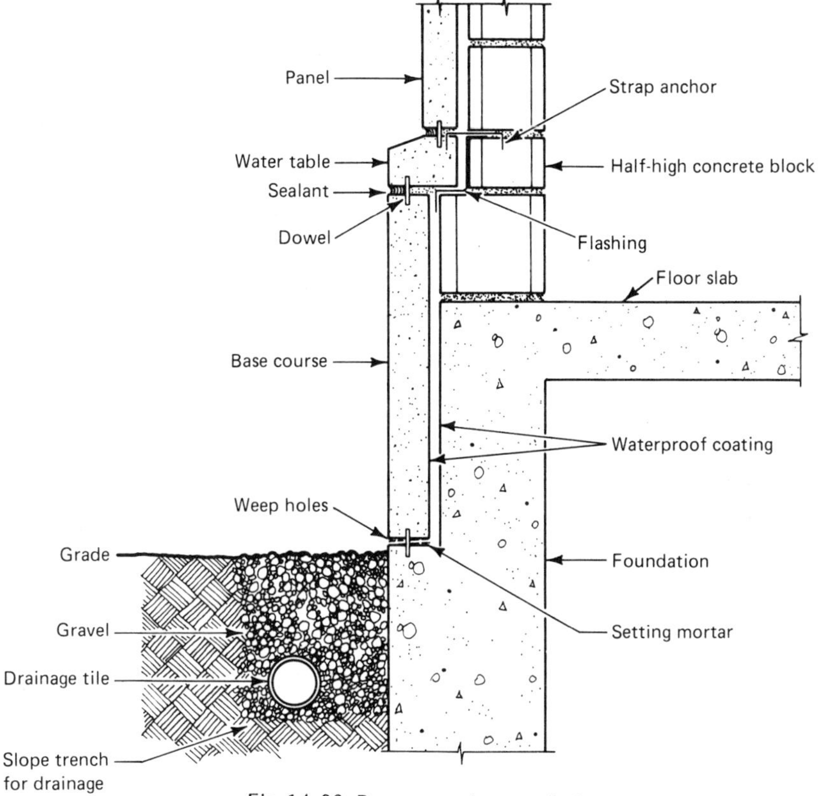

Fig. 14–83. Base course in stone facing.

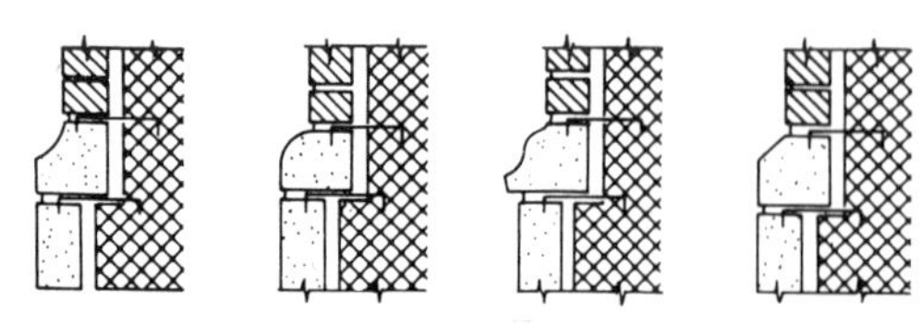

Fig. 14-84. Watertables.

given number of brick courses. Figure 14-85 illustrates some typical quoin details.

Cornices and **belt courses** are decorative courses of stone that may be used either with a stone-faced or a brick wall. The cornice is normally laid at the roof level, while a belt course may be used at any desired level, often in line with the window lintels, and is intended primarily to add some variety to an otherwise plain masonry wall. Belt course stones are similar to those used for watertable, and methods of setting and anchoring are also similar. Typical cornice details are shown in Figure 14-86.

The bed joint immediately below the heavy cornice stone should be raked back far enough to remove any compressive stress that would have a tendency to break off the edge of the stone below.

A good grade of nonstaining, waterproof paper should be laid between a copper gutter and the stone to prevent staining due to condensation. (See plan view, Fig. 14-86.)

Coping stones may be fabricated to any desired size and profile but should be sloped towards the roof. Typical flashing and anchoring methods are illustrated in Figure 14-87. Flashing and head joints require special attention because they are important factors in preventing water from entering the wall and causing extensive damage.

Continuous flashings are one good method of protecting the vertical joints, but if such flashing is not used, backer rod and sealant, properly installed, is preferable to mortar for the head and top joints between stones.

Copings meeting at the corners of walls that are subjected to possible expansion should be dowelled to the wall below, and the joint between stones should be sealed with a backer rod and sealant, making the joint an expansion joint.

The trim for doors and windows consists of the **lintel** or **head, jambs** and **sill.** Sills are of two general types–*slip sill* and *lug sill.* A slip sill is set into an opening and carries no load except the weight of the window itself, whereas the ends of a lug sill are built into the wall at the sides of the opening and support some of the wall load above them. (See Fig. 14-88.) Stone sills may be used under a variety of conditions; for example, with wood sash and with metal sash as Figure 14-88 indicates. It

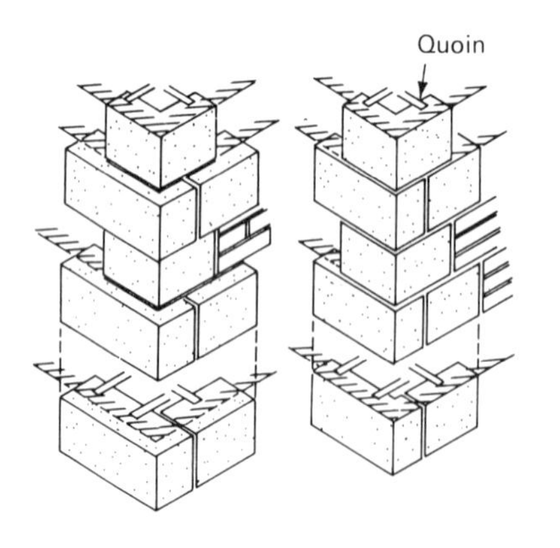

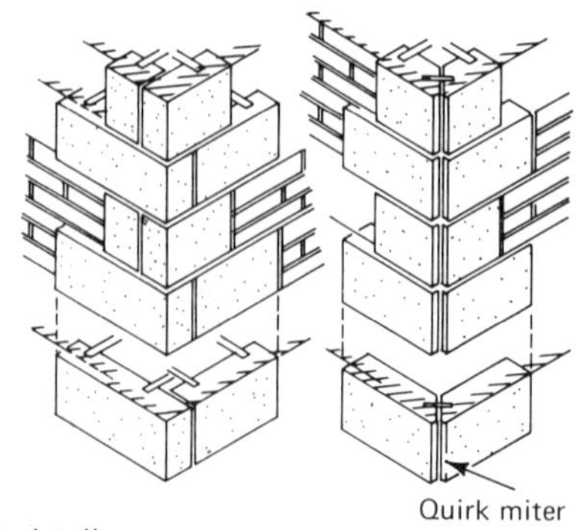

Fig. 14-85. Quoin details.

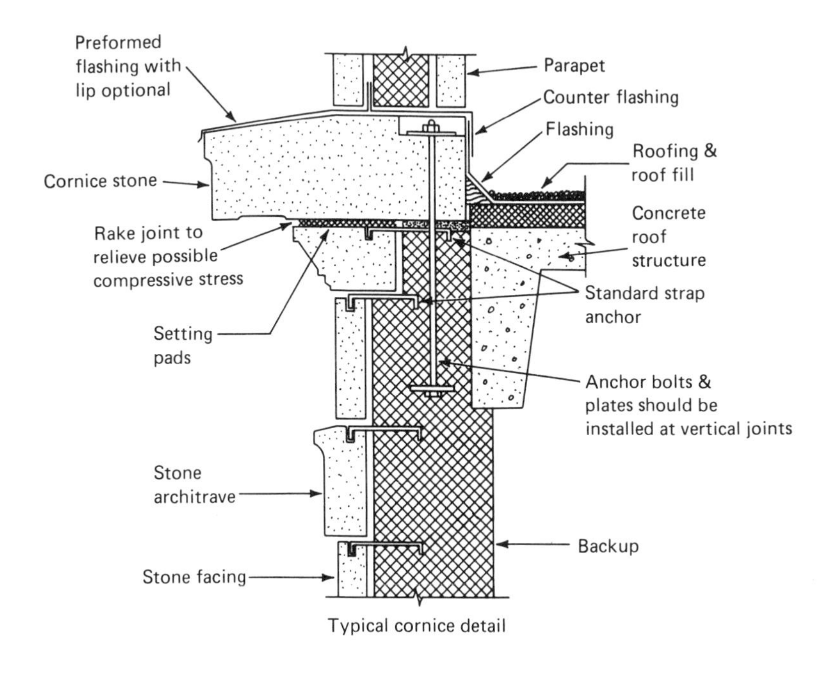

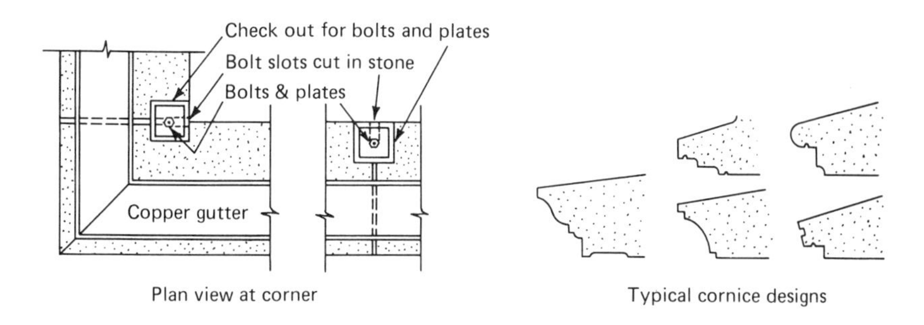

Fig. 14-86. Stone cornice details.

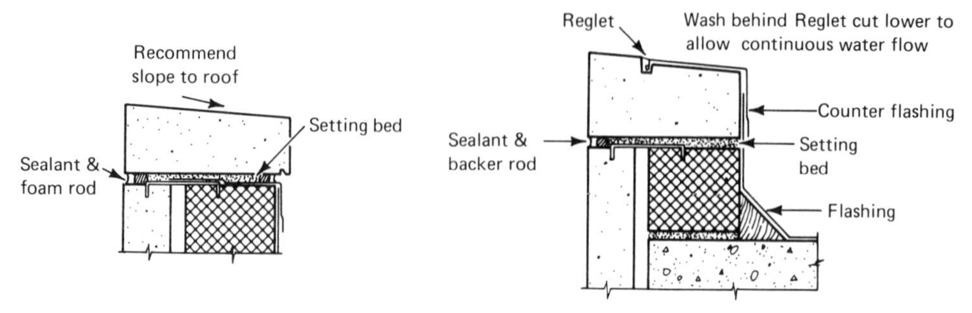

Flashing for stone copings

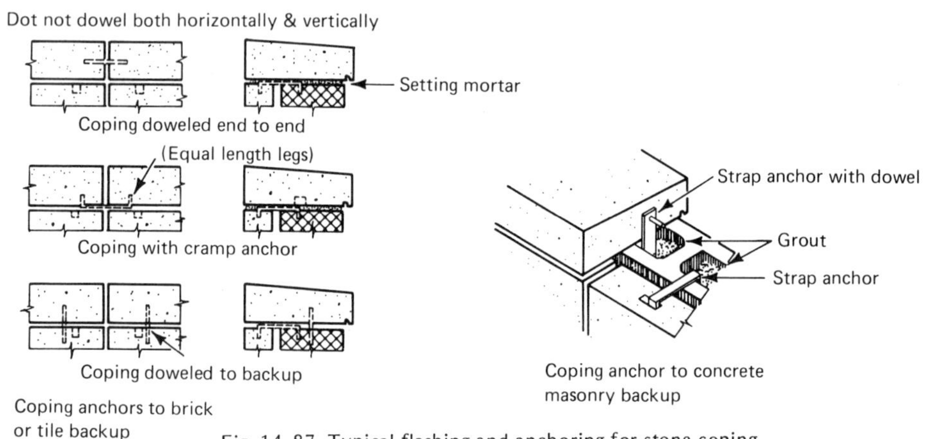

Fig. 14-87. Typical flashing and anchoring for stone coping.

should be noted that the lip of the sill should not have a height greater than its thickness.

An unlimited variety of side jamb designs are possible. Side jambs will rest either on the ends of a lug sill or on the wall at the edges of the opening. (See Fig. 14-88.) Dowels are used to anchor the bottom end of the jambs to the sill or wall, and cramp anchors tie the top ends to the backup wall.

A window or door lintel may be a single-stone unit spanning the opening, of such a depth and thickness as to adequately support the load above it or a steel structural member, faced with one or more pieces of stone. The alternative to a lintel is to use a stone arch over an opening, fabricated in much the same way as a brick arch.

Stone treads are used for both interior and exterior stairs and, in both cases, the basic stair is cast in concrete or framed from steel and the treads applied as a finish material.

Several types of safety nosing may be used with stone treads, as indicated in Figure 14-89.

USE OF SCAFFOLDING

Scaffolds and hoisting equipment have a very important place in the masonry construction industry. At heights greater than a mason can reach from the ground, some type of platform must be provided from which he can work safely and on which the materials that he is using may be placed. In order to get the materials to that platform level, some type of hoisting equipment must be used.

Most of today's scaffolds are made from

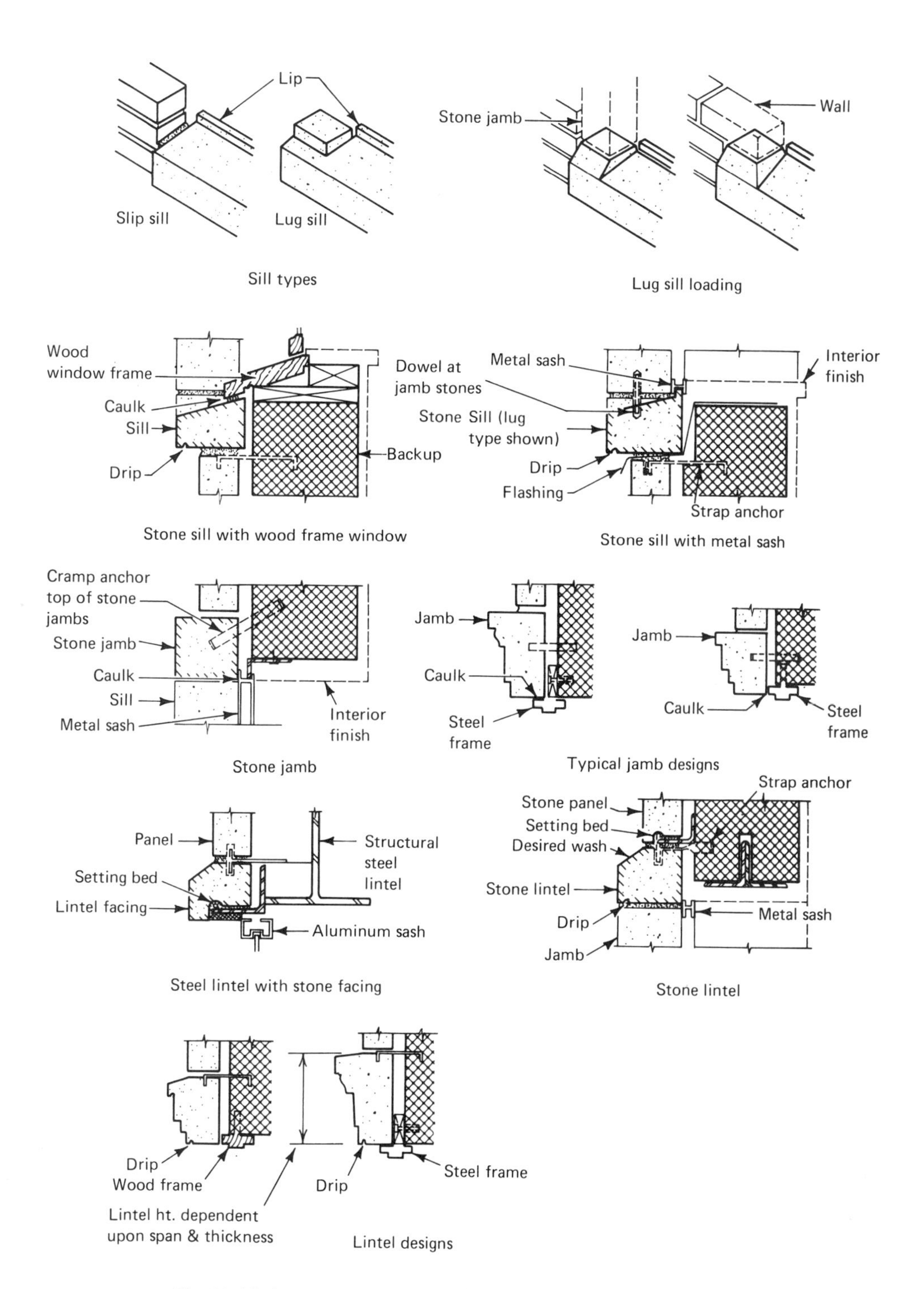

Fig. 14-88. Door and window trim. (*Courtesy Indiana Limestone Co.*)

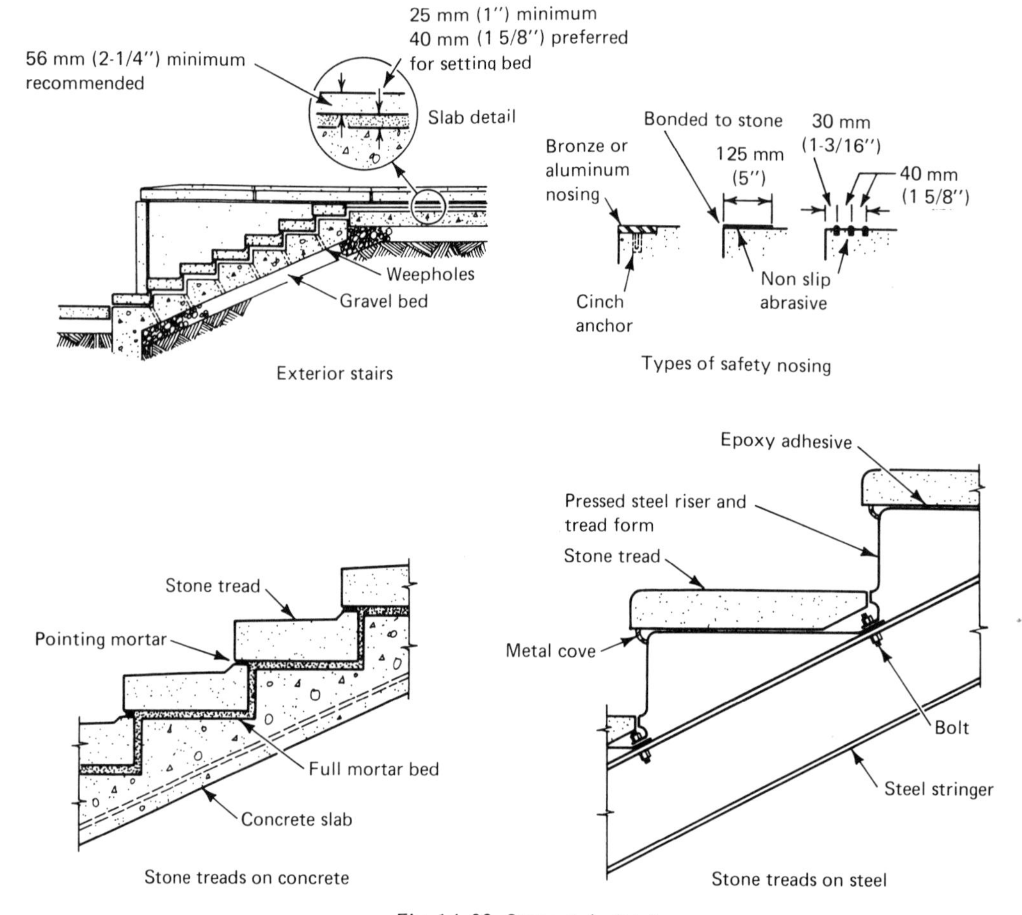

Fig. 14-89. Stone stair details.

steel, in a variety of types and designs. Two of the typical ones are *sectional* scaffolding and tower scaffolds. The platforms that support workers and materials will usually be *wood* or *metal scaffold planks.*

Sectional Scaffolding

Sectional scaffolding is made up of tubular steel in *double-pole* style, as shown in Figure 14-90. A standard unit consists of two end frames and two crossed braces, the ends of which fit over studs on the inside of the end-frame legs, where they are held in place by wing nuts. A heavy-duty type of tubular scaffold is made with double legs and top rail, as illustrated in Figure 14-91.

The height of such scaffolds is increased by setting one unit on top of another, with a *coupling pin* set into the top end of one set of legs and the bottom end of the other. In this way scaffolding properly braced and tied to the building, can be raised to any reasonable height. (See Fig. 14-92.)

Base plates, casters, screw jacks and *guard rails* are accessories that are available with scaffolds of this type.

Fig. 14-90. Tubular scaffolding, one section high. (*Courtesy Portland Cement Assn.*)

Fig. 14-91. Heavy-duty tubular scaffolding. (*Courtesy Portland Cement Assn.*)

Fig. 14-92. Tubular scaffolding used on high-rise building.

Fig. 14-93. Tower scaffold in use. (*Courtesy Morgen Mfg. Co.*)

Tower Scaffolds

Another type of scaffold used by masons is the *tower* scaffold, as shown in Figure 14-93. A standard unit consists of a pair of towers tied together with *stringer braces* and *crossed braces.* (See Fig. 14-94.) Pairs of towers are connected by stringer braces to make continuous scaffolding of any desired length. Each tower supports a *carriage* that carries planks to form a mason's platform and a laborers' and material platform. On each tower there is also a *winch,* used to raise the platform to the desired height.

The height of these scaffolds can be increased by adding long sections, called *inserts,* to the ends of towers to produce scaffolds that will reach to any required height. (See Fig. 14-95).

Steel scaffolding is made with the utmost consideration for safety, but of course care must be exercised in its erection and use. The following rules should be observed by erectors and workmen:

1. Provide sufficient sills or underpinning, in addition to standard base plates, for all scaffolds to be erected on fill or other soft ground.

2. Compensate for unevenness of ground by using adjusting screws, rather than blocking.

3. Be sure that all scaffolds are plumb and level at all times.

4. Anchor scaffold to the wall at least every 8 m (26 ft) horizontally and every 6 m (20 ft) vertically.

5. Adjust scaffolds until braces fit into place with ease.

6. Use guard rails on all scaffolds, regardless of height.

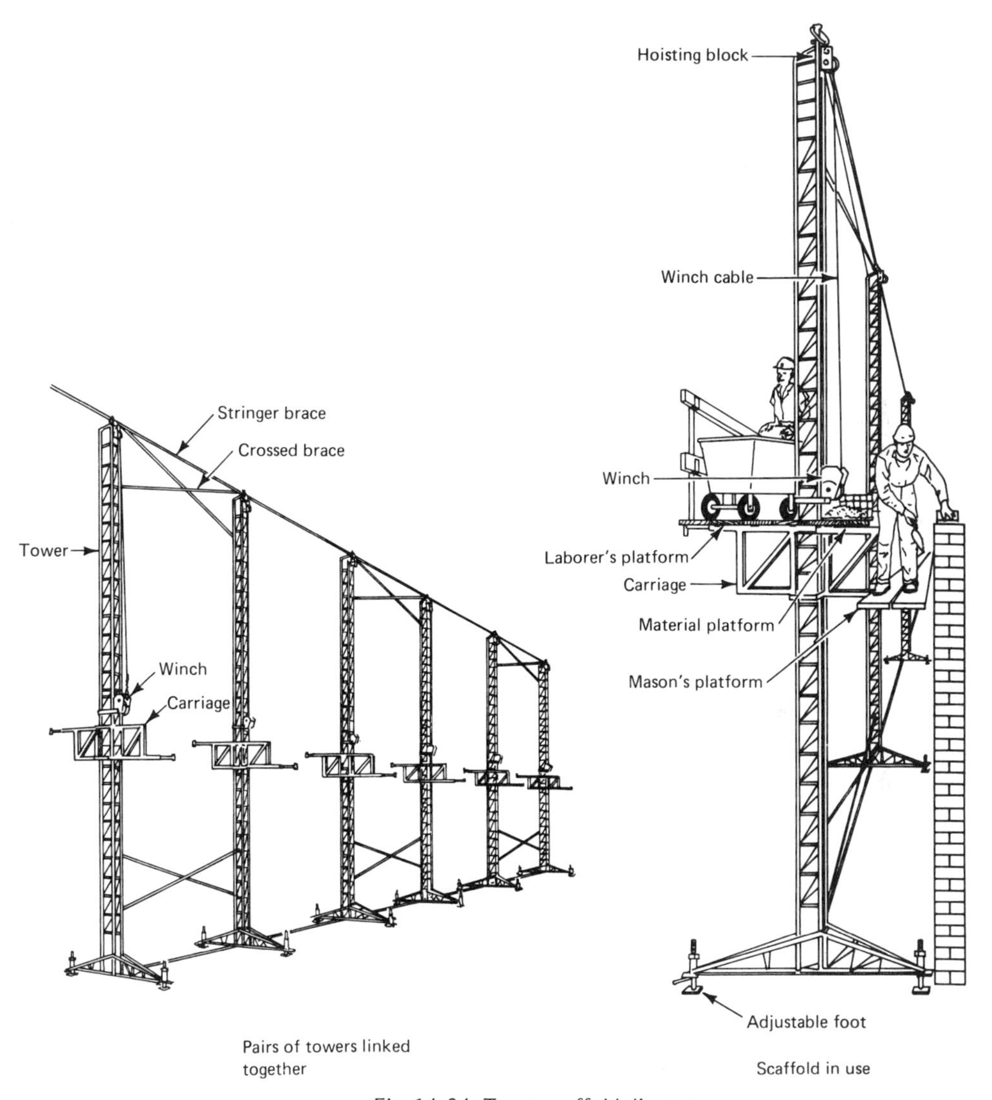

Fig. 14-94. Tower scaffold diagram.

7. Use ladders–not the cross braces–to climb scaffolds.

8. Tighten all scaffold bolts and wing nuts.

9. All wood planking used on a scaffold should be of sound quality, straight-grained and free from knots.

10. When using steel planks, always fill the space available because steel planks tend to slide sidewise easily.

11. Handle all rolling scaffolds with additional care.

12. Horizontal bracing should be used at the bottom, the top, and at intermediate levels of 6 m (20 ft).

Fig. 14-95. Tower scaffold on tall building. (*Courtesy Morgen Mfg. Co.*)

Hoists

A great variety of hoisting equipment is available, the choice at least partially depending on the total amount of material to be hoisted, the rate at which it must be hoisted and the height involved.

For relatively low lifts, typical equipment which might be used would include a *fork lift,* similar to the type shown in Figure 3-3, or lifts of the type illustrated in Figure 14-96(a) and (b), which will lift about 225 kg (500 lb) to a height of about 6 m (20 ft). For greater heights, an *elevator-type materials lift,* a *tower crane* mounted on the building or beside it or a *mobile crane* (see Fig. 14-92) may be used.

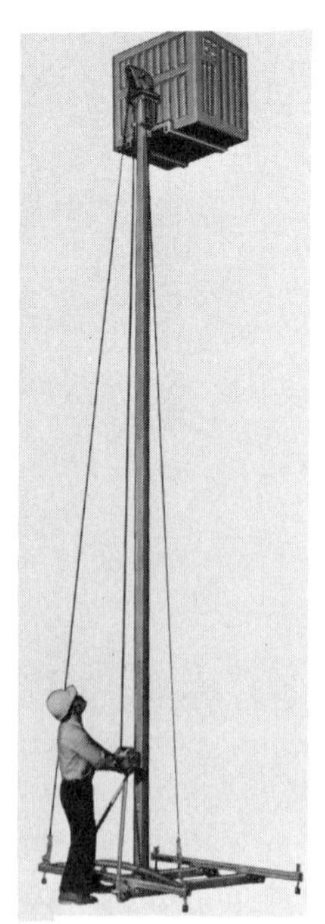

Fig. 14-96. (left) Materials hoist attached to scaffold; (right) free-standing materials hoist. (*Courtesy Vermette Machine Co.*)

Glossary of Terms

Adhesion The sticking together of substances in contact with one another.

Caulk To seal up crevices with some flexible material.

Chase A groove or channel.

Contemporary Of the same age.

Cornice A decorative, horizontal, projecting masonry course near the top of a wall.

Double-Pole Scaffold A scaffold with two legs at each support location.

Jamb Blocks Concrete blocks, made to accommodate door or window jambs.

Lime Putty A paste resulting from the slaking, by soaking, of quicklime in water.

Lintel A horizontal member spanning the top of an opening in a wall.

Neoprene Synthetic, thermo-setting polymer, resembling rubber.

Rack-Back Lead A brick lead sloped in two directions.

Reglet A narrow channel.

Soffit The flat underside of a part of a building such as an eave, stairway or archway.

Turnbuckle An elongated ring, with threads at both ends, one right-handed and the other left-handed.

Wash A sloping surface, designed to shead water away from a wall plane.

Review Questions

1. Explain the difference between a *load-bearing* wall, a *curtain* wall, and a *veneered* wall.
2. Explain what is meant by:
   (a) a collar joint
   (b) initial rate of absorption
   (c) mortar bond.
   (d) header course
   (e) bedding to the line
   (f) a corner pole
   (g) a line block
3. Explain what is meant by *structural bonding* in masonry work and outline three ways by which structural bonding is accomplished.
4. List all the different locations in a building where you would use flashing and explain its purpose in each case.
5. With the aid of a carefully drawn diagram, outline how you would introduce a control joint into a long composite wall with concrete masonry backup and brick face.
6. Describe the specific function of each of the following structural elements found in a building:
   (a) Pier (d) Bond beam
   (b) Lintel (e) Buttress
   (c) Pilaster (f) Coping

7. In stone construction, outline the difference between a *base* course, a *water-table* course and a *belt* course.
8. Outline the specific function of:
   (a) a *weep hole* in a masonry wall
   (b) a *bondstone* in a stone wall
   (c) a *sealant backer* in a masonry joint
   (d) *pointing mortar* in stone finishing
   (e) *lime putty* in a mortar mix
   (f) a *bond breaker* at a control joint

Selected Sources of Information

Alberta Masonry Institute, Calgary, Alberta.
Brick Institute of America, McLean, Virginia.
Clayburn Brick Products, Vancouver, British Columbia.
Dresser Industries Canada Ltd., Vancouver, British Columbia.
Garson Limestone Company Ltd., Winnipeg, Manitoba.
Gillis Quarries Ltd., Winnipeg, Manitoba.
Indiana Limestone Company, Bedford, Indiana.
I.X.L. Industries Ltd., Medicine Hat, Alberta.

Chapter 15

ALL-WEATHER CONSTRUCTION

The techniques used to achieve good masonry construction will vary with the weather. Some advance planning and preparation is required so that changes in weather will not affect the quality or progress of the structure. The planning should take into consideration the effects of *hot weather, cold weather, wet weather* and *windy weather.*

HOT WEATHER CONSTRUCTION

Building with masonry in hot weather [30°C (86°F) and over] can cause special problems. High temperatures can cause the materials to become very warm, affecting their *performance.* Rapid evaporation will also occur that will have an effect on hydration and curing. Special consideration must be given to the *handling and selection of materials* and to *construction procedures* during hot weather.

Performance

The physical properties of masonry will change with an increase in temperature:

1. Bond strength can be poor as units are hotter and drier, causing an increase in suction rate.

2. The compressive strength of the mortar and grout will be weaker as water is quickly evaporated, leaving little for hydration.

3. Workability is affected, as more water is required in the mortar for constant consistency and in grout to make filling of spaces possible.

4. Heat will affect the amount of air entraining used, since more is required in hot weather.

5. The initial and final sets of mortar will occur faster.

6. Water will evaporate quickly on a joint's exterior surfaces, causing a decrease in strength and durability.

7. The initial water content of mortar will be higher, but the placing will be difficult and the time period will be short.

Handling and Selection of Materials

When hot weather is expected, the materials should be stored in a cool place. Increasing the cement content will cause the mortar and grout to gain strength quickly. The amount of lime can be increased giving the mortar a higher water retentivity. (See Fig. 5-9.) The use of an air-entraining agent will also assist the retentivity of the mortar.

Covering the aggregate with a plastic sheet will retard the evaporation of water. Adding extra water will help keep the aggregate cool as evaporation has a cooling effect.

The units used should be stored in the shade and covered. The use of cold water on the units, especially block, for cooling purposes is not recommended. The use of cold water or ice water as mixing water will lower the temperature of the mortar.

Construction Procedures

When placing masonry units in hot weather, special consideration should be given to all equipment that comes in contact with the mortar. Flushing the mixers, tools and mortar boards occasionally with cold water helps keep temperature to a minimum. Units should be laid in the shade of walls or screens, moving as necessary to keep out of direct sunlight. Cover the wall and damp cure as soon as completed. (See Fig. 15-1)

Mortar should not be mixed too far ahead, and when mixed, should be stored in a cool shady place. Avoid placing long mortar beds ahead of the units as bond is reduced. When extremely high temperatures are expected, consideration should be given to stopping placement of masonry during the hottest times of day.

COLD WEATHER CONSTRUCTION

Masonry construction, once only carried on during the warm months of the year, now is continued for the entire year. On large projects, it is common to see masons working in favorable conditions within heated enclosures. On projects where enclosures are not available the use of warm mortar and dry masonry units are effective techniques. The *performance* of materials will vary with the temperature range. Special consideration must be given to the *selection and handling* of materials in cold weather [less than 5°C (41°F)]. Productivity, workmanship, and methods depend on whether or not *protection* is provided.

Performance

The performance of masonry *mortar* and *units* vary with a drop in temperature. The amount of water in the mortar when it freezes is significant. The presence of wind affects the temperature, as the wind chill factor creates temperatures less than measured air temperatures as shown in Table 15-1.

Fig. 15-1. Wall covered and damp cured. (*Courtesy Portland Cement Assn.*)

Table 15-1. Wind Chill Factor

Wind Speed, Kilometers Per Hour													
100	-21	-27	-34	-40	-47	-53	-60	-66	-73	-79	-86	-92	-99
90	-21	-27	-34	-40	-47	-53	-60	-66	-73	-79	-86	-92	-99
80	-21	-27	-34	-40	-47	-53	-60	-66	-73	-79	-86	-92	-99
70	-20	-27	-33	-40	-46	-52	-59	-65	-72	-78	-85	-91	-98
60	-19	-26	-32	-39	-45	-51	-58	-64	-70	-77	-83	-89	-96
50	-18	-25	-31	-37	-43	-49	-56	-62	-68	-74	-80	-87	-93
40	-17	-23	-29	-35	-41	-47	-53	-59	-65	-71	-77	-83	-89
30	-14	-20	-25	-31	-37	-43	-48	-54	-60	-65	-71	-77	-82
20	-10	-15	-21	-26	-31	-36	-42	-47	-52	-57	-63	-68	-73
10	-4	-8	-13	-17	-22	-26	-31	-35	-40	-44	-49	-53	-58
Air Temperature, Degrees Celsius	0	-4	-8	-12	-16	-20	-24	-28	-32	-36	-40	-44	-48

°F	mph					
Air Temperature	5	10	15	20	25	30
+40	37	28	22	18	16	13
+30	27	16	9	4	0	-2
+20	16	4	-5	-10	-15	-18
+10	6	-9	-18	-25	-29	-33
0	-5	-21	-36	-39	-44	-48
-10	-15	-33	-45	-53	-59	-63
-20	-26	-46	-58	-67	-74	-79
-30	-36	-58	-72	-82	-88	-94
-40	-47	-70	-88	-96	-104	-109

Mortar. Mortar will hydrate and increase in strength only when it is kept above freezing. Damage to mortar subjected to freezing can be avoided if:

1. The water content of the mortar is below 6 percent at the time of freezing.

2. The mortar has hardened enough before freezing (kept above freezing for 48 hours).

3. A masonry wall, built with cold units, freezes within two hours of placing. This method is not recommended, due to the compressive load that occurs on the mortar as it thaws.

Mortar that is frozen and dried will be reduced in strength unless it is exposed to water in the curing process. In order to achieve the desired consistency, water requirements are lower in cold weather, and less air-entraining agent is needed to get prescribed air content. The strength gain of mortar is slower, but the final strength may be higher.

Masonry construction may proceed in cold weather providing the mortar ingredients are heated and unit and structure temperature are maintained.

Units. Units are generally smaller when cold, but they perform in the same way as warm units. If the units are frozen wet, they will have less absorptive quality, creating a reduction in bond. Preheated units may have a more absorptive effect on mortar than normal units as a cooling body tends to absorb more moisture.

Selection and Handling of Materials

The masonry units used in cold weather should have a higher I.R.A., be dry at the time of laying, and be warm enough to stop freezing. The mortar used can be made of high early strength cement and a smaller amount of hydrated lime. Admixtures in general should be avoided, though some air may be considered.

The materials should be covered and stored on a platform that is elevated above the ground to cut down on the heat loss to the frozen ground. (See Fig. 15-2.) Materials should be heated according to the recommendations in Table 15-2. Avoid overheating materials as a flash set can occur if the temperature of the mortar exceeds 50°C (122°F).

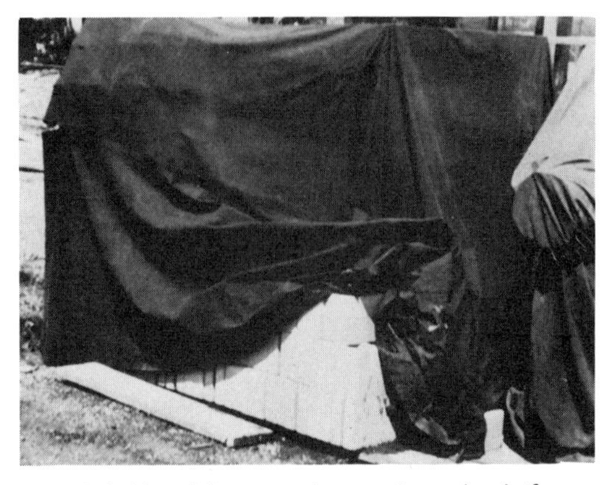

Fig. 15-2. Material covered on elevated platform. (*Courtesy Portland Cement Assn.*)

Protection

The use of protection for masonry construction in cold weather is more for the comfort of the masons than for the protection of the masonry. Productivity of the mason is generally increased in comfortable surroundings. The need for a wind break or a heated enclosure will depend on the amount of work to be done, the size of the project and the location. (See Fig. 15-3.)

Enclosures are generally made of a light framework of wood or metal, covered with polyethylene or canvas. They must be sufficiently strong to withstand the weight of snow

Table 15-2. Heating and Protection Recommendations

Air Temperature	Heating	Protection
Above 5°C (41°F)	Normal procedures	Cover walls at end of day to prevent moisture entering masonry
Below 5°C (41°F)	Heat water to maintain mortar temperature from 5°C to 50°C (41°F to 122°F)	Cover walls and materials with plastic or canvas to prevent wetting and freezing
Below 0°C (32°F)	In addition to above, heat sand. Frozen sand and units thawed	With wind velocities over 20 km/h (13 mph) provide wind breaks during the day and cover at night, keep above 0°C by using insulated blankets for 16 hrs
Below -5°C (23°F)	In addition to above, dry units heated to -5°C	Provide enclosure and heat to maintain 0°C for 24 hrs

Fig. 15-3. Enclosure. (*Courtesy Portland Cement Assn.*)

and be tied so they will not blow away in wind.

Enclosures can be heated with a variety of heaters (gas or oil), but care must be taken to insure sufficient ventilation so fumes will not accumulate. (See Fig. 15-4.) The cost of providing enclosures for cold weather construction is not as large as might be expected. It only adds about one to two percent of the total contract price to the cost.

Fig. 15-4. Heater. (*Courtesy Portland Cement Assn.*)

WET WEATHER

Building with masonry in rainy weather is possible if some type of shelter or covering is provided. The *performance* of the units and the mortar varies in wet conditions and some precautions must be taken in the *construction procedure*.

Performance

Rain can cause excessive wetting of materials, affecting their performance. The change in unit moisture content can cause considerable change in dimension. The amount will vary with the type of material used as shown in Table 15-3. Moisture will also reduce the absorptive quality of the units so that poor bond occurs between the units and the mortar. Water will evaporate more slowly so less mixing water need be added.

If it rains on a building before the mortar is set, the cementious material can be washed out, reducing strength and resulting in the possible collapse of the joint. The mortar will be washed

Table 15-3. Moisture Expansion of Various Materials

	Expansion on Wetting Percent Change in Length
Limestones	0.002 to 0.01
Clay and shale bricks	0.002 to 0.01*
Concrete	0.01 to 0.2**
Portland cement mortar	0.005 to 0.03
Lime mortar	0.001 to 0.02

(Courtesy National Research Council, Ottawa, Ontario)

*Highest expansions with soft burned bricks.
**Lightweight aggregates give higher expansions.

over the faces of the units causing a staining effect.

Construction Procedures

Building can be done in wet weather providing rain does not fall on the masonry materials or on the freshly laid walls. The cement, units and sand should be covered to keep them dry. (See Fig. 15-5.) They should also be stored off the wet ground so there is no migration of moisture from the ground to the materials.

A masonry wall, built in rainy conditions, should be built under or behind a shelter. This can be in the form of a roof or floor slab or inside an enclosure similar to the type used in cold weather. Walls should be protected from rain for 24 to 48 hours, depending on the temperature, so that the mortar is fully set and the bond has occurred. (See Fig. 15-6.)

Fig. 15-6. Wall covered to keep dry. (*Courtesy Portland Cement Assn.*)

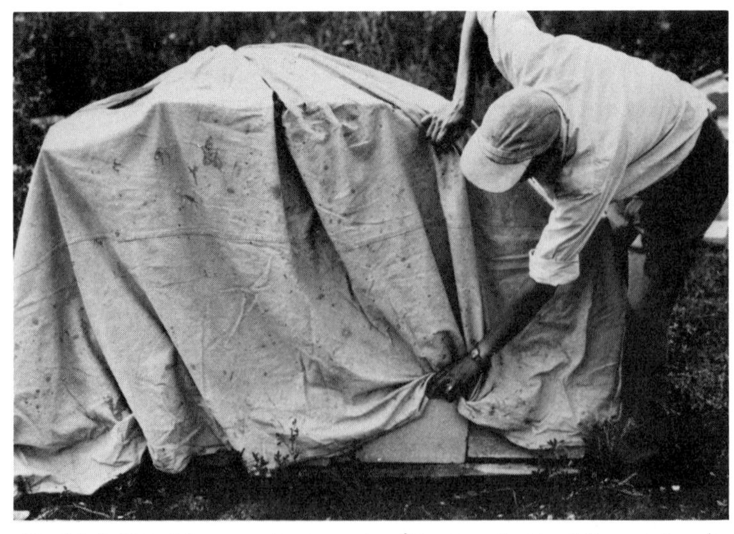

Fig. 15-5. Materials covered to keep dry. (*Courtesy Portland Cement Assn.*)

WINDY CONDITIONS

Building in windy conditions can proceed providing care is taken. The *mortar* must be adjusted for those conditions, and some means of *lateral support* should be provided for walls during construction.

Mortar

Mortar will react differently when building in windy conditions. If wind combines with heat, the water will evaporate very quickly, so the wall should be protected by wind breaks. Mortar should not be allowed to sit for long periods of time before the units are placed in the row, to avoid rapid evaporation. The amount of mixing water should be increased to allow for evaporation. The wall can be covered after placing is complete, and then damp cured.

Lateral Support

Masonry walls are subjected to wind loads while being erected. These winds may have sufficient velocity to cause the wall or a portion of it to fall, resulting in personal injury, significant damage and expense. The maximum safe unsupported height of a masonry wall will depend on:

1. Wall weight and thickness.
2. Wind velocity.
3. Wall exposure conditions.
4. Whether the wall is at ground level or on a high rise. (See Fig. 15-7.)

Grouting and reinforcing will increase the height of a masonry wall that can be unsupported. It seems that regardless of the circumstances, some sort of temporary bracing is required during construction.

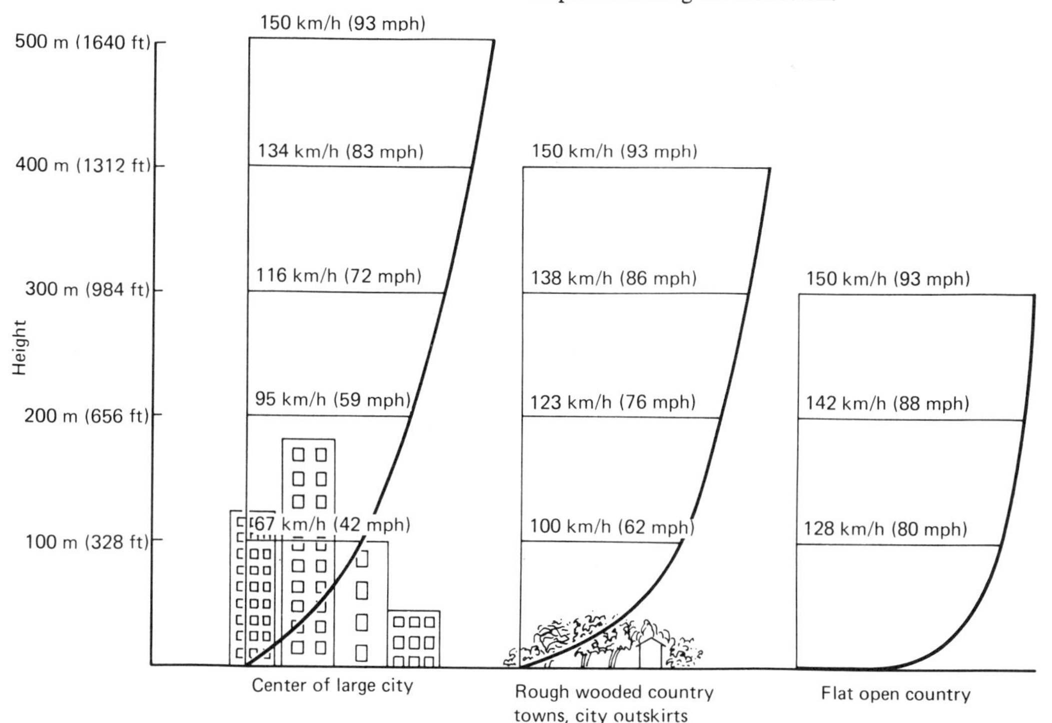

Fig. 15-7. Difference in wind velocity at ground level.

The pressure exerted on a masonry wall can be calculated by the following formula:

$$P = 0.05\ V^2\ (0.0027\ V^2)$$

where P = Pressure on the wall in pascals (lb/ft²).
V = Wind velocity in km/h (mph).

A wall will resist lateral loads in the horizontal and vertical spans depending on the height of the wall and the distance between cross walls. Bending due to wind is mainly in the vertical span. The bending stress in a free-standing wall is four times the stress in a supported wall and occurs at the bottom. (See Fig. 15-8.) Most bracing is placed in the vertical direction and should not be fastened only at the top of the wall.

The following is an example of the minimum bracing that would be needed for a free-standing wall, calculated on the basis of 640 Pa (13.4 lb) pressure exerted by a wind velocity of 100 km/h (62 mph).

1. At least two braces on each side of the wall.
2. Maximum horizontal interval of braces:
   –3600mm (12 ft) for a 200mm (8 in) thick block wall,
   –5400mm (18 ft) for a 300mm (12 in) thick block wall.

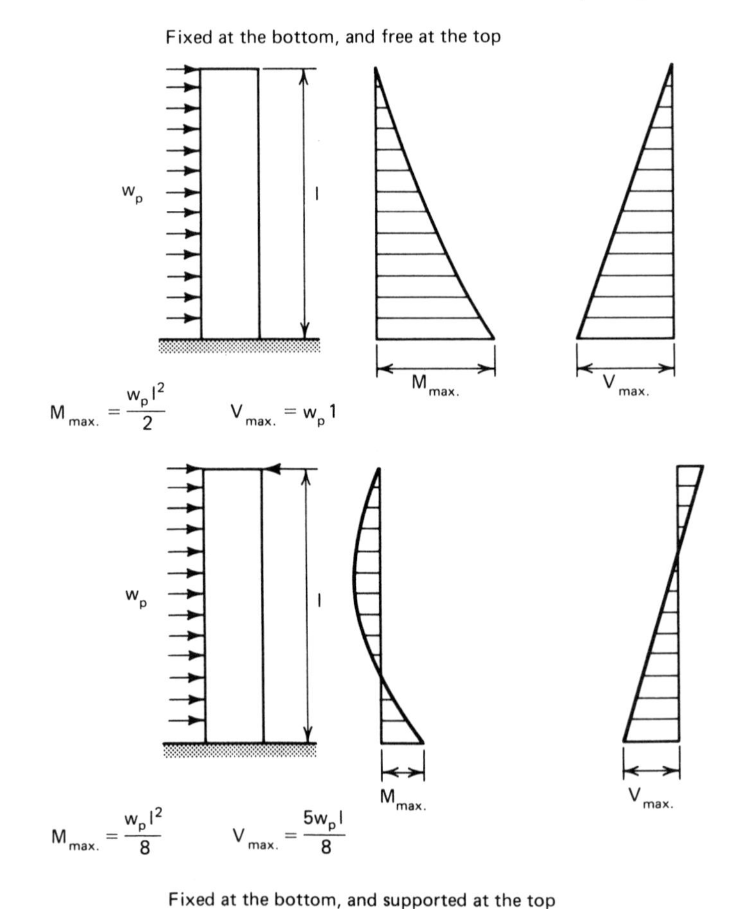

Fig. 15-8. Maximum moment and shear. (*Courtesy National Concrete Masonry Assn.*)

Cavity wall thicknesses are considered as two-thirds of the sum of the wythes.

3. Maximum height unsupported:
During erection, a masonry wall should not be higher than 10 times the thickness unless braced. Minimum size of lumber for adequate braces—38 × 140 mm (2 × 6 in) and 38 × 184 mm (2 × 8 in).

4. Braces should be inclined at 45° to 60° to the horizontal and be adequately fastened.

The following is an example of a method of finding the number of braces for wind. Generally impact forces are neglected.

Example
Panel size is 2400 mm (8 ft) high by 6000 mm (20 ft) long. Wind force is 140 km/h (87 mph). Force created by mass of wall = 218 kg, mass of wall/m^2 surface area (45 lb/ft^2), × 2.4 m × 6 m × 9.8 ÷ 1000 = 30.7 kN (7200 lb). Safe working load of brace = 18 kN (4000 lb).

Determine: R = axial load on brace
V = vertical load at bottom of brace
H = horizontal load at bottom of brace
Number of braces

1. Force created by the wind by using formula

$$P = 0.05\ V^2$$

where P = Wind force in Pa
V = Velocity of the wind in km/h

$P = 0.05 \times 140 \times 140 = 980$ Pa (20 lb/ft^2)

2. Length of brace = $\sqrt{1500^2 + 2000^2}$ = 2500 mm (100 in)

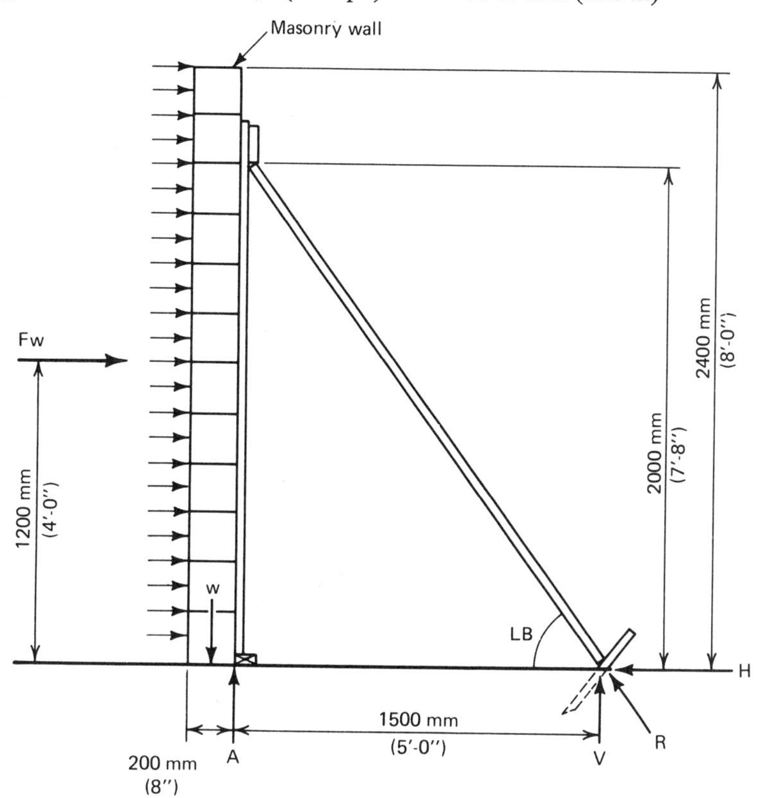

Fig. 15-9. Brace example.

3. Sine ∠B = 2000/2500 = 0.8
4. Cosine ∠B = 1500/2500 = 0.6
5. Force on entire wall = F_w = 2.4 m × 6 m × 980 = 14.11 kN (3200 lb)
6. Take moments about A
   $1200\,(F_w) = 1500(V) + 100(W)$
   $1200 \times 14.11 = 1500 \times V + 100 \times 30.7$
   $V = 9.24$ kN (2077 lb)
7. Sine ∠B = H/R
   $R = 9.24/0.8 = 11.55$ kN (2600 lb)
8. Cosine ∠B = H/R
   $H = 11.55 \times 0.6 = 6.93$ kN (1560 lb)
9. One brace should carry the load but two are required for panel stability. 18 kN > 11.55 kN (4000 > 2600)

A brace that is constructed for a temporary support for a wall should be adequately supported. A long brace will tend to vibrate with wind gusts and allow movement as shown in

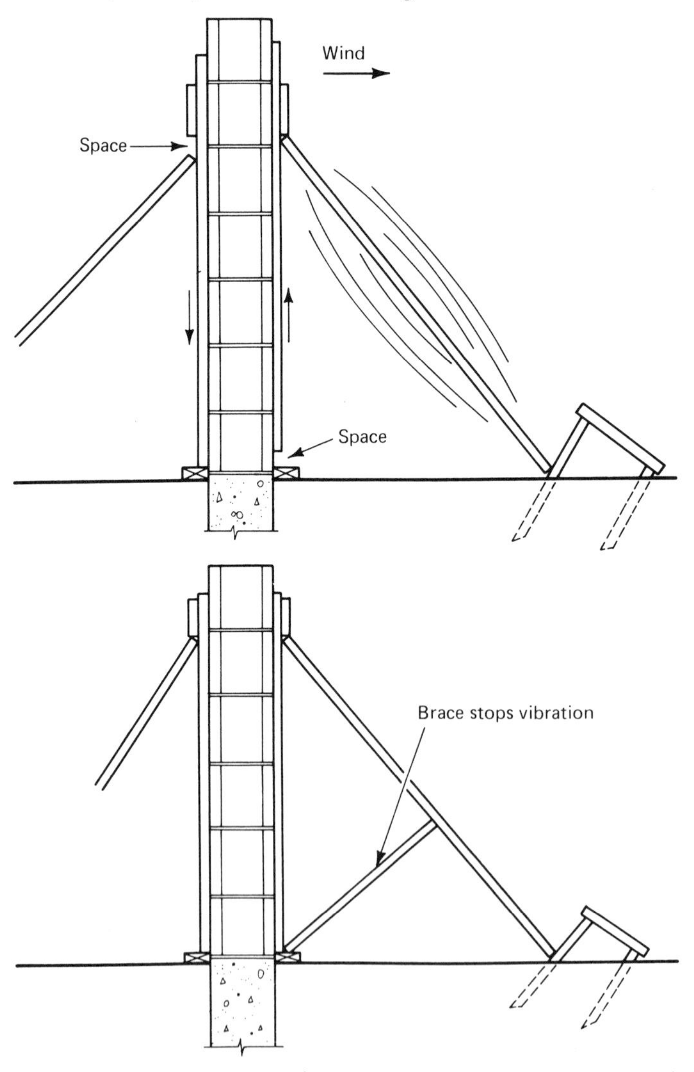

Fig. 15-10. Brace movement and control. (*Courtesy Worker's Compensation Board, Alberta*)

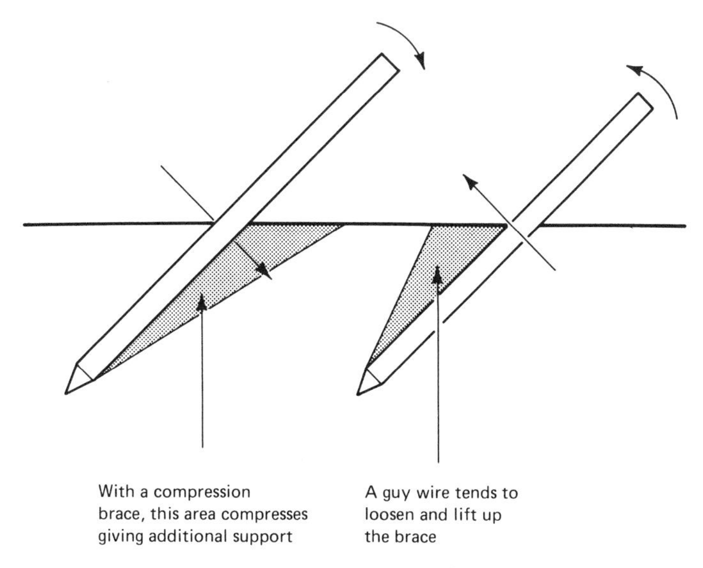

Fig. 15-11. Forces on a stake.

Figure 15-10. A brace support will stiffen the brace and reduce the movement.

Stakes that are driven into the ground to support the bottom end of the brace should be placed at an angle. (See Fig. 15-11.) A stake works quite well as a support for a compression brace but for a tension brace a deadman is more satisfactory. (See Fig. 15-12.) When the ground is very soft, the stake can be reinforced by bracing to a second stake. (See Fig. 15-10.) In frozen ground, a 25 mm (1 in) metal rod can be used. To reduce movement, it can be driven

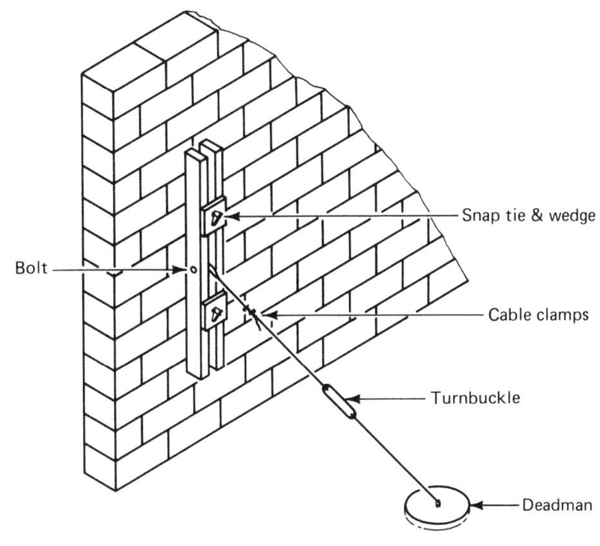

Fig. 15-12. Deadman.

through a piece of plywood, which will insulate against rising temperatures and resist movement. (See Fig. 15-13.)

A brace is securely fastened at the top if a piece of wood is nailed above the brace on the member against the wall. (See Fig. 15-10.) When bracing a wall on a high rise, the bottom of the brace can be nailed to a piece of material placed on a concrete floor held by weights. The top of the brace is supported by nailing it to pieces of wood that are bolted or fastened to the wall by ties. (See Fig. 15-14.)

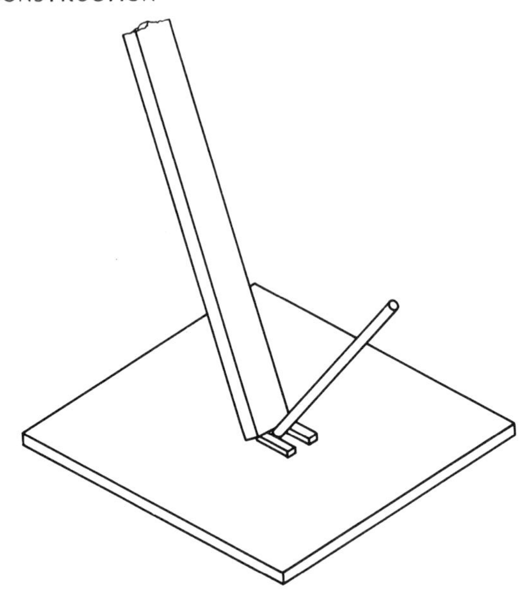

Fig. 15-13. Rod driven through plywood.

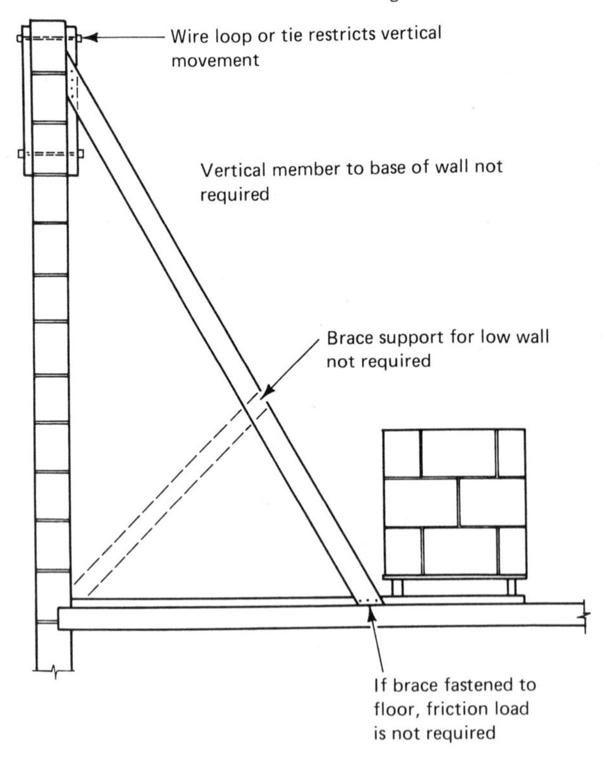

Fig. 15-14. Double-acting brace. (*Courtesy Worker's Compensation Board, Alberta*)

Glossary of Terms

Enclosure A temporary building or cover usually built of polyethylene or tarpaulin to protect against elements.

Flash Set The initial set in concrete or mortar that occurs very quickly after mixing, usually due to excess heat.

Hydrated Lime A dry powder obtained by treating quicklime (CaO) with enough water to satisfy its chemical affinity for water due to hydration.

Wind Break A wall or shield that is used to deflect the force of the wind and protect the element.

Wind Chill Factor The amount of heat lost by a surface due to varying wind and temperature conditions.

Review Questions

1. Why can an increase in suction rate of masonry units cause a decrease in bond strength?
2. Why should the use of water as a cooling method for block be avoided?
3. Why will mortar that is frozen with a water content of below six percent not be damaged?
4. What is meant by flash set of mortar?
5. Discuss the advantages and disadvantages of using heated enclosures during cold weather construction.
6. How many braces would be required to support the wall in the following illustration?

   Panel size 3600 mm (12 ft) high by 8000 mm (26 ft–8 in) long
   Wind force of 125 km/h (78 mph)
   Safe working load of brace is 12 kN (2700 lb)
   Mass of wall is 195 kg/m^2 (40 lb/ft^2)

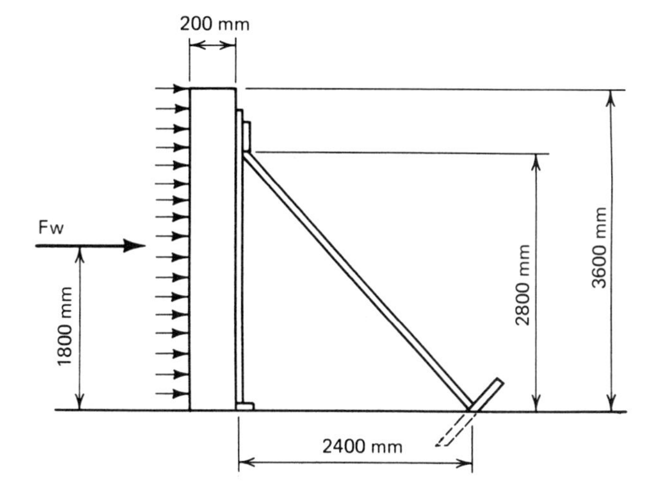

Selected Sources of Information

First Canadian Masonry Symposium, Calgary, Alberta.
National Concrete Masonry Association, McLean, Virginia.
National Research Council, Ottawa, Ontario.
Portland Cement Association, Skokie, Illinois.
Worker's Compensation Board—Alberta, Edmonton, Alberta.

Chapter 16

MAINTENANCE AND REPAIR

We have previously discussed some causes of failure that occur in masonry walls and what can be done in the detailing stage to avoid these problems. This is not very helpful, however, in cases of existing buildings where breakdown occurs. We will deal with the *repair* of problems that occur due to compression, thermal movement, unit shrinkage, water and vapor leakage and cracking.

Masonry walls require repair due to reasons mentioned above, but they must also be *maintained* and problems such as the natural weathering of units, and the failure of joints, and caulking beads must be checked. Removal of stains that occur on masonry walls must also be considered.

REPAIR

Failures due to compression appear in the form of spalling under bearing (see Fig. 16-1), walls bowing in or out, and as cracking (see Fig. 16-2). The failure shown in Figure 16-1 can be repaired by patching the area with non-shrink pressure grout. The main concern is to relieve the pressure points and then rebuild the affected area. This can be done by using a mortar mix preferably made with masonry cement, sand, and water, a little drier than normal mortar. The area should be wetted before forcing the new mortar into place. The placement of a metal plate or angle under the pressure point can be considered if it is possible to insert. In the case of bowed walls the main concern is to relieve the pressure on the wall. This can be done by removing the mortar at the angle iron support and replacing it with caulking. If this does not give enough room, then part of the masonry unit may also be removed.

The failure shown in Figure 16-2 will require the removal of all damaged units after temporary bracing has been installed where necessary. New units should then be placed in the damaged area. As the failure is due to inadequate grouting of the wall, the new units require a grout fill. The top units need to be cut to allow space for the grout to be placed. When the void is filled this opening is closed.

Failures due to thermal movement usually show up in the form of cracking, spalling or cracks up intersecting corners. The vertical cracking can be relieved by installing a control joint in the distressed area. The damaged area can then be repaired in the normal way. Horizontal cracks can be repaired by providing vertical reinforcing or a slip plane in the horizontal distress area. To correct the failure in Figure 16-3 would require allowing room for the horizontal movement of the concrete

Fig. 16-1. Wall spalling under bearing. (*Courtesy Alberta Masonry Institute*)

Fig. 16-2. Cracking. (*Courtesy Alberta Masonry Institute*)

Fig. 16-3. Thermal expansion. (*Courtesy Alberta Masonry Institute*)

beams. The broken block is removed and replaced by a new facing piece allowing room behind with drainage for the space.

Unit shrinkage shows up in the form of cracks at the mortar joint-unit interface, or in some cases across the unit at vertical mortar joints. Joint reinforcing tends to diminish these cracks. After failure, rake out the affected mortar joint and replace with new mortar. This area should not be allowed to dry too quickly, so it should be covered with polyethylene. Where a unit has cracked due to shrinkage, the entire unit is replaced. In some cases when the cracks are very small, a paint with sufficient body to span the crack can be used.

Walls that have been affected by water or vapor will have stains and efflorescence on the surface. In some cases there will be displacement of units as shown in Figure 16-4. The stains are usually caused by water entering the wall from the outside due to rain and from the inside due to vapor movement where the moisture comes in contact with salts. When the water carrying the salts reaches the surface, it evaporates and leaves efflorescent salts as a residue. (See Fig. 6-2.) The main sources of efflorescent salts are the masonry units and mortar, though some amounts can result from soil, fuel, and water. It can be prevented by using salt-free water and nonstaining cement and keeping moisture out of the wall. Stains can also occur on masonry surfaces from water that has come in contact with flashings, angle irons or mortar droppings. Check the wall for points of water entry such as improper caulking, inadequate flashing and poor vapor barriers. Clogged weep holes will immediately cause a band of staining but this can be rectified by clearing the debris.

Fig. 16-4. Displacement of units. (*Courtesy Alberta Masonry Institute*)

To stop vapor movement, reseal from within if the seal is inadequate. Silicone and paint can be used as a seal, but any movement of the wall will cause cracks thereby allowing water to pass through. If this does not solve the problem, then remove the outer skin, repair or install proper flashings, seal the inner wall and rebuild.

Cracking in masonry walls can be caused by vibration, an incompatible frame, deflection and settlement. These cracks can often be avoided by allowing for movement, by using flexible anchors and ties, a low-strength high-bond and water-retentive mortar or control joints. When repairing these cracks, flexible anchors and ties can be installed. Reinforcing, grouting and patching after failure will reinforce the particular area. Strengthening of foundations and installing control joints in the failure area can help solve the problem. The repair of cracks will depend on whether they are in the units or in the mortar. Failures in the units will usually require replacement of the unit while in the mortar, *tuck pointing* is a common practice.

MAINTENANCE

Maintenance involves the regular or continuous repair or renewal of deteriorated materials. Masonry is generally free of most maintenance problems. In some cases, a masonry unit may need to be replaced, a mortar joint retucked, cracks filled, caulking replaced or stains removed in order to maintain the building in excellent condition.

The replacement of a unit will require careful removal of the old unit or debris of a broken one and replacement with a new one. The connecting mortar should be scraped, and the surface cleaned with a wire brush. The surface is then wetted, and mortar is applied to the bed and the head joints of the space. Mortar is placed on top of the new unit and is pushed into the space. (See Fig. 16-5.) The joint around the unit is completely filled with mortar and the excess is removed. When the joint is sufficiently stiffened (thumbprint hard), the joint is tooled and then kept moist or covered with polyethylene to prevent too rapid drying.

Tuck pointing involves the removal of the loose mortar by scraping it with a screwdriver or chisel and cleaning it with a wire brush. Care should be taken not to damage the unit edge. The area is then wetted and packed with a low-slump mortar. When the mortar is stiffened, it is tooled to match existing joints. The area is then kept moist by applying water or by covering it with a layer of polyethylene. (See Fig. 16-6.)

Cracks that cross units cannot be readily obscured. If the unit is not replaced, a method that can be used to repair cracks is to widen and undercut cracks to 10 to 15 mm (3/8 to 5/8 in) at the surface, 15 to 20 mm (5/8 to 3/4 in) at the back, and 10 to 15 mm deep. The notch is then filled with a low-slump mortar or an epoxy resin. If mortar is used the crack should be dampened prior to filling. A fine crack may be bridged with a heavy body paint. A mortar mixture may be worked in prior to painting if paint will not bridge the gap.

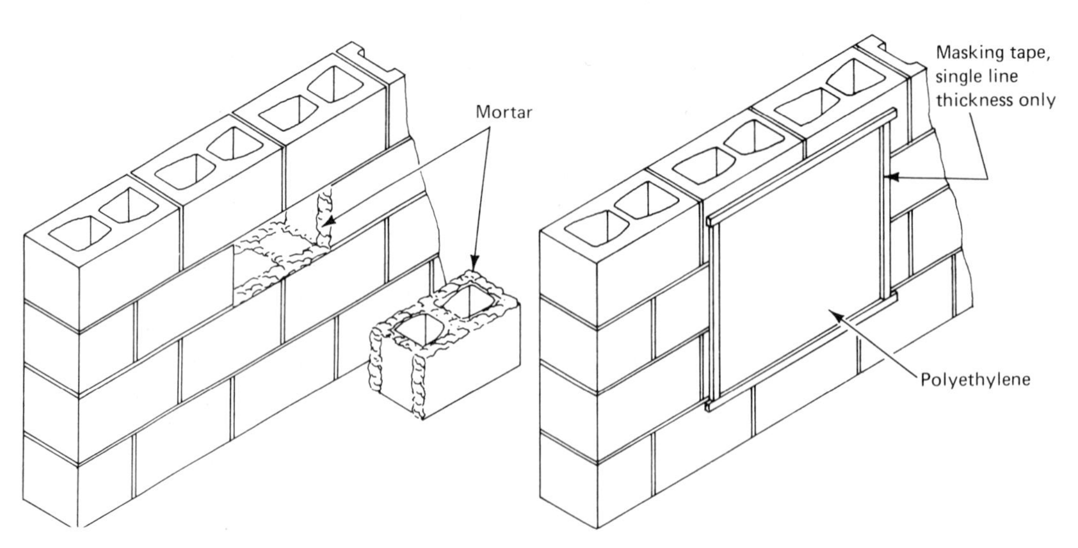

Fig. 16-5. Replacing masonry unit.

Fig. 16-6. Covering replaced unit.

Table 16-1. Selection Guide For Use of Sealants

Type of Sealant	Use	Water Immersion	Range of Movement
Butyl-Polyisobutylene base Solvent curing	Interior and exterior	Dry	±5%
Acrylic base Solvent curing	Interior and exterior	Dry	±7-1/2%
Acrylic emulsion base	Interior	Dry	±10%
Polysulfide base Chemical curing	Interior and exterior	Wet	±25%
Silicone base Chemical and solvent curing	Interior and exterior	Dry	±25%
Polyurethane base Chemical curing	Interior and exterior	Wet	±25%

Periodic checks should be made on the sealant for deterioration and performance. The success of a joint between two building components is determined by the effectiveness of the seal. A sealant is commonly thought of as a plug for cracks and joints without consideration to the conditions they must withstand. The type of sealant used should be determined by the specific properties that are required. The sealant should withstand the passage of moisture, resist solar deterioration and allow for anticipated movement. The joint movement should be estimated and the width of the joint should be matched with sealants' movement capabilities as shown in Table 16-1. If failure recurs, the joint size may be increased or sealant may be moved to an inner locale where it is protected from solar and physical abuse and where less stress occurs due to movement.

When sealants are installed, they should be placed in such a way and under conditions that will lessen the stress on them. The ideal temperature for placing sealants is the mean between maximum high and low temperature exposure. This will tend to reduce maximum tensile and compressive stresses on the sealant. Figure 16-7 illustrates the effect that installation at varying temperatures will have. Installation at low temperatures will cause excessive compression which is not recommended by some manufacturers, while installation at high temperatures will cause high tensile stress in cold weather when the sealant is not as flexible. A lap joint as shown in Figure 16-8 should be considered, as a given deformation will cause a stress imposed in shear that is less than that caused in tension.

The shape and size of a sealant bead will affect its durability. The bead should be narrower in the middle so there is less stress on the outer edge of the joint. The overall joint thickness should not be too great as this will cause greater stress on the sealant. (See Fig. 16-9.) Figure 16-10 illustrates the proper method of installing sealants to provide a joint that will adhere well to the side of the joint and be durable.

Stains occasionally occur on masonry walls, and care should be exercised in their removal. Once the stain has been identified, select the best means for removal and attend to the problem as soon as possible. There are two general

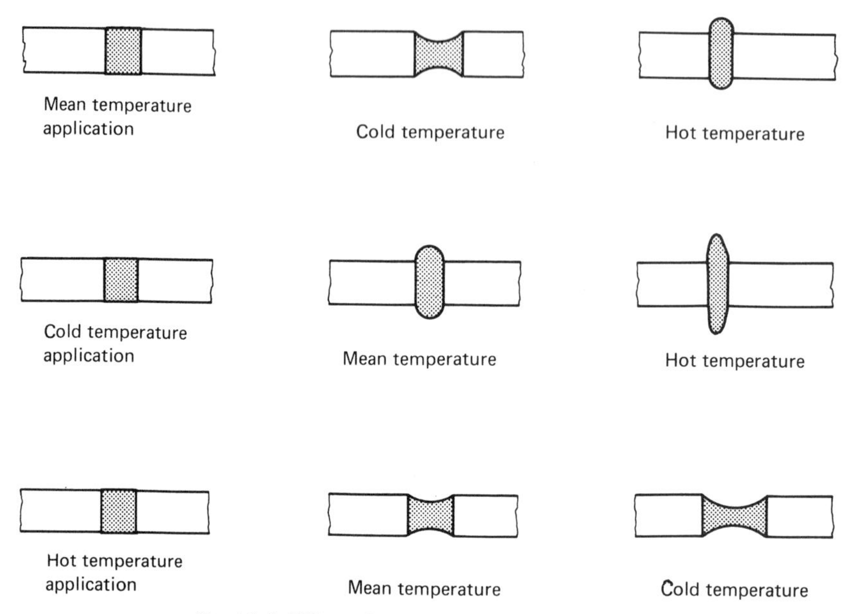

Fig. 16-7. Effect of varying temperature application.

methods of removing stains; that is, by *mechanical means* or by *chemical means* or a combination of the two.

Mechanical means include *high pressure water cleaning, steam cleaning, brushing, grinding* and *sand blasting.* High pressure cleaning utilizes water sprayed under pressure resulting in the removal of dirt and grime with no harm to the units, though there may be some effect on the mortar joints. Steam cleaning is commonly and effectively used by skilled personnel with no damage to the materials. Special equipment is required to produce the steam and would be difficult to maneuver on the exterior of a high-rise building. Brushing with a stiff bristled brush is effective in the removal of loose material on the surface of the masonry units. Cleaning with a sander or grinder can occur when the surfaces are flat and the finish is relatively smooth. Sand blasting is an effective method to remove all the dirt and discoloration from the surface. This method has a disadvantage in that it also removes some of the surface, marring the finish.

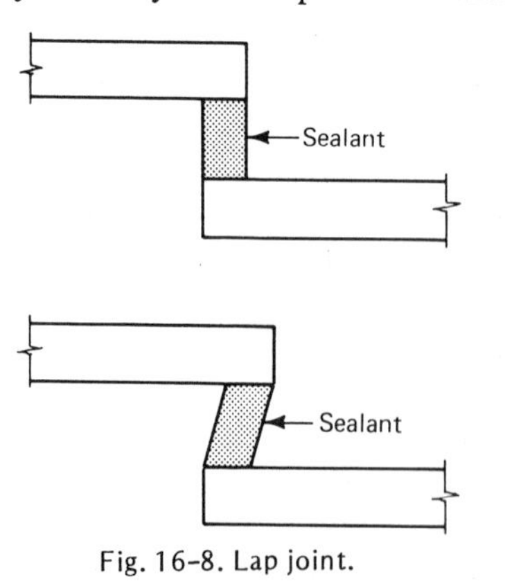

Fig. 16-8. Lap joint.

Chemical means involve materials that act as *solvents* to dissolve the stains or react with them to form a compound that will not show as a stain. When used as solvents, these materials are applied directly on the surface or with a cloth that has been soaked in the cleaning agent; or a *poultice* is made of the solvent and a fine powder such as whiting (calcium carbonate), hydrated lime (calcium hydroxide) or talc. Solvents applied on the surface will dissolve or react with the stain and, if the area

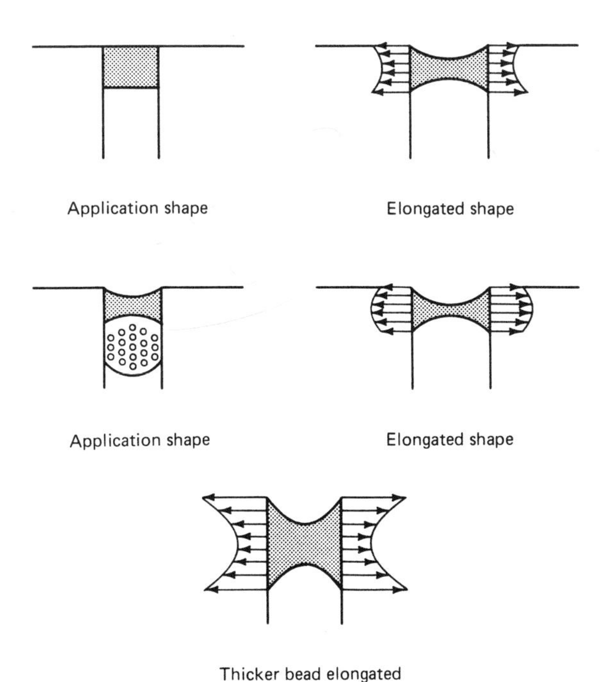

Fig. 16-9. Stress caused by shape. (*Courtesy National Research Council*)

is flushed with water before the solvent evaporates, there is less tendency for the stain to reappear. If the solvent evaporates too quickly it will be more effective to use the cloth or poultice method. The cloth will keep enough solvent at the surface to complete the reaction. The poultice method is much better for drawing the stain out of the pores of the material.

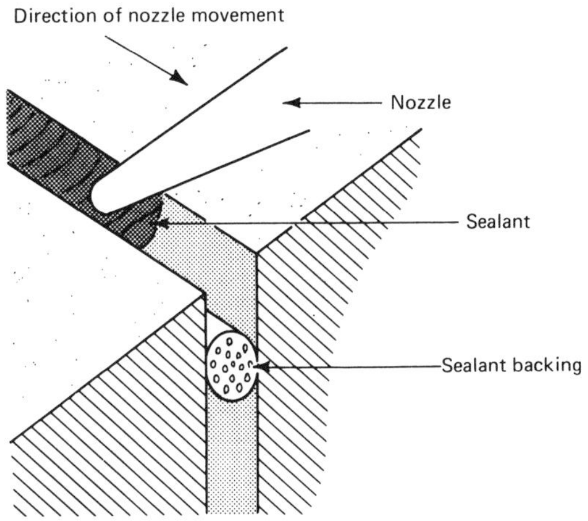

Fig. 16-10. Method of installing sealant. (*Courtesy National Research Council*)

Types of Stains

Metallic stains include those caused by iron rust, copper and bronze, and aluminum. Iron rust stains can be removed by:

1. Mopping the area with a solution of oxalic acid and water, (1 kg per 10 L or 1 lb/gal) and cleansing with water.

2. A ten percent solution of muriatic acid can be effective but it can cause a slight roughness on the wall surface.

3. A poultice made with one part of sodium or ammonium citrate, six parts of water and six parts glycerol. The poultice is placed on the stained area for a few days with a continual supply of liquid to keep the mix wet.

Copper and bronze stains can be removed by using a poultice made of one part ammonium or aluminum chloride, and three to four parts talc mixed with ammonium hydroxide. The poultice is applied to the stain and allowed to dry. The area is then scrubbed with clean water. Aluminum stains are removed by scrubbing the area with a ten percent solution of muriatic acid and then rinsing with clean water.

Bitumen stains include those caused by bitumen or oil-based products such as asphalts, lubricating oils, grease and paint. Asphalt and bitumen stains are removed by:

1. Scrubbing the area with kerosene followed by carbon tetrachloride and then scouring powder and water.

2. Cut-back asphalt may require a poultice of toluene or benzol followed by scrubbing with scouring powder and water.

Lubricating oils can be removed by:

1. Wiping the surface immediately with a cloth and then applying a layer of hydrated lime and allowing to set.

2. Scrubbing the surface with scouring powder, soap, trisodium phosphate followed by a poultice of a five-percent solution of sodium hydroxide.

Grease can be removed by scraping and scrubbing and then following the method used for lubricating oils. Paint can be removed by scraping and then applying a poultice impregnated with paint remover or lacquer thinner. The poultice is allowed to stand for twenty to thirty minutes and is then scrubbed with scouring powder and water.

Miscellaneous stains such as efflorescence, vanadium and manganese stains, smoke and ink can be removed by using the following methods. Efflorescence can be removed by:

1. Brushing with a stiff-bristled brush and then scouring with water.

2. Washing and scouring surface with a diluted solution of muriatic acid followed by a clear water rinse.

A vanadium stain can be removed by washing with water, then with a solution of potassium or sodium hydroxide and allowing it to dry on the wall for two to three days and then washing with water. A manganese stain is removed from a wet wall with a solution composed of one part acetic acid (80 percent), one part hydrogen peroxide (30 to 35 percent) and six parts water. A water rinse should follow. Smoke can be removed by:

1. Scrubbing with pumice or scouring powder and water.

2. Soaking with a cloth impregnated with a mixture of 1 kg trisodium phosphate and 350 ml chlorinated lime.

3. A poultice made with trichloroethylene.

Ink stains can be removed by:

4. A ten-percent solution of oxalic acid.

5. A poultice made with ammonium hydroxide and whiting.

There are numerous commercial stain removers that are available for the removal of stains previously mentioned.

Glossary of Terms

Epoxy Resin A thick pastelike synthetic adhesive.

Manganese Stain A brown oily stain caused by the manganese used to color brick.

Poultice A paste made with solvents and some inert material.

Sand Blasting Using air pressure, sand is forced through a nozzle against a masonry surface.

Sealant A weather-resistant compound used to seal a space and also to allow for movement in it.

Solar Deterioration Breakdown resulting from the action of the sun's rays.

Spalling Surface piece popping out due to pressure on the edge of the surface.

Vanadium Stain A green stain caused by salts of vanadium contained in the brick.

Weathering Gradual breakdown of a surface due to the action of the elements.

Review Questions

1. What is the reasoning behind using a nonshrink pressure grout when patching a pressure point spalled area?
2. How does placing a metal plate under the end of a beam relieve the pressure at that point?
3. What type of paint is better for sealing a small crack in a masonry wall?
4. Discuss some methods that will control efflorescent stains.
5. Why should the mortar used for repairing cracks in a masonry wall have a high water retention?
6. What effect does the type of sealant and the shape of the bead in a control joint have on its lasting ability?

Selected Sources of Information

Brick Institute of America, McLean, Virginia.
Clay Brick Association of Canada, Willowdale, Ontario.
National Concrete Masonry Association, McLean, Virginia.
National Research Council, Ottawa, Ontario.
Portland Cement Association, Skokie, Illinois.

Appendix

DESIGN AIDS

With permission of the Associate Committee on the National Building Code, National Research Council of Canada, the following material is reproduced from Supplement No. 4 to the National Building Code of Canada 1975.

Metric conversions have been made by the Building Standards Branch, General Services Division, Alberta Labour.

Minor changes have been made to some of the material by the authors to facilitate its use as a reference within the text.

Comments or questions on the material should be directed to

The Secretary
Associate Committee on the National Building Code,
National Research Council of Canada
Ottawa, Ontario
Canada
K1A 0R6

SUBSECTION 4.4.3. DESIGN OF PLAIN AND REINFORCED MASONRY BASED ON ENGINEERING ANALYSIS

4.4.3.1. In this Subsection

A_g = gross cross-sectional area
A_n = net cross-sectional area
A_s = effective cross-sectional area of reinforcement
A_v = cross-sectional area of web reinforcement
a = angle between inclined web bars and axis of beam
b = width of rectangular beam or column or width of flange of T beam
b' = width of web of a T beam
C_b = bending coefficient
C_e = eccentricity coefficient
C_s = slenderness coefficient
d = effective depth of flexural members
E_m = modulus of elasticity of masonry in compression
E_s = modulus of elasticity of steel
E_v = modulus of rigidity of masonry
e = virtual eccentricity
e_1 = the smaller virtual eccentricity occurring at the top or bottom of a vertical member at lateral supports
e_2 = the larger virtual eccentricity occurring at

the top or bottom of a vertical member at lateral supports

e_b = virtual eccentricity about the principal axis which is normal to the width, b, of the member

e_t = virtual eccentricity about the principal axis which is normal to the effective thickness, t, of the member

f_m = compressive stress in masonry

f_b = bearing stress on masonry

f_{cs} = compressive stress due to dead loads

f_t = tensile stress in masonry

f'_m = ultimate compressive strength of masonry at 28 days

f_s = stress in reinforcement

f_v = stress in web reinforcement

f_y = yield strength of reinforcement

h = effective height of a wall or column

j = ratio of distance between centroid of compression and centroid of tension to the effective depth, d

P = allowable vertical load

p_n = ratio of the area of tensile reinforcement to the net cross-sectional area, A_n, of the masonry

p = ratio of the area of tensile reinforcement to effective masonry area, bd

p_g = ratio of effective cross-sectional area of reinforcement, A_s, to the gross cross-sectional area, A_g

r = radius of gyration

r_b = ratio of area of bars cut off to the total area of bars at the section

s = spacing of stirrups parallel to direction of main reinforcement

t = effective thickness of a wall or column

u = bond stress per unit of surface area of bar

V = total shear

v = shear stress in masonry having shear reinforcement

v_m = shear stress in masonry having no shear reinforcement

v_{sw} = shear stress in a shear wall

Σo = sum of perimeters of bars.

GENERAL REQUIREMENTS

4.4.3.2. (1) This Subsection applies to the design and construction of plain masonry and reinforced masonry where the design is based on engineering analysis of the structural effects of the loads and forces acting on the structure.

(2) Engineering inspection of masonry construction shall be carried out to ensure that the construction is consistent with the design by the person responsible for its design or by another person qualified in the inspection of masonry construction.

4.4.3.3. Mortar shall be of Type M, S or N conforming to Sentence 4.4.2.15.(1).

4.4.3.4. The allowable stresses in masonry shall be based on its compressive strength, f'_m, as established in Article 4.4.3.6.

4.4.3.5. The actual dimensions of masonry shall be used in stress calculations.

DETERMINATION OF f'_m FOR THE PURPOSE OF DESIGN

4.4.3.6. (1) Except as provided in Sentence (4), the compressive strength, f'_m, shall be established in advance of design by tests of specimens which

(a) are built of the same type of materials under the same conditions, and insofar as possible, of the same thickness and bonding arrangements as for the structure,
(b) if of hollow masonry have unfilled cores and if of solid filled construction have solid filled cores,
(c) are constructed so that the moisture content of the units, mortar consistency, mortar joint thickness and workmanship are the same as will be used in the structure,
(d) if of brick masonry, are not less than 300 mm (0.98 ft) in height and have a height-to-thickness ratio of (h/t) not less than 2 nor more than 5,
(e) if of concrete block or structural clay tile, are not less than 400 mm (1.31 ft) in height and have a height-to-thickness ratio, h/t, not less than 1.5 nor more than 3, and
(f) are stored in air at a temperature not less than 20°C (68.00°F) and are tested after 28 days in conformance with CSA A23.2.13–1973, "Test for Compressive Strength of Moulded Concrete Cylinders."

(2) The compressive strength of each specimen in Sentence (1) shall be calculated by dividing its ultimate test load by its net cross-sectional area and the result multiplied by the appropriate correction factor in Table 1.

Table 1. Forming Part of Sentence 4.4.3.6.(2)

Ratio of Height-to-Thickness h/t	Correction Factor[1]	
	Brick Masonry	*Concrete Block or Structural Clay Tile*
1.5	NA	0.86
2.0	0.73	1.00
2.5	0.80	1.11
3.0	0.86	1.20
3.5	0.91	NA
4.0	0.95	NA
4.5	0.98	NA
5.0	1.00	NA
Column 1	2	3

Note to Table 1:
(1) Correction factors for values of h/t not listed may be interpolated from the values shown.

(3) At least 5 specimens described in Sentence (1) shall be tested and the compressive strength, f'_m, shall be obtained by multiplying the average compressive strength determined in conformance with Sentence (2) by

$$1 - \frac{1.5}{\bar{x}}\sqrt{\frac{\Sigma(x - \bar{x})^2}{n - 1}}$$

where x = an individual test result,
$\bar{x}$ = average of individual test results,
n = number of specimens.

(4) Where the value of the compressive strength of masonry, f'_m, is not determined in accordance with Sentences (1) to (3), it shall be based on tests of the masonry units and mortar in conformance with Sentences (5) to (11).

(5) Compressive strength test of clay or shale brick shall be conducted in conformance with CSA A82.2-1967, "Methods of Sampling and Testing Brick," and for concrete brick in conformance with CSA A165.2-M1977, "Concrete Brick Masonry Units."

(6) Compressive strength tests shall be made in conformance with

(a) CSA A165.1-M1977, "Concrete Masonry Units," and
(b) CSA A82.6-1954, "Standard Methods for Sampling and Testing Structural Clay Tile," for structural clay tile.

(7) At least 5 units shall be tested as described in Sentence (5) or (6) and the compressive strength shall be obtained by multiplying the average compressive strength of the specimens by

$$1 - \frac{1.5}{\bar{\bar{x}}}\sqrt{\frac{\Sigma(x - \bar{x})^2}{n - 1}}$$

where x = an individual test result,
$\bar{x}$ = average of individual test results,
n = number of specimens.

(8) At least six 50 mm (1.97 in.) mortar cubes shall be prepared from the same materials and in the same proportions as those to be used in the masonry, cured and tested in accordance with CSA A179M-1976, "Mortar and Grout for Unit Masonry," including Erratum published September 1977. The average strength determined from these tests shall conform to Article 4.4.3.3. for the type of mortar specified.

(9) The value of the compressive strength, f'_m, to be used in design of brick masonry shall conform to Table 2.

(10) The value of compressive strength, f'_m, to be used in the design of masonry constructed with solid concrete units, hollow concrete or structural clay tile units, or hollow units filled with concrete or grout having a compressive strength at least equal to that of the units, shall conform to Table 3.

Table 2. Forming Part of Sentence 4.4.3.6.(9)

Compressive Strength of Units		Ultimate Compressive Strength of Brick Masonry (f'_m)(1)					
		Type M Mortar		*Type S Mortar*		*Type N Mortar*	
MPa	*psi*	*MPa*	*psi*	*MPa*	*psi*	*MPa*	*psi*
90 plus	13 053.40 plus	30	4 351.13	25	3 625.94	21	3 045.79
80	11 603.02	27	3 916.02	23	3 335.87	19	2 755.72
70	10 152.64	24	3 480.91	20	2 900.75	17	2 465.64
55	7 977.08	19	2 755.72	16	2 320.60	14	2 030.53
40	5 801.51	15	2 175.57	13	1 885.49	11	1 595.42
25	3 625.94	10	1 450.38	8.8	1 276.32	7.5	1 087.78
15	2 175.57	7.5	1 087.78	6.8	986.26	6.0	870.23
Col. 1	1a	2	2a	3	3a	4	4a

Note to Table 2:
(1) Linear interpolation is permitted.

(11) The compressive strength of the units referred to in Sentence (10) shall be based on

(a) the net cross-sectional area of units with voids,
(b) the gross cross-sectional area of units without voids, and
(c) the gross cross-sectional area of filled hollow units.

(12) In composite faced walls, cavity walls or other structural members constructed of different kinds or grades of units or mortars, the value of f'_m used in design shall correspond to the weakest combination of units and mortars of which the member is constructed, except that in a cavity wall where only 1 wythe supports the vertical load, the value of f'_m shall be appropriate for the materials in the loaded wythe.

FIELD CONTROL TESTS

4.4.3.7. (1) Where the value of f'_m used in design is determined in accordance with Sentences 4.4.3.6.(1) to (3)

(a) at least 3 test specimens shall be made on

Table 3. Forming Part of Sentence 4.4.3.6.(10)

Compressive Strength of Units(1)		Ultimate Compressive Strength of Concrete Block Masonry or Structural Clay Tile Masonry (f'_m)			
		Types M and S Mortar		*Type N Mortar*	
MPa	*psi*	*MPa*	*psi*	*MPa*	*psi*
40 plus	5 801.51	16	2 320.60	8.6	1 247.32
30	4 351.13	15	2 175.57	8.6	1 247.32
20	2 900.75	12	1 740.45	8.4	1 218.32
15	2 175.57	10	1 450.38	7.1	1 029.77
10	1 450.38	7.7	1 116.79	5.4	783.20
Col. 1	1a	2	2a	3	3a

Note to Table 3:
(1) Linear interpolation is permitted.

site for each 500 m^2 (5 381.96 sq ft) or portion thereof of wall constructed, but not less than 3 test specimens per storey,
(b) at least 5 such test specimens shall be made for each type and strength of masonry used in any building,
(c) the field control test specimens shall be constructed on the site without using a jig near the walls being built, and using the same materials and workmanship as the site work and of a size conforming to clause 4.4.3.6.(1)(d) or 4.4.3.6.(1)(e),
(d) field control test specimens shall be wrapped in polyethylene and stored at the site for 24 hr and stored in air temperatures not less than 20°C (68.00°F) thereafter,
(e) except as provided in (f), the test specimens shall be tested 28 days after being constructed,
(f) field control test specimens may be tested at 7 days provided that the relationship between 7 and 28-day strengths of the masonry has been established by previous tests, or the compression strengths obtained from 7-day test results shall be assumed to be 90 per cent of the 28-day value, and
(g) the compressive strength of every test specimen shall be calculated in conformance with Sentence 4.4.3.6.(2), and the average compression strength from any 5 consecutive 28-day field control tests or from the 28-day strengths predicted from 7-day tests in accordance with Clause (f) shall exceed the value of f'_m used in the design, and no individual test result shall have a value less than 0.80 f'_m.

(2) If the requirements in Clause (1)(g) are not met, the authority having jurisdiction shall require proof that the strength of the structure is adequate.

4.4.3.8. (1) Where the value of f'_m used in design is determined in accordance with Sentences 4.4.3.6.(4) to (11), at least 5 masonry units and six 50 mm (1.97 in.) mortar cube specimens (3 from each of 2 different locations) shall be made for each 500 m^2 (5 381.96 sq ft) of wall or for each storey height, whichever required the greatest number of tests.

(2) For tests of units referred to in Sentence (1), units shall be selected and tested in conformance with

(a) CSA A82.2–1967, "Methods of Sampling and Testing Brick," for clay or shale brick units,
(b) CSA A165.2–M1977, "Concrete Brick Masonry Units,"
(c) CSA A165.1–M1977, "Concrete Masonry Units," and
(d) CSA A82.6–1954, "Methods for Sampling and Testing Structural Clay Tile," for tile units.

(3) The average of any 5 consecutive compressive test results for units referred to in Sentence (1) shall exceed the compressive strength of the units used in the selection of f'_m as provided in Sentence 4.4.3.6.(9) or (10), and no individual test result shall be less than 0.80 of that compressive strength.

(4) For tests of mortar cubes referred to in Sentence (1)

(a) the mortar shall be taken at random from the mortar boards currently in use, but care shall be taken that no old mortar from the edges of the boards is included,
(b) mortar test cubes shall be made, cured and tested in accordance with CSA A179M–1976, "Mortar and Grout for Unit Masonry," including Erratum published September 1977,
(c) except as provided in (d), compression strength tests of mortar cubes shall be made at an age of 28 days, and
(d) tests may be made after 7 days on mortar test cubes provided that the relationship between 7- and 28-day strength of the mortar has been established by previous tests, or the compression strengths obtained from 7-day test results may be assumed to be 90 per cent of the 28-day value.

(5) The average compression strength of 3 mortar cubes obtained from any 3 consecutive 28-day field control tests or from the 28-day strengths predicted from 7-day tests in accordance with Clause (d) referred to in Sentence (4) shall be at least 0.80 of the compressive strength determined in accordance with Article 4.4.3.3. for the type of mortar used, and no individual test result shall have a value less than 0.67 of that strength.

(6) If the requirements in Sentence (3) or (5) are not met, the authority having jurisdiction shall require proof that the strength of the structure is adequate.

Table 4. Forming Part of Article 4.4.3.11.

Maximum Allowable Stresses in Plain Brick Masonry

Type of Stress or Modulus	*Designation*	*Maximum Allowable Stress or Modulus*	
		MPa	*psi*
Compressive, axial[1]			
Walls	f_m	$0.25\ f'_m$	$0.25\ f'_m$
Columns	f_m	$0.20\ f'_m$	$0.20\ f'_m$
Compressive, flexural[1]			
Walls	f_m	$0.32\ f'_m$	$0.32\ f'_m$
Columns	f_m	$0.26\ f'_m$	$0.26\ f'_m$
Tensile, flexural[4][5]			
Normal to bed joints[1]			
M or S mortar	f_t	0.25	36.26
N mortar	f_t	0.19	27.56
Parallel to bed joints[2]			
M or S mortar	f_t	0.50	72.52
N mortar	f_t	0.39	56.56
Shear[6]			
M or S mortar	v_m	$0.083\sqrt{f'_m}$ but not to exceed 0.35	$\sqrt{f'_m}$ but not to exceed 50.76
N mortar	v_m	$0.083\sqrt{f'_m}$ but not to exceed 0.24	$\sqrt{f'_m}$ but not to exceed 34.81
Bearing on masonry	f_b	$0.25\ f'_m$ [3]	$0.25\ f'_m$ [3]
Modulus of elasticity	E_m	$1\ 000\ f'_m$ but not to exceed 20 000	$1\ 000\ f'_m$ but not to exceed 2 900 754.88
Modulus of rigidity	E_v	$400\ f'_m$ but not to exceed 8 000	$400\ f'_m$ but not to exceed 1 160 301.95
Column 1	2	3	3a

Notes to Table 4:

(1) Direction of stress is normal to bed joints.

(2) Tensile stresses parallel to bed joints are not permitted in stack bond masonry.

(3) Where a vertical load is supported on a masonry surface and the ratio of the loaded surface to the total surface is not more than 1:3, f_b may be increased to $0.375\ f'_m$ provided the least distance between the edges of the loaded and unloaded surfaces is at least 1/4 of the length of the edge of the loaded area perpendicular to such least distance. The allowable bearing stress on a reasonably concentric area greater than 1/3 the full area may be interpolated between the values given.

(4) For computing flexural stresses, the section modulus of a cavity wall shall be assumed to be equal to the sum of the section moduli of the wythes.

(5) Allowance shall be made for unusual vibration and impact forces.

(6) See also Article 4.4.3.33. for shear walls.

4.4.3.9. Loads and associated reduction factors shall conform to Section 4.1 of the Alberta Building Code 1978, except as provided in Article 4.4.3.33.

4.4.3.10. Where reinforcement is required by Sentence 4.1.9.3.(4), such reinforcement shall not be less than that required to resist the effects of seismic forces and shall not be less than that required in Articles 4.4.3.30, and 4.4.3.32.

ALLOWABLE STRESSES

4.4.3.11. The allowable stresses in plain masonry of brick shall conform to Table 4.

4.4.3.12. The allowable stresses in plain masonry of concrete block or structural clay tile shall conform to Table 5.

4.4.3.13. The allowable stresses in reinforced masonry of brick shall conform to Table 6.

4.4.3.14. The allowable stresses in reinforced masonry of concrete block or structural clay tile shall conform to Table 7.

4.4.3.15. (1) The allowable tensile stress in reinforcement shall not exceed

(a) 125 MPa (18 129.72 psi) for billet-steel or axle-steel reinforcing bars of structural grade.
(b) 165 MPa (23 931.23 psi) for deformed bars with a yield strength of at least 400 MPa (58 015.10 psi) and not exceeding 35 mm (1.38 in.) diameter, and
(c) 140 MPa (20 305.28 psi) for all other reinforcement.

(2) The allowable compressive stress in vertical column reinforcement shall not exceed 40 per cent of the yield strength of the steel and shall be not greater than 165 MPa (23 931.23 psi).

(3) The allowable compressive stress for compression reinforcement in flexural members shall be not greater than the allowable tensile stress shown in Sentence (1).

4.4.3.16. The modulus of elasticity of steel reinforcement shall be assumed as 200 000 MPa (29 007 548.78 psi).

4.4.3.17. The allowable shear on steel bolts and anchors shall conform to Table 4.4.3.A.

Table 4.4.3.A. Forming Part of Article 4.4.3.17.

Maximum Allowable Shear on Bolts and Anchors[1]

Diameter of Bolt or Anchor		*Minimum Embedment*[2]		*Maximum Allowable shear*	
mm	*in.*	*mm*	*in.*	*kN*	*lb.*
6.3	0.25	100	3.94	1.2	269.77
9.5	0.38	100	3.94	1.8	404.66
12.7	0.5	100	3.94	2.4	539.54
15.9	0.63	100	3.94	3.3	741.87
19.0	0.75	125	4.92	4.8	1 079.08
22.2	0.87	150	5.91	6.6	1 483.74
25.4	1.00	175	6.89	8.2	1 843.43
28.6	1.13	200	7.87	10.0	2 248.09
Column 1	1a	2	2a	3	3a

Notes to Table 4.4.3.A.:

(1) In determining the stress in masonry, the eccentricity due to loaded bolts and anchors shall be considered.

(2) Bolts and anchors shall be solidly embedded in mortar or grout to develop adequate resistance to the design shear forces, except that the embedment shall not be less than given in Table 4.4.3.H.

Table 5. Forming Part of Article 4.4.3.12.

Maximum Allowable Stresses and Moduli for Plain Concrete Block Masonry and Structural Clay Tile Masonry[1]

		Maximum Allowable Stress or Modulus			
		Units Without Voids or Filled Hollow Units[7] Based on Gross Cross-Sectional Area		*Units with Voids Based on Net Cross-Sectional Area*	
Type of Stress or Modulus	Designation	*MPa*	*psi*	*MPa*	*psi*
Compressive, axial[2]					
Walls	f_m	$0.20\ f'_m$	$0.20\ f'_m$	$0.225\ f'_m$	$0.225\ f'_m$
Columns	f_m	$0.18\ f'_m$	$0.18\ f'_m$	$0.20\ f'_m$	$0.20\ f'_m$
Compressive, flexural[2]					
Walls	f_m	$0.30\ f'_m$	$0.30\ f'_m$	$0.30\ f'_m$ [1]	$0.30\ f'_m$ [1]
Columns	f_m	$0.24\ f'_m$	$0.24\ f'_m$	$0.24\ f'_m$ [1]	$0.24\ f'_m$ [1]
Tensile, flexural[5][6]					
Normal to bed joints[2]					
M or S mortar	f_t	0.25	36.26	0.16[1]	23.21[1]
N mortar	f_t	0.19	27.56	0.11[1]	15.95[1]
Parallel to bed joints[3]					
M or S Mortar	f_t	0.50	72.52	0.32[1]	46.41[1]
N mortar	f_t	0.39	56.56	0.22[1]	31.91[1]

Shear[8]					
M or S mortar	v_m	0.23	33.36	0.23[1]	33.26[1]
N mortar	v_m	0.16	23.21	0.16[1]	23.21[1]
Bearing on masonry[4]	f_b	$0.25\ f'_m$	$0.25\ f'_m$	$0.25\ f'_m$	$0.25\ f'_m$
Modulus of elasticity	E_m	1 000 f'_m but not to exceed 20 000	1 000 f'_m but not to exceed 2 900 754.88	1 000 f'_m but not to exceed 20 000	1 000 f'_m but not to exceed 2 900 754.88
Modulus of rigidity	E_v	400 f'_m but not to exceed 8 000	400 f'_m but not to exceed 1 160 301.95	400 f'_m but not to exceed 8 000	400 f'_m but not to exceed 1 160 301.95
Column 1	2	3	3a	4	4a

Notes to Table 5:

(1) Shear and flexural calculations shall be based on net mortar bedded area.

(2) Direction of stress normal to bed joints.

(3) Direction of stress parallel to bed joints. Tensile stresses in horizontal planes are not permitted in stack bond masonry.

(4) Where a vertical load is supported on a masonry surface and the ratio of the loaded surface to the total surface is not more than 1:3, f_b may be increased to 0.375 f'_m provided the least distance between the edges of the loaded and unloaded surfaces is at least 1/4 of the length of the loaded area perpendicular so such least distance. The allowable bearing stress on a reasonably concentric area greater than 1/3 the full area may be interpolated between the values given.

(5) For computing flexural stresses, the section modulus of a cavity wall shall be assumed to be equal to the sum of the section moduli of the wythes.

(6) Allowance shall be made for unusual vibration and impact forces.

(7) For filled-hollow units the strength of the concrete or grout fill shall be at least equal to that of the units.

(8) See also Article 4.4.3.33. for shear walls.

Table 6. Forming Part of Article 4.4.3.13.

Maximum Allowable Stresses in Reinforced Brick Masonry

Type of Stress or Modulus	*Designation*	*Maximum Allowable Stress or Modulus*	
		MPa	*psi*
Compressive, axial			
Walls	f_m	$0.25\ f'_m$	$0.25\ f'_m$
Columns[1]	f_m	$0.20\ f'_m$	$0.20\ f'_m$
Compressive, flexural			
Walls and beams	f_m	$0.40\ f'_m$	$0.40\ f'_m$
Columns[1]	f_m	$0.32\ f'_m$	$0.32\ f'_m$
Shear			
No shear reinforcement			
Flexural members	v_m	$0.06\sqrt{f'_m}$ but not to exceed 0.35	$0.7\sqrt{f'_m}$ but not to exceed 50.76
Shear walls[3]	v_m	$0.04\sqrt{f'_m}$ but not to exceed 0.69	$0.5\sqrt{f'_m}$ but not to exceed 100.08
With shear reinforcement taking entire shear			
Flexural members	v	$0.16\sqrt{f'_m}$ but not to exceed 0.83	$2.0\sqrt{f'_m}$ but not to exceed 120.38
Shear walls[3]	v	$0.12\sqrt{f'_m}$ but not to exceed 1.03	$1.5\sqrt{f'_m}$ but not to exceed 149.39
Bond			
Plain bars	u	0.55	79.77
Deformed bars	u	1.10	159.54
Bearing[2]	f_b	$0.25\ f'_m$	$0.25\ f'_m$
Modulus of elasticity	E_m	$1\ 000\ f'_m$ but not to exceed 20 000	$1\ 000\ f'_m$ but not to exceed 2 900 754.88
Modulus of rigidity	E_v	$400\ f'_m$ but not to exceed 8 000	$400\ f'_m$ but not to exceed 1 160 301.95
Column 1	2	3	3a

Notes to Table 6:

(1) See Article 4.4.3.31.

(2) Where a vertical load is supported on a masonry surface and the ratio of the loaded surface to the total surface is not more than 1:3, f_b may be increased to $0.375\ f'_m$ provided the least distance between the edges of the loaded and unloaded surfaces is at least 1/4 of the length of the edge of the loaded area perpendicular to such least distance. The allowable bearing stress on a reasonably concentric area greater than 1/3 the full area may be interpolated between the values given.

(3) See also Article 4.4.3.33. for shear walls.

Table 7. Forming Part of Article 4.4.3.14.

Maximum Allowable Stresses in Reinforced Concrete Block and Structural Clay Tile Masonry

Type of Stress or Modulus	*Designation*	*Maximum Allowable Stress or Modulus*	
		MPa	*psi*
Compressive, axial			
Walls	f_m	$0.225\ f'_m$	$0.225\ f'_m$
Columns(1)	f_m	$0.20\ f'_m$	$0.20\ f'_m$
Compressive, flexural			
Walls and beams	f_m	$0.33\ f'_m$	$0.33\ f'_m$
Columns(2)	f_m	$0.28\ f'_m$	$0.28\ f'_m$
Shear			
No shear reinforcement			
Flexural members	v_m	$0.02\ f'_m$ but not to exceed 0.35	$0.02\ f'_m$ but not to exceed 50.76
Shear walls(2)	v_m	$0.015\ f'_m$ but not to exceed 0.35	$0.015\ f'_m$ but not to exceed 50.76
With shear reinforcement taking entire shear			
Flexural members	v	$0.05\ f'_m$ but not to exceed 1.03	$0.05\ f'_m$ but not to exceed 149.39
Shear walls(2)	v	$0.04\ f'_m$ but not to exceed 0.52	$0.04\ f'_m$ but not to exceed 75.42
Bond			
Plain bars	u	0.55	79.77
Deformed bars	u	1.10	159.54
Bearing on masonry(3)	f_b	$0.25\ f'_m$	$0.25\ f'_m$
Modulus of elasticity	E_m	$1\ 000\ f'_m$ but not to exceed 20 000	$1\ 000\ f'_m$ but not to exceed 2 900 754.88
Modulus of rigidity	E_v	$400\ f'_m$ but not to exceed 8 000	$400\ f'_m$ but not to exceed 1 160 301.95
Column 1	2	3	3a

Notes to Table 7:

(1) See Article 4.4.3.31.

(2) See also Article 4.4.3.33. for shear walls.

(3) Where a vertical load is supported on a masonry surface and the ratio of the loaded surface to the total surface is not more than 1:3, f_b may be increased to $0.375\ f'_m$ provided the least distance between the edges of the loaded and unloaded surfaces is at least 1/4 of the length of the edge of the loaded area perpendicular to such least distance. The allowable bearing stress on a reasonably concentric area greater than 1/3 the full area may be interpolated between the values given.

Table 8. Forming Part of Sentence 4.4.3.20.

Slenderness Coefficient (C_s)*

h/t	e_1/e_2 = -1	-3/4	-1/2	-1/4	0	+1/4	+1/2	+3/4	+1
5.0	1.00	1.00	1.00	1.00	1.00	1.00	1.00	1.00	1.00
6.0	1.00	1.00	1.00	1.00	1.00	0.99	0.98	0.97	0.96
7.0	1.00	1.00	1.00	1.00	1.00	0.98	0.97	0.94	0.92
8.0	1.00	1.00	1.00	0.99	0.99	0.97	0.94	0.91	0.88
9.0	1.00	0.99	0.99	0.98	0.96	0.94	0.91	0.88	0.84
10.0	1.00	0.99	0.98	0.96	0.93	0.91	0.88	0.84	0.80
11.0	0.98	0.97	0.96	0.94	0.91	0.88	0.84	0.80	0.76
12.0	0.96	0.95	0.93	0.91	0.88	0.85	0.81	0.77	0.72
13.0	0.94	0.93	0.91	0.89	0.86	0.82	0.77	0.73	0.68
14.0	0.92	0.92	0.88	0.86	0.83	0.79	0.74	0.70	0.64
15.0	0.90	0.88	0.86	0.83	0.80	0.76	0.71	0.66	0.60
16.0	0.88	0.86	0.84	0.81	0.77	0.73	0.68	0.62	0.56
17.0	0.86	0.84	0.82	0.78	0.74	0.70	0.65	0.59	0.52
18.0	0.84	0.82	0.80	0.76	0.72	0.67	0.61	0.55	0.48
19.0	0.82	0.80	0.77	0.73	0.69	0.64	0.58	0.52	0.44
20.0	0.80	0.78	0.75	0.71	0.67	0.61	0.55	0.48	0.40
21.0	0.78	0.76	0.74	0.69	0.64	0.58	0.52	0.44	
22.0	0.76	0.74	0.71	0.66	0.61	0.55	0.49	0.41	
23.0	0.74	0.72	0.68	0.64	0.59	0.52	0.45		
24.0	0.72	0.70	0.66	0.61	0.56	0.50	0.42		
25.0	0.70	0.67	0.64	0.59	0.53	0.47	0.39		
26.0	0.68	0.65	0.62	0.57	0.51	0.44			
27.0	0.66	0.63	0.59	0.54	0.48	0.41			
28.0	0.64	0.61	0.57	0.52	0.46				
29.0	0.62	0.59	0.54	0.49	0.43				
30.0	0.60	0.57	0.52	0.47	0.40				
31.0	0.58	0.55	0.50	0.45					
32.0	0.56	0.53	0.48	0.42					
33.0	0.54	0.51	0.46						
34.0	0.52	0.48	0.43						
35.0	0.50	0.46	0.41						
36.0	0.48	0.44							
37.0	0.46	0.42							
38.0	0.44								
39.0	0.42								
40.0	0.40								

$$*C_s = 1.20 - \frac{h/t}{300}\left[5.75 + \left(1.5 + \frac{e_1}{e_2}\right)^2\right] \leqslant 1.0$$

DESIGN OF MASONRY WALLS AND COLUMNS

4.4.3.18. (1) The slenderness ratio of a load-bearing masonry wall (the ratio of its effective height, h, to the effective thickness, t, shall not exceed

$$10\,(3 - (e_1)/(e_2))$$

(2) The values e_1/e_2 in Sentence (1) shall be assumed to be positive where the wall is bent in single curvature and negative where the wall is bent in double curvature.

4.4.3.19. (1) The slenderness ratio of a load-bearing masonry column (the greatest value obtained by dividing the effective height, h, by the effective thickness, t, shall not exceed

$$5\,(4 - (e_1)/(e_2))$$

(2) The value e_1/e_2 in Sentence (1) shall be assumed to be positive where the column is bent in single curvature and negative where the column is bent in double curvature.

4.4.3.20. The slenderness coefficient, C_s, shall conform to Table 8.

4.4.3.21. (1) Where a wall is laterally supported at more than 1 level, the effective height, h, between supports shall be assumed as the clear height between such supports.

(2) Where a wall is not laterally supported at the top, its effective height, h, shall be assumed as twice the height of the wall above the lateral support.

4.4.3.22. (1) Where a column is laterally supported at more than 1 level in the directions of both principal axes, the effective height, h, in relation to any axis shall be assumed as the clear distance between such supports.

(2) Where a column is provided with lateral support in the directions of both principal axes at the bottom and 1 principal axis at the top, its effective height in relation to the axis about which the column has support top and bottom shall be assumed as the distance between such supports, and its effective height at right angles, to this axis shall be assumed as twice this distance.

(3) Where a column is not provided with lateral support at the top, its effective height relative to 2 principal axes shall be assumed as twice its height above the lower support.

4.4.3.23. (1) Except as provided in Article 4.4.3.25., for all solid masonry walls the effective thickness, t, shall be assumed as the actual thickness.

(2) Except as provided in Article 4.4.3.25., for cavity walls loaded on not more than 1 wythe, the effective thickness shall be assumed as the actual thickness of the loaded wythe.

(3) Except as provided in Article 4.4.3.25., for cavity walls loaded on both wythes, each wythe shall be considered to act independently and the effective thickness of each wythe shall be assumed as its actual thickness.

4.4.3.24. (1) Except as provided in Article 4.4.3.25. for rectangular columns, the effective thickness in the direction of each principal axis shall be assumed as the actual thickness in that direction.

(2) Except as provided in Article 4.4.3.25. for non-rectangular columns, the effective thickness, t, in relation to each principal axis shall be assumed as 3.5 times its radius of gyration about the axis considered.

4.4.3.25. Where raked mortar joints are used, the effective thickness shall be assumed as the effective thickness in Articles 4.4.3.23. and 4.4.3.24. reduced by the depth of the raking.

4.4.3.26. (1) Lateral movements due to loads, thermal effects and other causes shall be taken into account in calculating the virtual eccentricity of loads on walls or columns.

(2) Where members are constructed of different kinds or grades of units or mortar, the variation in the moduli of elasticity shall be taken into account and the eccentricity of the load shall be measured from the centroid of the transformed section of the member.

(3) Where a cavity wall is loaded on 1 wythe, the eccentricity of the load shall be measured from the centroid of the loaded wythe.

(4) Where a cavity wall is loaded on both wythes, the load shall be distributed to each wythe according to the eccentricity of the load from the centroidal axis of the wall.

(5) For walls or columns of solid masonry subject to bending about not more than 1 principal axis

Table 9. Forming Part of Sentence 4.4.3.26

Eccentricity Coefficient (C_e)*

e/t or $\frac{e_t b + e_b t}{bt}$	e_1/e_2								
	-1	*-3/4*	*-1/2*	*-1/4*	*0*	*+1/4*	*+1/2*	*+3/4*	*+1*
0.05 or less	1.00	1.00	1.00	1.00	1.00	1.00	1.00	1.00	1.00
0.10	0.86	0.86	0.85	0.84	0.84	0.83	0.82	0.81	0.80
0.15	0.79	0.77	0.76	0.75	0.73	0.72	0.71	0.70	0.68
0.20	0.74	0.71	0.70	0.67	0.66	0.64	0.62	0.60	0.58
0.25	0.69	0.66	0.64	0.61	0.59	0.56	0.54	0.51	0.49
0.30	0.64	0.60	0.58	0.54	0.51	0.48	0.45	0.42	0.39
0.33	0.61	0.57	0.54	0.50	0.47	0.43	0.40	0.36	0.32

*Where the maximum virtual eccentricity (e) exceeds t/20 but is equal to or less than t/6,

$$C_e = \frac{1.3}{1 + 6\frac{e}{t}} + \frac{1}{2}\left(\frac{e}{t} - \frac{1}{20}\right)\left(1 - \frac{e_1}{e_2}\right)$$

Where the maximum virtual eccentricity (e) exceeds t/6 but is equal to or less than t/3,

$$C_e = 1.95\left(\frac{1}{2} - \frac{e}{t}\right) + \frac{1}{2}\left(\frac{e}{t} - \frac{1}{20}\right)\left(1 - \frac{e_1}{e_2}\right)$$

(a) the eccentricity of any load shall be measured from the centroid of the member, and
(b) the eccentricity coefficient, C_e, shall conform to Table 9.

(6) Where walls and columns are subject to bending about both principal axes, the eccentricity coefficient, C_e, shall conform to Table 9 with reference to $e_t b + e_b t/bxt$.

4.4.3.27. (1) Except as provided in Sentence (2), for cavity walls loaded on both wythes A_g shall be assumed as the gross cross-sectional area of the wythe under consideration.

(2) Where raked mortar joints are used, the thickness used in determining A_g shall be the actual thickness of the member reduced by the depth of the raking.

4.4.3.28. (1) Except as provided in Sentence (2), the allowable vertical load, P, on a plain masonry wall or column subject to bending about not more than 1 principal axis shall be computed by

(a) $P = C_s f_m A_n$ where the virtual eccentricity is less than t/20 and f_m is the allowable axial compressive stress, or

(b) $P = C_e C_s f_m A_n$ where the virtual eccentricity is at least t/20 but does not exceed t/3 and f_m is the allowable axial compressive stress.

(2) Where the virtual eccentricity exceeds t/3, P in Sentence (1) shall be computed in accordance with Clause (1)(b), except that the allowable flexural tensile stress, f_t in Articles 4.4.3.11. and 4.4.3.12. shall be substituted for f_m.

(3) Except as provided in Sentence (4), the allowable vertical loads on rectangular plain masonry walls and columns subject to bending about both principal axes shall be calculated in conformance with

(a) Clause (1)(a) where ($e_t b + e_b t$) is less than bt/20, or
(b) Clause (1)(b) where ($e_t b + e_b t$) is at least equal to bt/20 but does not exceed bt/3.

(4) Where ($e_t b + e_b t$) exceeds bt/3, walls and columns subject to bending about 2 principal axes shall be reinforced and designed in accordance with Articles 4.4.3.29. and 4.4.3.31.

4.4.3.29. (1) Except as permitted in Sentences (2) and (4), the allowable load, P, on a reinforced masonry wall subject to bending about not more than 1 principal axis shall be

(a) $P = C_s f_m A_n$ where the virtual eccentricity is less than t/10 and f_m is the allowable axial compressive stress, or
(b) $P = C_e C_s f_m A_n$ where the virtual eccentricity is at least t/10 but does not exceed t/3 or a value which would produce tension in the reinforcement and f_m is the allowable axial compressive stress.

(2) Where the virtual eccentricity exceeds t/3 or a value which would produce tension in the reinforcement, P in Sentence (1) shall be determined on the basis of a transformed section and linear stress distribution. Reinforcement in compression shall be neglected except as provided in Sentence (4). The compressive stress in the masonry shall not exceed the allowable flexural compressive stress, f_m, and the tensile stress in the reinforcement shall conform to Article 4.4.3.15. The vertical load determined in accordance with this Sentence shall be modified by the slenderness coefficient in Article 4.4.3.20.

(3) Except as provided in Sentence (4), the allowable vertical load, P, on a reinforced masonry wall subject to bending about both principal axes shall be calculated in conformance with

(a) Clause (1)(a) where $(e_t b + e_b t)$ is less than bt/10,
(b) Clause (1)(b) where $(e_t b + e_b t)$ is at least equal to bt/10 but does not exceed bt/3 or a value which would produce tension in the reinforcement, or
(c) Sentence (2) where $(e_t b + e_b t)$ exceeds bt/3 or a value which would produce tension in the reinforcement.

(4) Where the reinforcement in bearing walls is designed, placed and tied in position as for columns, the walls may be designed as columns in accordance with Article 4.4.3.31. provided the length of the wall considered as a column does not exceed the center-to-center distance between concentrated loads nor exceed the width of the bearing plus 4 times the wall thickness.

4.4.3.30. (1) Reinforced masonry walls shall be reinforced horizontally and vertically with steel having a total cross-sectional area of not less than 0.002 times the gross cross-sectional area of the wall so that not less than 1/3 of the required steel is either vertical or horizontal.

(2) The principal reinforcing bars shall be spaced not more than 6 times the wall thickness nor more than 1.2 m (3.94 ft) apart.

(3) Horizontal reinforcement shall be provided in the wall immediately above every footing, at the bottom and top of every wall opening, at roof and floor level and at the top of every parapet wall.

(4) All required wall reinforcement in Sentences (1) to (3) shall be continuous or shall be spliced in accordance with Sentence 4.4.3.51.(4).

(5) In addition to the minimum reinforcement or that required by the structural design, there shall be not less than the equivalent of one 15 mm (0.59 in.) diameter bar around all window and door openings extending at least 0.6 m (1.97 ft) beyond the corners of the openings.

4.4.3.31. (1) Except as provided in Sentence (2), the allowable vertical load, P, on a reinforced masonry column subject to bending about not more than 1 principal axis, shall be

(a) $P = C_s(f_m + 0.80\ p_n f_s)\ A_n$ where the virtual eccentricity is less than t/10 and f_m is the allowable axial compressive stress, or
(b) $P = C_e C_s(f_m + 0.80\ p_n f_s)\ A_n$ where the virtual eccentricity is at least t/10 but does not exceed t/3 or a value which would produce tension in the reinforcement and f_m is the allowable flexural compressive stress.

(2) Where the virtual eccentricity exceeds t/3 or a value which would produce tension in the reinforcement, P in Sentence (1) shall be determined on the basis of a transformed section and linear stress distribution. The compressive stress in the masonry shall not exceed the allowable flexural compressive stress and the stresses in the reinforcement, f_s, shall conform to Article 4.4.3.15. The vertical load determined in accordance with this Sentence shall be modified by the slenderness coefficient in Article 4.4.3.20.

(3) Allowable vertical loads on rectangular reinforced masonry columns subject to bending about both principal axes shall be calculated in conformance with

(a) Clause (1)(a) where $(e_t b + e_b t)$ is less than bt/10,

(b) Clause (1)(b) where $(e_t b + e_b t)$ is at least equal to bt/10 but does not exceed bt/3 or a value which would produce tension in the reinforcement, or
(c) Sentence (2) where $(e_t b + e_b t)$ exceeds bt/3 or a value which would produce tension in the reinforcement.

4.4.3.32. (1) The cross-sectional area of vertical reinforcement in columns shall be at least 0.5 per cent and not more than 4 per cent of the gross cross-sectional area of the column, except that a column stressed to less than 1/2 of its allowable stress may have its reinforcement reduced to not less than 0.27 per cent.
(2) Lateral ties shall be not less than 3.8 mm (0.15 in.) diameter and the spacing shall not exceed 16-bar diameters, 48-tie diameters, nor the least dimension of the column, whichever gives the smallest spacing. Ties may be placed in horizontal mortar joints or in contact with the vertical steel.
(3) The ties shall be so arranged that every corner bar and intermediate bar is laterally supported by a tie forming an included angle of not more than 135 deg. at the bar, except that an intermediate bar that is not more than 150 mm (5.91 in.) from a laterally supported bar need not be supported, and where the bars are located around the periphery of a circle a circular tie may be used.

SHEAR WALLS

4.4.3.33. (1) A plain masonry shear wall shall be designed so that no part of the wall is in tension.
(2) Reinforced masonry shear walls may be designed in conformance with Article 4.4.3.29.
(3) The maximum horizontal shear stress in a shear wall, $_{sw}$, shall not exceed the value

$$(v \text{ or } v_m) + 0.3f_{cs}$$

where v or v_m = the allowable applicable shear stress.
(4) In computing the shear resistance of a shear wall, flanges or projections formed by intersecting walls shall be neglected.
(5) In calculations of shear stresses in masonry shear walls subjected to earthquake forces, the load probability combination factor in Section 4.1 of the Alberta Building Code 1978 shall not apply.

4.4.3.34. (1) Except as provided in Sentence (3), where masonry shear walls intersect a masonry wall or walls to form symmetrical T or I sections, the effective flange width shall not exceed 1/6 of the total wall height above the level being analyzed and its overhanging width on either side of the shear wall shall not exceed 6 times the thickness of the intersected wall.
(2) Except as provided in Sentence (3), where masonry shear walls intersect a masonry wall or walls to form L or C sections, the effective overhanging flange width shall not exceed 1/16 of the total wall height above the level being analyzed nor 6 times the thickness of the intersected wall.
(3) Limits on effective flange width in Sentences (1) and (2) may be increased where it can be shown that such increases are justified.
(4) The vertical shear stress at the intersection of masonry walls shall not exceed the allowable shear stress in Articles 4.4.3.11. to 4.4.3.14. for shear walls if the intersection is laid up in true masonry bond conforming to Clause 4.4.5.18. (1)(a) or shall not exceed the allowable shear values in Article 4.4.3.17. where metal bolts or anchors are provided. Metal anchors shall be embedded to the depth required to develop the tensile strength of the anchors.

4.4.3.35. (1) When floors or roofs are designed to transmit horizontal forces to walls, the anchorage of the floor or roof to the wall shall be designed to resist the horizontal force.
(2) Steel anchors to resist shear force shall be designed in conformance with Article 4.4.3.17.

FLEXURAL MEMBERS

4.4.3.36. (1) The design of flexural members of reinforced masonry shall be in accordance with the following assumptions:

(a) A section that is plane before bending remains plane after bending,
(b) Moduli of elasticity of the masonry and of the reinforcement remain constant,
(c) Tensile forces are resisted only by the tensile reinforcement, and
(d) Reinforcement is completely surrounded by and bonded to masonry material.

4.4.3.37. (1) All members shall be designed to resist at all sections the maximum bending moment and shears as determined by the principle of continuity and relative rigidity.

(2) The clear distance between lateral supports of a beam shall not exceed 32 times the least width of the compression flange or face.

(3) Where compression steel is required in beams, it shall be anchored by ties or stirrups not less than 6 mm (0.24 in.) in diameter, spaced not more than 16-bar diameters or 48-tie diameters apart, whichever is less.

(4) In computing flexural stresses in walls where reinforcement occurs, the effective width shall be not greater than 4 times the wall thickness.

4.4.3.38. Where tensile reinforcement at any section of a flexural member is required, the ratio, p, of the area of tensile reinforcement to effective masonry area shall be at least $0.55/f_y$, where f_y is in Megapascals ($79.77/f_y$, where f_y is in psi), unless the tensile reinforcement at every section, positive or negative, is at least 1/3 greater than that required by analysis.

4.4.3.39. (1) The shearing stress, v, as a measure of diagonal tension in reinforced masonry flexural members shall be calculated by

$$v = \frac{V}{bd}$$

except for members of I or T section where b′ shall be substituted for b.

(2) Except for corbels, brackets and other short cantilevers, the maximum shear in a flexural member shall be assumed as that occurring at a distance equal to the effective depth, d, of the member, from the face of the support.

(3) The effects of flexural compression in variable-depth members and the significant effects of torsion shall be included in calculating the shear stress.

(4) Where the value of the calculated shearing stress exceeds the allowable shearing stress permitted on masonry without web reinforcement, web reinforcement shall be provided to carry the entire shearing stress. Such reinforcement shall be continued for a distance equal to the depth, d, of the member beyond the point theoretically required.

4.4.3.40. (1) Web reinforcement shall consist of

(a) bars or stirrups perpendicular to or at an angle of at least 45 deg. with the longitudinal tension reinforcement,

(b) longitudinal bars bent so that the axis of the inclined portion of the bar makes an angle of at least 30 deg. with the axis of the longitudinal portion of the bar, or

(c) combination of (a) and (b).

4.4.3.41. (1) The area of steel, A_v, required in stirrups placed perpendicular to the longitudinal reinforcement shall be calculated by

$$A_v = \frac{Vs}{f_v d}$$

4.4.3.42. (1) The required area, A_v, of inclined stirrups or parallel bars bent up at different distances from the support shall be calculated by

$$A_v = \frac{Vs}{f_v d\,(\sin a + \cos a)}$$

(2) When the web reinforcement consists of a single bent bar or of a single group of parallel bars all bent up at the same distance from the support, the required area, A_v, of such bar or bars, shall be calculated by

$$A_v = \frac{V}{f_v \sin a}$$

(3) Only the center 3/4 of the inclined portion of a bent bar shall be considered effective as web reinforcement.

4.4.3.43. Where web reinforcement is required, it shall be spaced so that every 45-deg. line, representing a potential diagnonal crack, extending from the mid-depth, d/2, of the beam to the longitudinal tension bars shall be crossed by at least 1 line of effective web reinforcement.

4.4.3.44. (1) In flexural members in which tensile reinforcement is parallel to the compressive face, the bond stress, u, shall be calculated by

$$u = \frac{V}{\Sigma o j d}$$

(2) The tension or compression in any bar at any section shall be developed on each side of that section by adequate embedment length,

end anchorage or hooks. A tension bar may be anchored by bending it across the web at an angle of not less than 15 deg. with the longitudinal portion of the bar and making it continuous with the reinforcement on the opposite side of the member.

(3) Except at supports, every reinforcing bar shall be continued beyond the point at which it is no longer needed to resist flexural stress, for a distance of not less than the effective depth of the member but not less than 12 bar diameters.

(4) Tension bars shall not be terminated in a tension zone except where

(a) the shear is not over 1/2 that permitted,
(b) additional stirrups in excess of those required are provided each way from the termination point, a distance equal to the depth of the beam. The stirrup spacing shall not exceed $d/8r_b$ where r_b is the ratio of the area of bars terminated to the total area of bars at the section, or
(c) the continuing bars provide double the area required for moment resistance at the termination point or double the perimeter required for bond.

(5) Tensile reinforcement for negative moment in any span of a continuous, restrained or cantilever beam, or in any member of a rigid frame shall be adequately anchored by bond, hooks or mechanical anchors in or through the supporting member.

(6) At least 1/3 of the total reinforcement required for negative moment at a support shall be extended beyond the extreme position of the point of inflection a distance at least 1/16 of the clear span but not less than the effective depth of the member.

(7) At least 1/3 of the total reinforcement required for positive moment in simple beams or at the simply supported end of continuous beams shall extend along the same face of the beam at least 150 mm (5.91 in.) past the edge of the support. At least 1/4 of the total reinforcing required for positive moment in a continuous beam shall extend along the same face of the beam past the face of intermediate supports at least 150 mm (5.91 in.).

(8) Plain bars in tension shall terminate in standard hooks, except that hooks shall not be required on the positive reinforcement at interior supports of continuous members.

4.4.3.45. (1) Single separate bars used as web reinforcement shall be anchored at each end by

(a) welding to longitudinal reinforcement,
(b) hooking tightly around the longitudinal reinforcement through 180 deg.,
(c) embedment above or below the mid-depth of the beam on the compression side a distance sufficient to develop by bond the stress in the bar, or
(d) standard hook as specified in Article 4.4.3.46. developing 50 MPa (7 251.89 psi), plus embedment sufficient to develop by bond the remainder of the stress in the bar; the effective embedded length shall be assumed not to exceed the distance between the mid-depth of the beam and the tangent of the hook.

(2) The ends of bars forming single U-stirrups or multiple U-stirrups shall be anchored by one of the methods of Sentence (1) or shall be bent through an angle of at least 90 deg. tightly around longitudinal reinforcing bars not less in diameter than the stirrup bar, and shall project beyond the bend at least 12 diam. of the stirrup bar.

(3) The loops or closed ends of single U-stirrups or multiple U-stirrups shall be anchored by bending around the longitudinal reinforcement through an angle of at least 90 deg., or by being welded or otherwise rigidly attached to such reinforcement.

(4) Hooking or bending stirrups or separate web reinforcing bars around the longitudinal reinforcement shall be considered effective only when these bars are perpendicular to the longitudinal reinforcement.

(5) Longitudinal bars bent to act as web reinforcement in tension zones shall be continuous with the longitudinal reinforcement. The tensile stress in each bar shall be fully developed in both the upper and lower half of the beam by anchorage through bond or hooks.

4.4.3.46. (1) A hook used for anchoring reinforcement shall have

(a) a complete semicircular bend with a radius on the axis of the bar of at least 3 and not more than 6 bar diameters, plus an extension at the free end of the bar equal to at least 4 bar diameters,

(b) a 90-deg. bend having a radius of at least 4 bar diameters plus an extension beyond the bend equal to at least 12 bar diameters, or
(c) for stirrup anchorage only, a 135-deg. turn with a radius on the axis of the bar of 3 diameters plus an extension at the free end of the bar of at least 6 bar diameters.

(2) Hooks having a radius of bend of more than 6 bar diameters shall be considered merely as extensions to the bars.
(3) Hooks shall not be assumed to carry a load which would produce a tensile stress in the bar greater than 50 MPa (7 251.89 psi).
(4) Hooks shall not be considered effective in anchoring bars in compression.
(5) Any mechanical device capable of developing the strength of the bar without damage to the masonry may be used in lieu of a hook provided test evidence is submitted to show the adequacy of such device.

GROUTED REINFORCED MASONRY

4.4.3.47. (1) Grouted reinforced masonry shall be constructed so that

(a) at the time of laying, all masonry units are free of excessive dust or dirt,
(b) Type S mortar is used,
(c) the proportions of materials in fine or coarse grout conform to Article 4.4.2.18.,
(d) fine grout is used in grout spaces, except that coarse grout may be used in grout spaces 50 mm (1.97 in.) or more in least horizontal dimension,
(e) the grout completely fills all spaces intended to receive grout,
(f) grout is used before it has begun to set but not more than 1 1/2 hr after initial mixing, and
(g) the units in all wythes are laid with full head and bed mortar joints.

4.4.3.48. (1) Where grouted masonry is grouted in low lifts

(a) masonry headers shall not project into the grout space,
(b) all spaces to be grouted shall be not less than 20 mm (0.79 in.) in width,
(c) grout shall be puddled immediately after pouring,
(d) wythes shall be carried up to a height not greater than that required to accommodate 1 grout lift, except that 1 wythe may be carried to a height of not more than 400 mm (1.31 ft) before grouting,
(e) grout shall be placed in lifts of not more than 200 mm (7.87 in.) but not more than 6 times the width of the grout space, and
(f) the grout shall be stopped 25 mm (0.98 in.) below the top of the lowest wythe where the work may be stopped for 1 hr or longer.

4.4.3.49. (1) Where grouted masonry is grouted in high lifts, the wythes may be constructed to the full wall height and grouting carried out in conformance with Sentences (2) to (6) after the mortar has set.
(2) The outer wythes of grouted masonry in Sentence (1) shall be bonded together with wall ties of not less than 3.8 mm (0.15 in.) diameter corrosion-resistant wire bent into rectangles 100 mm (3.94 in.) wide and 50 mm (1.97 in.) less in length than the over-all wall thickness, or other ties providing equivalent strength, stiffness and bond. Kinks or other deformations in the ties shall not be permitted. One wythe of the wall shall be built up not higher than 400 mm (1.31 ft) above the other wythe. Ties shall be laid not more than 600 mm (1.97 ft) o.c. horizontally and 400 mm (1.31 ft) o.c. vertically for running bond, and not more than 600 mm (1.97 ft) o.c. horizontally and 300 mm (0.98 ft) o.c. vertically for stack bond.
(3) Cleanouts shall be provided for each lift in grouted masonry in Sentence (1) by omitting every second unit in the bottom course of the section being placed. Mortar fins and other foreign matter shall be removed from the grout space by a high pressure jet of water or air. Such cleanouts shall be sealed after inspection and before grouting.
(4) The grout space in grouted masonry in Sentence (1) shall be not less than 75 mm (2.95 in.) in width, and vertical grout barriers of solid masonry not more than 8 m (26.25 ft) apart shall be built across the grout space the entire height of the wall.
(5) Grout used in grouted masonry in Sentence (1) shall be mixed thoroughly to a consistency suitable for pumping without segregation, and placed by pumping or other approved method.
(6) Grouting of grouted masonry in Sentence (1) shall be done in a continuous pour in lifts

of not more than 1.5 m (4.92 ft). It shall be consolidated by puddling or vibrating during pouring and again after excess moisture has been absorbed and while the grout is plastic. The grouting of any section between vertical grout barriers shall be completed in 1 day with no interruptions greater than 1 hr.

REINFORCED MASONRY OF HOLLOW UNITS

4.4.3.50. (1) Reinforced masonry of hollow unit construction shall be constructed of hollow masonry units in which certain cells contain reinforcement and are filled with concrete or grout.

(2) All reinforced masonry of hollow units shall be built so that walls and cross webs forming cells to be filled shall be fully bedded in mortar to prevent leakage of grout. All head joints shall be filled with mortar for a distance in from the face of the wall or unit not less than the thickness of the face shells. Bond shall be provided by lapping units in successive vertical courses or by equivalent mechanical anchorage.

(3) Vertical cells of hollow units to be filled in Sentence (1) shall have vertical alignment sufficient to maintain an unobstructed continuous cell of at least 50 mm (1.97 in.) by 75 mm (2.95 in.), except that where the total grout pour exceeds 3 m (9.84 ft) such cells shall be at least 75 mm (2.95 in.) by 75 mm (2.95 in.).

(4) Cleanout openings shall be provided at the bottoms of all cells to be filled at each lift or where such lift or pour of grout exceeds 1.2 m (3.94 ft) in height. Any overhanging mortar or other obstruction or debris shall be removed from the insides of such walls. The cleanouts shall be inspected before being sealed.

(5) Vertical reinforcement shall be held in position at top and bottom and at intervals not exceeding 192 diam. of the reinforcement.

(6) All cells containing reinforcement shall be completely filled with grout in lifts not exceeding 3 m (9.84 ft), except that where the total grout pour exceeds 3 m (9.84 ft) in height the grout shall be placed in lifts not exceeding 1.5 m (4.92 ft). Grout shall be consolidated at the time of pouring by puddling or vibrating during pouring and again after excess moisture has been absorbed and while the grout is plastic.

(7) When the grouting is stopped for more than 1 hr, horizontal construction joints shall be formed by stopping the pour of grout 40 mm (1.57 in.) below the top unit.

(8) The porportions of materials in fine or coarse grout shall conform to Article 4.4.2.18.

(9) All grout shall be used within 1 1/2 hr of initial mixing but before it has begun to set.

PLACING REINFORCEMENT

4.4.3.51. (1) The thickness of grout or mortar between masonry units and reinforcement shall be not less than 6 mm (0.24 in.), except that 6 mm (0.24 in.) bars may be laid in not less than 10 mm (0.39 in.) horizontal joints. Spaces containing both horizontal and vertical reinforcement shall be not less than 12 mm (0.47 in.) larger than the sum of the diameters of such horizontal and vertical reinforcement.

(2) Except in columns the clear distance between parallel bars shall be at least equal to the diameter of the bar.

(3) Reinforcement shall be accurately placed and fixed rigidly in position during grouting, except that horizontal reinforcement may be placed as the work progresses.

(4) Splices shall be made so that the structural strength of the member is not reduced. Lapped splices shall provide sufficient lap to develop by bond the working stress of the reinforcement. Mechanical connections shall develop the strength of the reinforcement and welded connections shall conform to CSA W186-1970, "Welding of Reinforcing Bars in Reinforced Concrete Construction."

4.4.3.52. (1) Except as provided in Sentence (2), all reinforcing bars shall be completely embedded in mortar or grout and have a coverage of masonry not less than

(a) 75 mm (2.95 in.) at the tops and bottoms of footings and masonry in contact with soil,
(b) 50 mm (1.97 in.) over bars in masonry exposed to weather, except that 40 mm (1.57 in.) shall be permitted over bars 15 mm (0.59 in.) diameter or less and not located in the upper face of the masonry,
(c) 40 mm (1.57 in.) over reinforcement in columns not exposed to weather or soil,
(d) 40 mm (1.57 in.) on the bottom and sides of beams or girders not exposed to weather or soil.

(e) 20 mm (0.97 in.) from the face of all walls not exposed to weather or soil,
(f) 20 mm (0.97 in.) at the upper face of any member not exposed to weather or soil, and
(g) one bar diameter over all bars.

(2) Reinforcement consisting of bars or wire 6 mm (0.24 in.) or less in diameter embedded in the horizontal mortar joints shall have not less than 15 mm (0.59 in.) mortar coverage from the exposed face.

SUBSECTION 4.4.4. CONVENTIONAL DESIGN OF PLAIN MASONRY

General

4.4.4.1. This Subsection applies to the design and construction of plain masonry except for plain masonry designed in accordance with Subsection 4.4.3.

4.4.4.2. (1) Dimensions of masonry units or masonry in this Subsection are actual.
(2) Minimum actual dimensions of masonry units or masonry shall be determined in accordance with CAN3-A31-M75, "Series of Standards for Metric Dimensional Coordination in Buildings."

4.4.4.3. (1) The compressive stresses in plain masonry shall conform to Table 10.
(2) Mortar joints shall not exceed 12 mm (0.47 in.) in thickness.
(3) Where a type of masonry unit or type of mortar is not provided for in Sentence (1), the maximum allowable compressive stress of the masonry shall be 15 per cent of the ultimate compressive strength of the masonry as determined by tests performed in accordance with ASTM E72-77, "Standard Methods of Conducting Strength Tests of Panels for Building Construction."

4.4.4.4. (1) Where a masonry unit of natural stone directly supports a concentrated load, the maximum allowable compressive stress for that unit shall be 10 per cent of its compressive strength.
(2) The maximum allowable flexural stress for natural stone shall be 1/6 its modulus of rupture.

4.4.4.5. The thickness of every masonry wall shall conform to the appropriate requirements in Aritcles 4.4.4.10. to 4.4.4.18., and shall have a bearing capacity conforming to Article 4.4.4.7.

4.4.4.6. Every masonry partition or wall, including panel walls and curtain walls, shall be laterally supported in conformance with the appropriate requirements in Articles 4.4.4.8., 4.4.4.9., 4.4.4.23., 4.4.5.8. and 4.4.5.10.

BEARING CAPACITY

4.4.4.7. (1) The maximum allowable bearing capacity of masonry shall be the product of its maximum allowable stress provided for in Article 4.4.4.3. and

(a) its gross cross-sectional area when it is solid masonry, or
(b) its gross cross-sectional area minus the area of space between the wythes when it is a cavity wall.

(2) For the purpose of calculating the area of masonry the actual dimensions of the cross-section of the masonry shall be used.
(3) Where masonry is constructed of more than 1 type of masonry unit, its maximum allowable bearing capacity shall be determined on the basis of the weakest unit.

LATERAL SUPPORT

4.4.4.8. (1) Except as provided in Sentence (2), a wall of masonry shall have lateral supports at either horizontal or vertical intervals spaced not more than

(a) 20 times the thickness of the wall where the wall is of solid masonry of solid units, or
(b) 18 times the thickness of the wall where the wall is of solid masonry of hollow units or a cavity wall.

(2) Every partition shall be supported laterally at either horizontal or vertical intervals of not more than 36 times the thickness of the wall.

4.4.4.9. (1) Except as provided in Sentence (2), where a wall of masonry does not have lateral support along its top, and if its height exceeds 4 times its thickness, it shall have vertical lateral supports at horizontal intervals spaced in accordance with Sentence 4.4.4.21.(1).

Table 10. Forming Part of Sentence 4.4.4.3.(1)

Type of Masonry	Type of Masonry Units	Maximum Allowable Compressive Stress(1) — Type of Mortar(2)									
		M		*S*		*N*		*O*		*K*	
		MPa	*psi*	*MPa*	*psi*	*MPa*	*psi*	*MPa*	*psi*	*MPa*	*psi*
Solid Masonry	Rubble stone	1.0	145.04	0.8	116.03	0.7	101.53	0.5	72.52	–	–
	Ashlar granite	5.5	797.71	5.0	725.19	4.4	638.17	3.5	507.63	–	–
	Ashlar limestone and marble	3.5	507.63	3.1	449.62	2.8	406.11	2.2	319.08	–	–
	Ashlar sandstone and cast-stone	2.8	406.11	2.5	362.59	2.2	319.08	1.7	246.56	–	–
	Solid units, except concrete block, with an ultimate compressive strength of										
	over 70 MPa (10 152.64 psi)	3.5	507.63	3.1	449.62	2.4	348.09	1.7	246.56	0.7	101.53
	55 MPa (7 977.08 psi) to 70 MPa (10 152.64 psi)	2.8	406.11	2.4	348.09	2.1	304.58	1.4	203.05	0.7	101.53
	42 MPa (6 091.59 psi) to 55 MPa (7 977.08 psi)	2.2	319.08	1.9	275.57	1.7	246.56	1.2	174.5	0.7	101.53
	30 MPa (4 351.13 psi) to 42 MPa (6 091.59 psi)	1.7	246.56	1.5	217.56	1.4	203.05	1.0	145.04	0.7	101.53
	18 MPa (2 610.68 psi) to 30 MPa (4 351.13 psi)	1.2	174.05	1.1	159.54	1.0	145.04	0.7	101.53	0.5	72.52
	10 MPa (1 450.38 psi) to 18 MPa (2 610.68 psi)	0.9	130.53	0.8	116.08	0.7	101.53	0.5	72.52	0.3	43.51

Col. 1	2	3	3a	4	4a	5	5a	6	6a	7	7a
Cavity Walls	Solid concrete block										
	over 20 MPa (2 900.75 psi) to 27.5 MPa (3 988.54 psi)	1.8	261.07	1.6	232.06	1.4	203.05	1.0	145.04	–	–
	12.5 MPa (1 812.97 psi) to 20 MPa (2 900.75 psi)	1.2	174.05	1.1	159.54	1.0	145.04	0.7	101.53	–	–
	8 MPa (1 160.30 psi) to 12.5 MPa (1 812.97 psi)	0.9	130.53	0.8	116.08	0.7	101.53	0.5	72.52	–	–
	Hollow load-bearing units										
	over 7.5 MPa (1 087.78 psi) to 10 MPa (1 450.38 psi)	0.8	116.08	0.7	101.53	0.6	87.02	–	–	–	–
	5 MPa (725.19 psi) to 7.5 MPa (1 087.78 psi)	0.6	87.02	0.5	72.52	0.5	72.52	–	–	–	–
	Solid units, except concrete block, with an ultimate compressive strength of										
	over 18 MPa (2 610.68 psi)	1.0	145.04	0.9	130.53	0.8	116.08	–	–	–	–
	10 MPa (1 450.38 psi) to 18 MPa (2 610.68 psi)	0.7	101.53	0.6	87.02	0.5	72.52	–	–	–	–
	Solid concrete block										
	over 12.5 MPa (1 812.97 psi)	1.0	145.04	0.9	130.53	0.8	116.08	–	–	–	–
	8 MPa (1 160.30 psi) to 12.5 MPa (1 812.97 psi)	0.7	101.53	0.6	87.02	0.5	72.52	–	–	–	–
	Hollow load-bearing units										
	over 7.5 MPa (1 087.78 psi)	0.5	72.52	0.5	72.52	0.4	58.02	–	–	–	–

Notes to Table 10:
(1) Based on gross cross-sectional area.
(2) See Article 4.4.2.15. for type of mortar.

(2) The portion of a wall extending from the sill of a window to the floor immediately below shall be laterally supported along its top or have vertical lateral supports at horizontal intervals spaced in accordance with Sentence 4.4.4.8.(1) where

(a) its height exceeds 3 times its thickness, and
(b) the length of the wall below the window exceeds the limits in Sentence 4.4.4.8.(1).

HEIGHT AND THICKNESS OF SOLID MASONRY

4.4.4.10. (1) Where a solid masonry wall is made up of 2 or more wythes, the thickness of the wall shall not include any wythe less than 90 mm (3.54 in.) thickness for load-bearing masonry walls or 75 mm (2.95 in.) thickness for panel walls or curtain walls.

(2) Veneer shall not be considered part of the wall when computing the required thickness of the wall.

4.4.4.11. (1) Except as provided in Sentences (3) and (4) and Articles 4.4.4.14. and 4.4.4.18., the thickness of a load-bearing wall of solid masonry above the top of a foundation wall, and not including basement and cellar walls, shall be at least 290 mm (11.42 in.)

(a) for wall up to 11 m (36.09 ft) in height if constructed of hollow units, and
(b) for wall up to 15 m (49.21 ft) in height if constructed of solid units.

(2) Where a solid masonry wall exceeds the height limits in Sentence (1), the thickness requirements for the top 11 m (36.09 ft) of walls made with hollow units or the top 15 m (49.21 ft) of walls made with solid units shall conform to Sentence (1), and the wall thickness below these heights shall be increased in increments of at least 100 mm (3.94 in.) for each increment of 11 m (36.09 ft) of height or part thereof measured down from the top.

(3) Where a load-bearing wall of solid masonry is not over 11 m (36.09 ft) in height above the top of the foundation wall, and where the design live load on any floor above the first storey does not exceed 2.4 kPa (50.13 psf), the minimum wall thickness shall be 190 mm (7.48 in.), except that for rubble stone the minimum thickness shall be 300 mm (11.81 in.).

(4) Load-bearing walls of solid units 140 mm (5.51 in.) in thickness may be built to a height not exceeding 2.8 m (9.19 ft) at the eave and 4.6 m (15.09 ft) at the peak of a gable

(a) in 1-storey buildings, and
(b) for the top storey of 2-storey buildings where the wall of the first storey is permitted to be 190 mm (7.48 in.) in Sentence (3).

(5) Masonry foundation walls shall be designed in accordance with Section 4.4.3.

(6) When a change in thickness due to minimum thickness requirements occurs between floor levels, the greater thickness shall be carried up to the next higher floor level.

(7) Where a change in thickness of a masonry wall occurs, the top 190 mm (7.48 in.) of the thicker portion shall be of solid units.

4.4.4.12. (1) Except as provided in Sentence (2) and Article 4.4.4.14., the thickness of a solid masonry panel wall shall be not less than 175 mm (6.89 in.) thickness.

(2) Panel walls of solid masonry of solid units not less than 140 mm (5.51 in.) in thickness may be built to a height not exceeding 3 m (9.84 ft) provided Type S mortar is used.

4.4.4.13. (1) The thickness of every solid masonry curtain wall shall be at least 175 mm (6.89 in.) thickness for walls up to 11 m (36.09 ft) in height above its bearing support.

(2) Where a solid masonry curtain wall exceeds 11 m (36.09 ft) in height, the top 11 m (36.09 ft) of the wall shall be at least 175 mm (6.89 in.) thickness, and the wall thickness below this height shall be increased in increments of at least 100 mm (3.94 in.) for each increment of 11 m (36.09 ft) of height or part thereof measured down from the top.

4.4.4.14. (1) Where a solid masonry wall is stiffened by pilasters of plain masonry, the wall thickness required in Articles 4.4.4.11. and 4.4.4.12. may be reduced between pilasters by 1/2 of the thickness added by the pilaster to the wall thickness required without pilasters provided

(a) no part of the wall after reduction is less than 190 mm (7.84 in.) in thickness,
(b) the center-to-center spacing of pilasters is not more than 25 times the reduced thickness of the stiffened wall, and
(c) the width of the pilasters is not less than 1/8 of their center-to-center spacing.

HEIGHT AND THICKNESS OF CAVITY WALLS

4.4.4.15. (1) A cavity wall shall not be built to a height greater than 11 m (36.09 ft) above its bearing support.

(2) The minimum thickness of wythe in a cavity wall shall be 90 mm (3.54 in.).

(3) The width of a cavity in a cavity wall shall be not less than 50 mm (1.97 in.) and not more than 75 mm (2.95 in.) when tied with metal ties.

(4) The width of a cavity in a cavity wall shall be not less than 75 mm (2.95 in.) and not more than 100 mm (3.94 in.) when tied with masonry bonding units.

4.4.4.16. (1) Where a cavity wall is load-bearing the total thickness of wythes and cavities shall be at least

(a) 230 mm (9.06 in.) for the top 3.6 m (11.81 ft),
(b) 290 mm (11.42 in.) for that portion more than 3.6 m (11.81 ft) but not more than 7.2 m (23.62 ft) from the top, and
(c) 330 mm (12.99 in.) for that portion more than 7.2 m (23.62 ft) from the top.

(2) Where a cavity wall is non-load-bearing the total thickness of wythes and cavities shall be at least 230 mm (9.06 in.).

HEIGHT AND THICKNESS OF PARTITIONS

4.4.4.17. (1) Except as provided in Sentence (2), the height of any masonry partition between horizontal lateral supports shall not exceed 36 times the partion thickness.

(2) Where lateral support of a partition is provided by walls or columns spaced at horizontal intervals not exceeding 36 times the partition thickness, the height of a partition shall not exceed 72 times its thickness.

HEIGHT AND THICKNESS OF SHAFT AND PENTHOUSE WALLS

4.4.4.18. (1) Except as provided in Sentence (4), every interior load-bearing wall that encloses a stair shaft, elevator shaft or other vertical shaft and does not exceed 6 m (19.69 ft) between vertical lateral supports shall be at least 190 mm (7.48 in.) thick for walls up to 2 storeys in height.

(2) Except as provided in Sentence (4), where interior load-bearing walls in Sentence (1) exceed 2 storeys in height, the top 2 storeys shall be at least 190 mm (7.48 in.) in thickness and the minimum thickness below this height shall be increased in increments of at least 100 mm (3.94 in.) for each increment of 3 storeys measured downward from the top.

(3) Load-bearing masonry walls not more than 3.6 m (11.81 ft) in height above the main roof level that enclose mechanical rooms or elevator or stairway penthouses having an aggregate area not exceeding 15 per cent of the roof area, but not exceeding 500 m^2 (5 381.96 sq ft) shall be not less than 190 mm (7.48 in.) in thickness, except that where such exterior walls support beams carrying elevator loads the wall thickness shall be at least 290 mm (11.42 in.) up to the underside of such beams.

(4) Where penthouse walls described in Sentence (3) are supported on interior masonry walls described in Sentences (1) and (2), they need not be considered in computing the allowable height and thickness of such interior walls except as provided in Article 4.4.4.19.

CHANGES IN THICKNESS

4.4.4.19. The thickness of a wall of masonry at any height shall not be greater than the thickness of the wall immediately below, except as provided in Article 4.4.5.8.

CHASES AND RECESSES

4.4.4.20. (1) Chases or recesses shall not be made in walls 190 mm (7.48 in.) or less in thickness.

(2) Chases or recesses shall not be closer than 600 mm (1.97 ft) to any structural member that provides lateral support for any wall.

(3) Except as provided in Article 4.4.4.22., the depth of any chase or recess in any wall shall not exceed 1/3 the thickness of the wall.

(4) The clear distance between chases in a wall shall be not less than 4 times the wall thickness.

4.4.4.21. (1) Except as provided in Sentence (2), every chase or recess or bearing in masonry walls of hollow units shall be built in as construction proceeds.

(2) Where necessary to cut a chase, dry pack

concrete shall be used to form a chase of required size in a wall of hollow masonry after the wall has been constructed.

4.4.4.22. (1) Every chase or recess having a width exceeding 500 mm (1.64 ft) or a depth exceeding 1/3 the thickness of the wall shall be considered as an opening, and any masonry above such chase or recess shall be supported on a lintel or arch.

(2) The width of any sloping or horizontal chase or recess shall be assumed to be the horizontal distance between the vertical lines through its extremities.

ALLOWABLE OPENINGS

4.4.4.23. Evidence shall be provided to show that the openings do not cause stresses in the wall greater than the values given in Article 4.4.4.3.

COLUMNS

4.4.4.24. (1) Every masonry column shall be constructed of solid masonry of solid units or hollow units filled with grout or concrete.

(2) Every masonry column shall have lateral supports spaced so that the vertical distance between supports is not greater than 10 times the least dimension of the column.

INDEX